Studies in Computational Intelligence

Volume 611

Series editor

Janusz Kacprzyk, Polish Academy of Sciences, Warsaw, Poland
e-mail: kacprzyk@ibspan.waw.pl

About this Series

The series "Studies in Computational Intelligence" (SCI) publishes new developments and advances in the various areas of computational intelligence—quickly and with a high quality. The intent is to cover the theory, applications, and design methods of computational intelligence, as embedded in the fields of engineering, computer science, physics and life sciences, as well as the methodologies behind them. The series contains monographs, lecture notes and edited volumes in computational intelligence spanning the areas of neural networks, connectionist systems, genetic algorithms, evolutionary computation, artificial intelligence, cellular automata, self-organizing systems, soft computing, fuzzy systems, and hybrid intelligent systems. Of particular value to both the contributors and the readership are the short publication timeframe and the worldwide distribution, which enable both wide and rapid dissemination of research output.

More information about this series at http://www.springer.com/series/7092

Siddhartha Bhattacharyya
Paramartha Dutta · Susanta Chakraborty
Editors

Hybrid Soft Computing Approaches

Research and Applications

 Springer

Editors

Siddhartha Bhattacharyya
Department of Information Technology
RCC Institute of Information Technology
Kolkata, West Bengal
India

Paramartha Dutta
Department of Computer and System
  Sciences
Visva-Bharati University
Santiniketan, West Bengal
India

Susanta Chakraborty
Department of Computer Science
  and Technology
Indian Institute of Engineering Science
  and Technology
Howrah
India

ISSN 1860-949X   ISSN 1860-9503   (electronic)
Studies in Computational Intelligence
ISBN 978-81-322-2978-0   ISBN 978-81-322-2544-7   (eBook)
DOI 10.1007/978-81-322-2544-7

Springer New Delhi Heidelberg New York Dordrecht London
© Springer India 2016
Softcover re-print of the Hardcover 1st edition 2016

Printed on acid-free paper

Springer (India) Pvt. Ltd. is part of Springer Science+Business Media (www.springer.com)

Dedicated to

My father Late Ajit Kumar Bhattacharyya
My mother Late Hashi Bhattacharyya

My paternal uncles Late Beni Madhab Bhattacharyya, Late Ramesh Chandra Bhattacharyya and Mr. Narayan Chandra Bhattacharyya

My aunts Late Uma Bhattacharyya, Late Arati Bhattacharyya and Mrs. Anita Bhattacharyya

Dr. Siddhartha Bhattacharyya

Dedicated to
My father Late Arun Kanti Dutta
My mother Mrs. Bandana Dutta

Dr. Paramartha Dutta

Dedicated to
My respected teachers and beloved students

Dr. Susanta Chakraborty

Preface

With the shortcoming and limitation of classical platforms of computation, particularly for tackling uncertainty and imprecision prevalent in our day-to-day life, Soft Computing as an alternative and extended computation paradigm has been making its presence felt, practically in problems being faced by the human civilization along all walks of life. Accordingly, a phenomenal growth of research initiative in this field is quite evident. Soft Computing techniques include (i) the elements of fuzzy mathematics, primarily used for handling various real-life problems engrossed with uncertainty, (ii) the ingredients of artificial neural network, usually applied for cognition, learning, and subsequent recognition by machine inducing thereby the flavor of intelligence in a machine through the process of its learning, and (iii) components of Evolutionary Computation mainly used for search, exploration, efficient exploitation of contextual information, and knowledge useful for optimization.

These techniques individually have got their points of strength as well as limitations. On several real-life contexts, it is being observed that they play a supplementary role to one another. Naturally, this has given rise to serious research initiative for exploring avenues of hybridization of the above-mentioned Soft Computing techniques. It is in this context that the editors of the present treatise aim at bringing out some latest findings in the field of hybrid Soft Computing.

- "A Hybrid CS–GSA Algorithm for Optimization" intends to enhance the exploration effectiveness of the Gravitational Search Algorithm (GSA) by computing with Cuckoo Search. Authors apply their hybridized technique on 23 different benchmark test functions to justify the supremacy of their technique over existing methods.
- Authors of "Study of Economic Load Dispatch by Various Hybrid Optimization Techniques" consider the Economic Load Dispatch (ELD) problem of electrical power system. Many techniques figure in the literature ranging from classical, linear, quadratic, nonlinear programming methods to various Soft Computing techniques including ABC, PSO, CSA, ACO, SA, GA, etc. Of late, hybridization of these Soft Computing techniques proves to be even more

encouraging. Authors provide a comprehensive study on such hybrid techniques applied on ELD in this chapter.

- As of now, Template-Based Modeling (TBM) algorithms applied for structural and functional characterizations of protein sequences in biological research suffer from drawbacks in respect of accuracy as indicated by the authors of "Unsolved Problems of Ambient Computationally Intelligent TBM Algorithms". In the present initiative, the authors try to come out with corrective measures in various steps of TBM algorithms.
- In "Hybridizing Differential Evolution Variants Through Heterogeneous Mixing in a Distributed Framework", authors try to identify the effectiveness as well as advantages of heterogeneous DE variants having diverse characteristics applied in a distributed framework over its homogeneous counterpart. The robustness of the proposed method is due to benchmarked comparison with existing state-of-the-art DE techniques available in the literature.
- "Collaborative Simulated Annealing Genetic Algorithm for Geometric Optimization of Thermo-electric Coolers" contains the impact of collaboration (hybridization) of Simulated Annealing and Genetic Algorithm. Authors earmark the rate of refrigeration (ROR) for indexing the performance of thermo-electric coolers (TEC) and justify the supremacy of such collaboration over the performance of Simulated Annealing (SA) or Genetic Algorithm (GA) individually.
- In course of "Color Magnetic Resonance Brain Image Segmentation by ParaOptiMUSIG Activation Function: An Application", authors establish how parallel optimized multilevel sigmoidal (ParaOptiMUSIG) activation function associated to Parallel Self-organizing Neural Network (PSONN) outperforms MUSIG activation function-based technique for segmenting color MR brain images.
- Authors of "Convergence Analysis of Backpropagation Algorithm for Designing an Intelligent System for Sensing Manhole Gases" carry out a comprehensive study and report the effectiveness of Back Propagation (BP) algorithm in the context of sensing the presence in excess proportion of hazardous gas components in manholes. They supplement their finding with theoretical justification of convergence of the BP algorithm they used for this purpose.
- In "REFII Model as a Base for Data Mining Techniques Hybridization with Purpose of Time Series Pattern Recognition", authors take up Raise-Equal-Fall model, version II, in short REF II which essentially is a transformation characteristic unique in nature and capable of automating time series analysis. The utility of REF II, coupled with other methods such as Self-organizing Maps or Frequent pattern trees offers a hybrid platform for efficient data mining.
- Authors have demonstrated in "A Soft Computing Approach for Targeted Product Promotion on Social Networks", as to how the Soft Computing paradigm may be useful for promoting different items on social networks.
- In the course of "Hybrid Rough-PSO Approach in Remote Sensing Imagery Analysis", authors combine the technical ingredients of Rough Set theory on

one hand and of Particle Swarm Optimization on the other to offer a hybridized platform for effective application in remotely sensed imagery along with analysis.

- An extensive study and comprehensive analysis of the usefulness of hybrid intelligent techniques for detection of breast cancer on the basis of breast thermograms are reported in "A Study and Analysis of Hybrid Intelligent Techniques for Breast Cancer Detection Using Breast Thermograms".
- Indian Summer Monsoon Rainfall (ISMR) prediction is reportedly inadequate with the use of Artificial Neural Network (ANN) alone. However, when ANN is hybridized with the power of Fuzzy Time Series analysis (FTS), the accuracy of prediction gets enhanced drastically, as demonstrated by the author in "Neuro-Fuzzy Hybridized Model for Seasonal Rainfall Forecasting: A Case Study in Stock Index Forecasting".
- Comprehensive overview of 3D face registration as a Computer Vision problem is reported in "Hybridization of 2D-3D Images for Human Face Recognition". There, the authors also make use of two supervised classifier techniques along with 2D and 3D hybrid face images for effective 3D face recognition purpose.
- In view of the imprecision inherent in the Business to Customer (B2C) E-commerce trust, authors of "Neutrosophic Trust Evaluation Model in B2C E-Commerce" attempt to offer a solution by exploiting the capability of handling uncertainty in the context of relevant trust models.
- Authors of "Immune-Based Feature Selection in Rigid Medical Image Registration Using Supervised Neural Network" exploit the power of Artificial Immune-based System for feature extraction. The extracted features are then fed into back propagation-based supervised Artificial Neural Network (ANN) for achieving the task medical image registration.

Last but not the least, the editors of this volume take this opportunity to express their sincere gratitude to the Series Editor Dr. Janusz Kacprzyk of Polish Academy of Sciences, Warsaw, Poland, Mr. Aninda Bose, Editor (Springer India), Ms. Kamiya Khatter, Editorial Senior Assistant (Springer India), and the entire editorial team of Springer India. It is a truth that but for their active involvement and effective guidance the present initiative would not have come in place in the present form.

Siddhartha Bhattacharyya
Paramartha Dutta
Susanta Chakraborty

Contents

About the Editors

Dr. Siddhartha Bhattacharyya did his Bachelors in Physics and Optics and Optoelectronics and his Masters in Optics and Optoelectronics from University of Calcutta, India in 1995, 1998, and 2000, respectively. He completed his Ph.D. in Computer Science and Engineering from Jadavpur University, India in 2008.

He is currently Associate Professor and Head of Information Technology and Dean (R&D) at RCC Institute of Information Technology, India. Prior to this, he was Assistant Professor in CS & IT of UIT, BU, India from 2005 to 2011. He was a Lecturer in IT of Kalyani Government Engineering College, India during 2001–2005. He is co-author of about 125 research publications in international journals and conference proceedings. He is the co-editor of the Handbook of Research on Computational Intelligence for Engineering, Science and Business by IGI Global, USA. He is the co-author of Soft Computing for Image and Multimedia Data Processing by Springer-Verlag, Germany.

Dr. Paramartha Dutta (born in 1966), M. Stat., M. Tech. (Computer Science) from Indian Statistical Institute Kolkata, and Ph.D. (Engineering) from presently Indian Institute of Engineering Science and Technology, is currently Professor in the Department of Computer and System Sciences, Visva-Bharati University, Santiniketan, India for more than eight years. He has about 180 papers in journal and conference proceedings, six books, and three edited books to his credit. He has the experience of successfully handling various projects funded by the Government of India. Moreover, he has supervised three scholars who have earned their Ph.D. Dr. Dutta is member/fellow of different professional bodies including IEEE and ACM.

Dr. Susanta Chakraborty received the Bachelor's (B.Tech.) and Master's (M.Tech.) degree in Technology from the University of Calcutta in 1983 and 1985, respectively, and the Ph.D. (Tech.) in Computer Science in 1999 from University of Calcutta and research work done at Advance Computing and Microelectronic Unit, Indian Statistical Institute, Kolkata. He is currently Professor in the Department of Computer Science and Technology of the Indian Institute of Engineering Science and Technology, West Bengal, India. Prior to this he served the University of

Kalyani as **Dean of Engineering, Technology and Management Faculty.** He has published around 60 research papers in reputed international journals including IEEE Transactions on CAD and refereed international conference proceedings of IEEE Computer Science Press.

Part I
Hybrid Soft Computing Approaches:
Research

A Hybrid CS–GSA Algorithm for Optimization

Manoj Kumar Naik, Leena Samantaray and Rutuparna Panda

Abstract The chapter presents a hybridized population-based Cuckoo search–Gravitational search algorithm (CS–GSA) for optimization. The central idea of this chapter is to increase the exploration capability of the Gravitational search algorithm in the Cuckoo search (CS) algorithm. The CS algorithm is common for its exploitation conduct. The other motivation behind this proposal is to obtain a quicker and stable solution. Twenty-three different kinds of standard test functions are considered here to compare the performance of our hybridized algorithm with both the CS and the GSA methods. Extensive simulation-based results are presented in the results section to show that the proposed algorithm outperforms both CS and GSA algorithms. We land up with a faster convergence than the CS and the GSA algorithms. Thus, best solutions are found with significantly less number of function evaluations. This chapter also explains how to handle the constrained optimization problems with suitable examples.

Keywords Cuckoo search · Gravitational search algorithm · Optimization

M.K. Naik (✉)
Department of Electronics & Instrumentation Engineering, Institute
of Technical Education and Research, Siksha 'O' Anusandhan University,
Bhubaneswar 751030, India
e-mail: manojnaik@soauniversity.ac.in

L. Samantaray
Department of Electronics & Instrumentation Engineering,
Ajaya Binaya Institute of Technology, Cuttack, India
e-mail: leena_sam@rediffmail.com

R. Panda
Department of Electronics & Telecommunication Engineering,
VSS University of Technology, Burla 768018, India
e-mail: r_ppanda@yahoo.co.in

© Springer India 2016
S. Bhattacharyya et al. (eds.), *Hybrid Soft Computing Approaches*,
Studies in Computational Intelligence 611,
DOI 10.1007/978-81-322-2544-7_1

1 Introduction

Over the decades, the evolutionary computation (EC) algorithms have been successfully smeared to solve the different practical computational problems like
optimization of the objective functions [1, 2], optimization of filter parameters [3,
4], optimization of different parameters for improvising image processing results [5,
6], optimal feature selection in pattern recognition [7–9], etc. Aforementioned
engineering applications are basically motivated by the near-global optimization
norm of the evolutionary computational methods. This permits those algorithms to
accomplish the task within a very big search space (of a given problem). However,
certain ECs are not yet investigated for solving a particular tricky efficiently. The
room of the proposed idea is to conduit the slit. It may generate concern among the
readers doing research in this area. There is a strong need to develop new hybrid
algorithms to suit a particular problem in hand. Further, by presenting a widespread
simulation results on the performance of 2 moderately firsthand ECs, we try to
convince the readers of their importance. In this chapter, an attempt is made to
attract the researchers regarding the aptness of the proposed technique for solving
the delinquent of the function optimization.

Many types of heuristic search ECs were proposed by investigators Genetic
algorithm (GA) [10], Ant colony algorithm (ACA) [11], Particle swarm optimization (PSO) algorithm [12], Bacterial foraging optimization (BFO) algorithm [13],
Cuckoo search (CS) algorithm [14–19], Gravitational search algorithm (GSA) [20],
etc. But a particular algorithm is not efficient to solve different types of optimization
problems. They never provide us with the best solutions. Certain algorithms offer
best solutions for particular (given) problems only. Thus, it is necessary to devise
hybridized heuristic population-based optimization methods for solving different
applications efficiently. In this chapter, a population-based hybridized optimization
algorithm is proposed. In this work, the thrust is mainly to cartel the social thinking
capability found in the Cuckoo birds and the local search capability observed in
Gravitational search method. Actually, an efficient optimization scheme is well
umpired by its two key features—(i) exploration and (ii) exploitation. Note that
exploration is the capability of an EC to explore the complete search space, and
exploitation is the skill to congregate to a better result. Combining these two
features, the best solutions can be obtained with less number of function evaluations. Therefore, here an attempt is made to hybridize CS with GSA in order to
provide equilibrium for both exploration and exploitation ability. This chapter
describes the use of both the methods in a balanced manner. Note that hybridized
approach provides significant improvements in the solutions. Interestingly, the CS
is popular for its simplicity and its ability to search for a near-global solution.
Further, GSA provides a better local search method with good initial estimates to
solve a particular problem.

In fact, EC techniques were initially proposed by different researchers as an
approach to the artificial intelligence. Later, it has become more popular and is used

directly to solve analytical and combinatorial optimization problems. However, the demerit of an EC method is its slow convergence in solving multimodal optimization problems, to find near (global) optimum values. In this context, various EC techniques are proposed. Recently, there has been a strong need to develop new hybridized EC techniques to find better results. This chapter discusses the application of a new hybrid EC method for optimization. Extensive empirical evaluations are important to measure the strength and weaknesses of a particular algorithm. In this connection, we consider 23 standard benchmark [21] functions. The proposed CS–GSA algorithm performs better than GSA and CS techniques for multimodal functions with many local minima. In addition, the proposed algorithm is faster than GSA. The levy flight incorporated in the algorithm helps us for a speed to reach the near (global) optimum very quickly. The idea behind this is simple and straightforward. Thinking ability of the Cuckoo birds is very useful for exploitation. In this study, we propose the hybridization of these two algorithms. Here, 23 standard benchmark functions [21] are utilized to relate the performance of our method. The results are compared to both the CS and GSA algorithms. In this work, the solutions presented in the chapter reveal the fact that our proposed algorithm is well suited for function minimization.

For this work, we consider 23 benchmark functions. These functions are given in Table 1. More details on such test functions are discussed in the Appendix. Note that the functions F_1–F_{13} refer to the high-dimensional problems, whereas the functions F_1–F_7 are known as the unimodal functions. The function F_5 is a step function. This has only one minimum. Further, the function is discontinuous. Here, the function F_7 is coined as the quartic function with noise. Here, rand $(0, 1)$ is basically a uniformly distributed random variable in the range $(0, 1)$. Here, we also consider multimodal functions F_8–F_{13} for our experimental study. For these functions, the amount of local minima surges exponentially with the surge in the problem size. It is important to solve these types of functions to validate our

Table 1 Unimodal benchmark functions

Benchmark function
$F_1(X) = \sum_{i=1}^{n} x_i^2$
$F_2(X) = \sum_{i=1}^{n} |x_i| + \prod_{i=1}^{n} |x_i|$
$F_3(X) = \sum_{i=1}^{n} \left(\sum_{j=1}^{i} x_j \right)^2$
$F_4(X) = \max_{i} \{|x_i|, 1 \leq i \leq n\}$
$F_5(X) = \sum_{i=1}^{n-1} \left[100\left(x_{i+1} - x_i^2\right)^2 + (x_i - 1)^2 \right]$
$F_6(X) = \sum_{i=1}^{n} \left(\lfloor x_i + 0.5 \rfloor \right)^2$
$F_7(X) = \sum_{i=1}^{n} i x_i^4 + \text{random}[0, 1)$

algorithm for optimization. The functions F_{14}–F_{23} are called as the functions with the low dimension. These functions have only a few local minima. It is relatively easy to optimize the unimodal functions. However, the optimization is much more significant in the case of multimodal functions. The ability of an algorithm is well judged by its behavior, capable of escaping from the poor local optima while localizing the best near (global) solution. Note that the functions F_{14}–F_{23} are multimodal in nature, but are fixed-dimensional functions. These functions are also very useful for the measure of the accuracy of the hybrid soft computing techniques.

The organization of the rest of the chapter is as follows: Sect. 1 is the introduction part. Related work is discussed in Sect. 2. The hybridized population-based CS–GSA algorithm is discussed in Sect. 3. Extensive results and discussions are presented in Sect. 4. In this chapter, conclusions are presented in Sect. 5.

2 Related Works

The main objective of an optimization algorithm is to find a near or near-global optimal solution. Several EC algorithms are presented in the literature, but the CS algorithm [14–16] has its own importance. Interestingly, it consists of less parameters for search, and so it is a faster optimization method. Recently, Chetty and Adewumi [22] have done a case study on an annual crop planning problem using a CS algorithm along with the GA and glow worm swarm optimization (GSO); the former has shown superior results. Chen and Do [23] used the CS to train the feed-forward neural networks for predicting the student academic performances. Swain et al. [24] have proposed neural network based on CS and apply the same for the noise cancellation. Khodier [25] used the CS algorithm for optimization of the antenna array. Although the CS gives better performance, some researcher adds some more characteristics to the search process. Zhao and Li [26] proposed opposition-based learning to upsurge the exploration proficiency of the CS algorithm.

The other non-evolutionary algorithms are also good at finding the near-global optima. The GSA [20] is founded on the rule of gravity. It has better convergence in the search space. Saha et al. [27]; Rashedi et al. [28] used the GSA for the filter design and modeling. Many authors proposed new schemes to enhance the performance of the GSA such as Disruption operator [29], black hole operator [30], Niche GSA [31], and binary GSA (BGSA) [32].

Every algorithm has its own merits and demerits. To improve search performance of one algorithm, some researcher proposed hybrid algorithm combining the features of more than one algorithm. Mirjalili and Hashim [33] proposed hybrid algorithm PSOGSA by integrating the ability of exploitation in PSO and the ability of exploration in GSA. Jiang et al. [34] proposed HPSO-GSA for solving economic emission load dispatch problems by updating the particle position with PSO velocity and GSA acceleration. Ghodrati and Lotfi [35] proposed hybrid algorithm CS/PSO by combining the idea of cuckoo birds awareness of each other position via swarm intelligence of PSO. A hybrid algorithm on GSA–ABC was proposed by Guo [36] to update the ant colony with the help of an artificial bee colony algorithm (ABC) and GSA. Sun and Zhang [37] have proposed GA–GSA hybrid algorithm for image segmentation based on the GA and GSA. Yin et al. [38] proposed a hybrid IGSAKHM approach for clustering by combining K-harmonic means (KHM) clustering technique and improved gravitational search algorithm (IGSA).

In this section, we discuss the potential features of two different algorithms used for optimization.

2.1 Cuckoo Search (CS) Algorithm

Cuckoo search (CS) method is basically a nature-inspired technique. This method is introduced by Yang and Deb [14–17]. The CS is inspired by an interesting event how the Cuckoo bird leaves eggs in the nest of another horde bird. The available host nests are fixed. The egg laid by the Cuckoo bird may be exposed by the horde bird with a probability $p_a \in [0, 1]$. Then, the horde birds either throw those eggs or abandon the present nest. Sometimes, the horde bird builds a new nest in a totally different location [18] to deceive the Cuckoo bird. Here, each egg in the nest represents a solution. It is interesting to note here that the CS has similarity to the well-known hill climbing algorithm. Here, in the Cuckoo search, note that a particular pattern corresponds to a nest, whereas an individual feature of that pattern resembles to that of an egg of the Cuckoo bird.

Interestingly, the CS method is founded on the succeeding three more idealized strategies:

   i. Every Cuckoo bird puts an egg at a time, it junk yards its egg in a randomly selected nest.
  ii. Nests having the best class of eggs are carried over to the subsequent generations.
 iii. Always the quantity of available horde nests is fixed. Note that the egg placed by a Cuckoo bird is exposed by the horde bird with a probability $p_a \in [0, 1]$. Then, the worst nests are revealed and discarded from future calculations.

The CS algorithm is mathematically modeled as follows: For a new search space $X_i(t + 1)$ for Cuckoo i (for $i = 1, 2, \ldots, N$) at time $t + 1$,

$$X_i^{t+1} = X_i^t + \alpha \oplus \text{Lévy}(\lambda), \tag{1}$$

where $X_i(t)$ is the current search space at time t, represented as $X_i = (x_i^1, \ldots, x_i^d, \ldots, x_i^n)$, $\alpha > 0$ is the step size connected to the range of the problem, $\oplus$ is the entry wise multiplication, and Lévy(λ) is the random walk through the Lévy flight. The Lévy flight [19] provides random walk for step size from the Lévy distribution $\left(\text{Lévy} \sim u = t^{-\lambda}\right)$ by considering λ such that it satisfies $1 < \lambda < 3$.

Here, time t denotes the number for a recent group ($t = 1, 2, 3, \ldots, t_{max}$) and t_{max} represents the pre-determined extreme cohort position. Here, the initial values of the dth attribute of the ith pattern are found by

$$x_i^d(t = 0) = \text{rand} \cdot \left(u^d x_i^d - l^d x_i^d\right) + l^d x_i^d, \tag{2}$$

where l^d and u^d are called as the lower and the upper search space limits of the dth attributes, respectively. These attributes are useful for implementations. This method is used to provide control over the boundary conditions in every calculation step. Note that the value for the interrelated attribute is restructured, when it exceeds the allowed search space limits. This is achieved by considering the value for the nearby limit corresponding to the linked trait. In this discussion, it is seen that the CS method identifies the best fruitful pattern as the X_{best} pattern. This process is accomplished before starting the iterative search method. Actually, here in the CS algorithm, the iterative growth part of the pattern matrix initiates by the discovery step of the Φ as

$$\Phi = \left(\frac{\Gamma(1 + \gamma) \cdot \sin(\pi \cdot \gamma/2)}{\Gamma\left(\left(\frac{1+\gamma}{2}\right) \cdot \gamma \cdot 2^{\frac{\gamma-1}{2}}\right)}\right)^{\frac{1}{\gamma}}. \tag{3}$$

The Γ in the above expression represents a gamma function and $\gamma - 1 = \lambda$.

The evolution phase of the X_i pattern is defined by the donor vector v, where $v = X_i$. The required step size value is calculated as

$$\text{stepsize}_d = 0.01 \cdot \left(\frac{u_d}{v_d}\right)^{\frac{1}{\gamma}} \cdot (v - X_{best}).$$

Note that $u = \Phi \, \text{randn}[n]$ and $v = \text{randn}[n]$.

Here, a donor pattern is randomly mutated as

$$v = v + \text{stepsize}_d \cdot \text{randn}[n]. \tag{4}$$

In this scheme, the update procedure of X_{best} in CS is stated as

$$X_{best} \leftarrow f(X_{best}) \leq f(X_i). \tag{5}$$

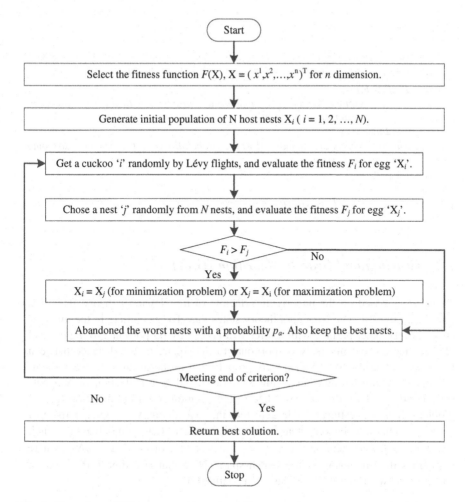

Fig. 1 Flow chart of CS

It is noteworthy to mention here that the controller constraints of the CS technique are coined as scale factor (λ) followed by the mutation probability (p_a). The flow chart for the CS algorithm is shown in Fig. 1.

PseudoCode for CS

First identifies the search space-like dimension of search problem 'n,' the range of the objective function, and objective function $F(X)$. Let us choose some important parameters N, p_a, α, λ, $t_{\max}$, and $t = 1$. Also, randomly initialize the population of N host nests $X_i(t) = \left(x_i^1, \ldots, x_i^d, \ldots, x_i^n\right)$ with n dimension for $i = 1, 2, \ldots, N$.

do {

   (a) *Get a cuckoo 'i' randomly by Lévy flights, and evaluate the fitness F_i for egg X_i.*
   (b) *Chose a nest 'j' randomly from N nests, and evaluate the fitness F_j for egg X_j.*
   (c) *If $(F_i > F_j)$*
              Replace X_i with X_j for minimization problem, or Replace X_j with X_i for
              maximization problem.
          End
   (d) *The worst nests are abandoned with a probability (p_a). The new ones are built*
        and keep the best ones.
   (e) *$t = t + 1$.*

} *while $(t < (t_{max} + 1))$ or End criterion not satisfied).*

2.2 Gravitational Search Algorithm (GSA)

The GSA is based on the underlying principle of the Newton's theory. This was introduced in [20]. The Newton's theory states that 'Every particle in the universe attracts every other particle with a force that is directly proportional to the product of their masses and inversely proportional to the square of the distance between them.' Note that the force is inversely proportional to the distance between them.

This algorithm is quite interesting and considered as a collection of N agents (masses) only. Here, the masses relate to the solution of an optimization (given) problem. It is interesting to note here that the heavier mass has greater force of attraction. This may be very near to the global optima. Let us initialize the search space of the ith agent as $X_i = (x_i^1, \ldots, x_i^d, \ldots, x_i^n)$ (for $i = 1, 2, \ldots, N$), where n signifies the dimension of the search space. Note that at a time t, the force of attraction between mass 'i' by mass 'j' is defined as

$$F_{ij}(t) = G(t)\frac{M_{pi}(t) \times M_{aj}(t)}{R_{ij}(t) + \varepsilon}\left(x_j(t) - x_i(t)\right), \tag{6}$$

where M_{aj} and M_{pj} are the active and passive gravitational masses connected to the agent i, $G(t)$ is coined as a gravitational constant at a particular time t, and $R_{ij}(t)$ is the Euclidian distance between agents 'i' and 'j,' which is written as

$$R_{ij} = \left\| X_i(t) \cdot X_j(t) \right\|_2. \tag{7}$$

It is noteworthy to mention here that the gravitational constant G gradually decreases with respect to the time. This event actually helps us to reach at the minima in the search space. Therefore, the gravitational constant G is considered as a function of the initial value G_0 together with the time t. The phenomenon can mathematically be modeled as

$$G(t) = G_0 \times e^{\left(-\beta \frac{t}{t_{\max}}\right)}, \tag{8}$$

where β is the descending coefficients. Note that $t_{\max}$ is the maximum number of iterations considered for the simulation work.

Thus, the total amount of the force that acts on the agent i is $F_i(t)$, which can be computed from Eq. (6) as

$$F_i(t) = \sum_{j=1, j\neq i}^{N} \text{rand}_j \, F_{ij}(t). \tag{9}$$

Here, different masses are computed from the fitness values. Further, the masses are updated by the following set of equations:

$$M_{\text{a}i} = M_{\text{p}i} = M_{ii} = M_i, i = 1, 2, \ldots, N, \tag{10}$$

$$m_i(t) = \frac{\text{fit}_i(t) - \text{worst}(t)}{\text{best}(t) - \text{worst}(t)}, \tag{11}$$

$$M_i(t) = \frac{m_i(t)}{\sum_{j=1}^{N} m_j(t)}, \tag{12}$$

where $\text{fit}_i(t)$ designates the fitness cost of an agent i at a particular time t. Here, best (t) signifies the best fitness value and worst(t) represents the worst fitness value among the N agents. Here, the acceleration of agent i at time t can be given by

$$a_i(t) = F_i(t) / M_i(t), \tag{13}$$

where $M_i(t)$ is coined as the mass of an agent i.

Finally, the velocity and the position of an agent in the search space are computed as given below:

$$x_i(t+1) = x_i(t) + v_i(t+1), \tag{14}$$

$$\text{and, } v_i(t+1) = \text{rand}_i \times v_i(t) + a_i(t). \tag{15}$$

Note that the positions are updated iteratively using the above equations till the GSA algorithm reaches the global or near-global minima. Actually, no further change in the mass should undergo after attaining the global or near-global minima. The flow chart for GSA is displayed in Fig. 2.

PseudoCode for GSA

In the beginning, it recognizes the search space, dimension of the search problem 'n,' the range of the objective function, and the objective function itself, i.e., $F(X)$.

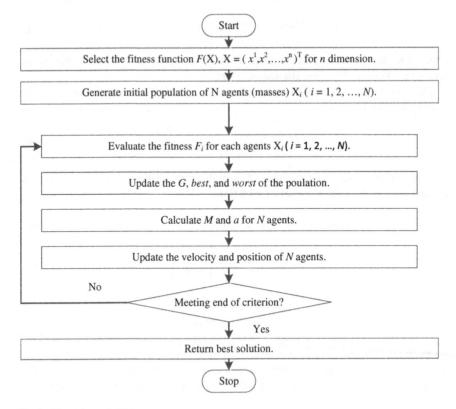

Fig. 2 Flow chart of GSA

Then choose some important parameters N, G_0, β, t_{max}, and $t = 1$. After choosing the parameters, randomly initialize the population of N agents (or masses) $X_i(t) = \left(x_i^1, \ldots, x_i^d, \ldots, x_i^n\right)$ with n dimension for $i = 1, 2, \ldots, N$.

> *do* {
>
>> (a) *Evaluate the objective function $F(X_i)$ for $i = 1,2,\ldots,N$.*
>> (b) *Update $G(t)$, best(t), and worst(t) of the current population.*
>> (c) *Calculate $M_i(t)$, $v_i(t)$ and $a_i(t)$ for all N agents.*
>> (d) *Then calculate the new position $X_i(t+1)$ of agents for $i = 1,2,\ldots,N$.*
>> (e) *$t = t + 1$.*
>
> *} while ($t < (t_{max} + 1)$) or End criterion not satisfied).*

3 A Hybrid CS–GSA Algorithm

It is well known from the literature that the CS is a heuristic search method based on evolutionary computational approach. The beauty of the CS algorithm is that it uses the randomized walk via a Lévy flight, as described in [14, 15]. Here, the Lévy flight is quite effectual in discovering the search space. Note that the step size is booked from the Lévy distribution as explained in [19].

Let us choose α as 1 (as $\alpha > 0$). So Eq. (1) is reduced to

$$X_i(t+1) = X_i(t) + \text{Lévy}(\lambda). \tag{16}$$

From the above Eq. (16), it is clearly showed that the new search space (the new solution) only rest on a Lévy distribution. Let us introduce a term l Best (t), which provides the best local solution among $i = 1, 2, \ldots, N$ at time t. Here, the lBest(t) can be expressed by

$$l\text{Best}(t) = X_j(t) | \forall j == i,$$
$$\text{for which } f(X_i(t)) \text{ is minimum,} \tag{17}$$
$$\text{for } i = 1, 2, \ldots, N \text{ at time } t.$$

Let us again incorporate an additional term (coined as the proportionate term) to the new solution, thereby incorporating the difference between the current solution and the local best solution at time t. Therefore, Eq. (16) can be expressed by

$$X_i(t+1) = X_i(t) + \text{Lévy}(\lambda) \times (l\text{Best}(t) - X_i(t)). \tag{18}$$

Further, let us see how every solution differs from the other at time t. In this sense, the acceleration of an agent i at a time t is used to provide enhancement to the local search in the GSA algorithm. Once more, we incorporate Eq. (13) in Eq. (18) as expressed by

$$X_i(t+1) = X_i(t) + \text{Lévy}(\lambda) \times (l\text{Best}(t) - X_i(t)) + a_i(t). \tag{19}$$

It is noteworthy to mention here that $a_i(t)$ is defined in Eq. (13). If we choose α as the proportionate measure of the step size, then Eq. (19) can be re-organized as

$$X_i(t+1) = X_i(t) + \alpha \times \text{Lévy}(\lambda) \times (l\text{Best}(t) - X_i(t)) + a_i(t). \tag{20}$$

Thus, Eq. (20) provides the new solution space for Cuckoo Search–Gravitational Search Algorithm (CS–GSA) from the list of current solutions obtained in this method. The flow chart for our new hybrid CS–GSA algorithm is shown in Fig. 3.

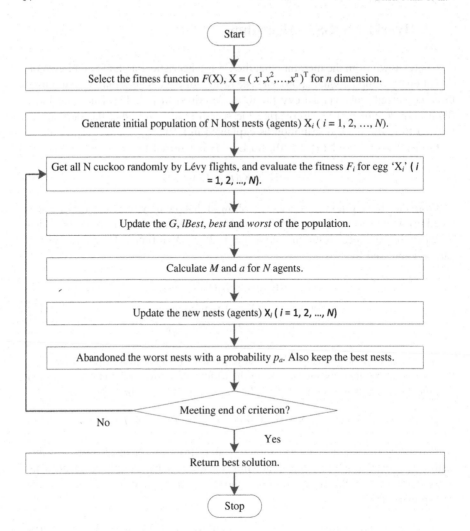

Fig. 3 Flow chart of CS–GSA

PseudoCode for CS–GSA

In the opening, the objective function $f(X)$ with the dimension n is recognized. Then choose the parameters N, p_a, G_0, α, λ, β, t_{max}, and $t = 1$ to control the algorithm while in iteration. Let us initialize randomly the population of N horde nests $X_i(t) = \left(x_i^1, \ldots, x_i^d, \ldots, x_i^n\right)$ with n dimension for $i = 1, 2, \ldots, N$ at $t = 1$.

do {

    (f) *Evaluate the objective function* $f(\mathrm{x}_i)$ *for i = 1,2,...,N.*

    (g) *Analyze all the fitness functions* $f(\mathrm{x}_i)$ *for i = 1,2,...,N. Then find the lBest(t) from the Eq.(17).*

    (h) *Update G(t) from the Eq. (8),* $M_i(t)$ *from the Eq. (12). Then compute acceleration* $a_i(t)$ *from the Eq. (13).*

    (i) *Then compute the new position of Cuckoo nests by using the Eq. (20).*

    (j) *The worst nests are abandoned with a probability* (p_a)*. The new ones are built. Then, keep the best ones.*

    (k) *t = t + 1.*

} while (t < (t_{max} +1) or End criterion not satisfied).

Finally, report the best $f(\mathrm{X}_i)$ with i = 1, 2, ..., N, also report the corresponding X_i.

4   Results and Discussions

In this work, the main thrust is to improve the CS algorithm in comparison to the standard CS methodology. Here, it is also important to bring some improvement over the GSA. In this context, a new CS–GSA algorithm has been developed in the previous Section.

4.1   *Performance Evaluation Using Standard Benchmark Functions*

In the performance evaluation of the proposed algorithm, we consider 23 standard benchmark functions [21] displayed in Tables 1, 2 and 3. Note that the convergence rate of the unimodal benchmark functions is important to validate an optimization algorithm. These useful functions are listed in Table 1.

It is noteworthy to mention here that the multimodal benchmark functions also have a significant role in validating optimization algorithms. These multimodal functions have many local minima, so it is difficult to optimize these functions. Such functions are displayed in Table 2 and are used for the performance measure.

Further, multimodal functions with fixed dimensions are also considered in this work. These types are displayed in Table 3. Generally, these functions have similar performances for all types of optimization algorithms.

Table 2 Multimodal benchmark functions

Benchmark function
$F_8(X) = \sum_{i=1}^{n} \left[x_i^2 - 10\cos(2\pi x_i) + 10 \right]$
$F_9(X) = -20\exp\left(-0.2\sqrt{\frac{1}{n}\sum_{i=1}^{n} x_i^2}\right) - \exp\left(\frac{1}{n}\sum_{i=1}^{n} \cos(2\pi x_i)\right) + 20 + e$
$F_{10}(X) = \frac{1}{4000}\sum_{i=1}^{n} x_i^2 - \prod_{i=1}^{n} \cos\left(\frac{x_i}{\sqrt{i}}\right) + 1$
$F_{11}(X) = \frac{\pi}{n}\left\{ 10\sin(\pi y_i) + \sum_{i=1}^{n} (y_i - 1)^2 \left[1 + 10\sin^2\left(\pi y_{i+1}\right) \right] + (y_n - 1)^2 \right\}$ $+ \sum_{i=1}^{n} u(x_i, 10, 100, 4)$
$F_{12}(X) = 0.1\left\{ \sin^2(3\pi x_1) + \sum_{i=1}^{n} (x_i - 1)^2 \left[1 + \sin^2(3\pi x_i + 1) \right] + (x_n - 1)^2 \cdot \left[1 + \sin^2(2\pi x_n) \right] \right\} + \sum_{i=1}^{n} u(x_i, 5, 100, 4)$

Table 3 Multimodal benchmark functions with fixed dimension

Benchmark function	n
$F_{13}(X) = \sum_{i=1}^{30} -x_i \sin\left(\sqrt{|x_i|}\right)$	30
$F_{14}(X) = \left(\frac{1}{500} + \sum_{j=1}^{25} \frac{1}{j + \sum_{i=1}^{2} (x_i - a_{ij})^6} \right)^{-1}$	2
$F_{15}(X) = \sum_{i=1}^{11} \left[a_i - \frac{x_1(b_i^2 + b_i x_2)}{b_i^2 + b_i x_3 + x_4} \right]^2$	4
$F_{16}(X) = 4x_1^2 - 2.1x_1^4 + \frac{1}{3}x_1^6 + x_1 x_2 - 4x_2^2 + 4x_2^4$	2
$F_{17}(X) = \left(x_2 - \frac{5.1}{4\pi^2}x_1^2 + \frac{5}{\pi}x_1 - 6 \right) + 10\left(1 - \frac{1}{8\pi}\right)\cos x_1 + 10$	2
$F_{18}(X) = \left[1 + (x_1 + x_2 + 1)^2 \left(19 - 14x_1 + 3x_1^2 -14x_2 + 6x_1 x_2\right) \right] \times \left[(2x_1 - 3x_2)^2 \times \left(18 - 32x_1 + 12x_1^2 + 48x_2 -36x_1 x_2 + 27x_2^2\right) + 30 \right]$	2
$F_{19}(X) = -\sum_{i=1}^{4} c_i \exp\left(-\sum_{j=1}^{3} a_{ij}\left(x_j - p_{ij}\right)^2 \right)$	3
$F_{20}(X) = -\sum_{i=1}^{4} c_i \exp\left(-\sum_{j=1}^{6} a_{ij}\left(x_j - p_{ij}\right)^2 \right)$	6
$F_{21}(X) = -\sum_{i=1}^{5} \left[(X - a_i)(X - a_i)^T + c_i \right]^{-1}$	4
$F_{22}(X) = -\sum_{i=1}^{7} \left[(X - a_i)(X - a_i)^T + c_i \right]^{-1}$	4
$F_{23}(X) = -\sum_{i=1}^{10} \left[(X - a_i)(X - a_i)^T + c_i \right]^{-1}$	4

Table 4 Parameter setting for GSA, CS, and CS–GSA for benchmark functions (F_1–F_{23})

Number of agents (masses or nests) $N = 50$
Number of maximum iteration $t_{max} = 1000$
Initial gravitational constant $G_0 = 100$
Mutation probability $p_a = 0.25$
Constant $\alpha = 1$, $\beta = 20$, $\lambda = 1.5$

In this simulation study, the range of x_i is different for different functions. The ranges of these functions are described in Appendix. The choice of parameters is important for evaluating the optimization algorithm. We have chosen best parameters for our study based on extensive simulation results. These parameters are used for all the three algorithms. The parameter setting for GSA, CS, and CS–GSA is shown in Table 4, for validating the benchmark functions given in Tables 1, 2 and 3.

The benchmark functions are categorized into three different tables as unimodal test functions (F_1–F_7), multimodal test functions (F_8–F_{12}), and multimodal test functions with fixed dimensions (F_{13}–F_{23}). The ranges of the objective functions and the global minima are given in the Appendix. The search dimension 'n' is taken as 10, 50 for the unimodal functions given in Table 1 and multimodal benchmark functions given in Table 2. The search dimension 'n' for the multimodal functions with fixed dimension is given in Table 3.

The performance evaluations of GSA, CS, and CS–GSA for the unimodal benchmark functions are presented in Tables 5 and 6.

From Table 5, it is observed that the performance of GSA for the unimodal benchmark functions F_1, F_2, and F_4 with $n = 10$ seems to be better than the other two algorithms. But for other functions, the performance of the CS–GSA algorithm is better. For all benchmark functions, final results are reflected as the 'Best,' 'Median,' and 'Ave' among 50 independent runs. Here, 'Best' implies the best fitness value obtained from 50 independent runs. 'Median' refers to the median of 50 fitness values obtained from 50 independent runs. The 'Ave' denotes the average value of 50 fitness values obtained from 50 independent runs. Within a function, the performance of GSA, CS, and CS–GSA is compared. The best solutions among all three algorithms are shown in boldface letters.

The performance evaluation of GSA, CS, and CS–GSA for the unimodal benchmark functions F_1 to F_7 with $n = 50$ is displayed in Table 6. Here, the proposed CS–GSA algorithm performs well as compared to other algorithms.

A comparison of these algorithms for the unimodal benchmark functions F_4 and F_7 with $n = 50$ is shown in Fig. 4. It is seen that CS–GSA offers us best values compared to other algorithms. Note that the maximum number of iterations considered here is 1000. Here, GSA is the second best.

The performance evaluation of GSA, CS, and CS–GSA for the multimodal benchmark functions F_8 to F_{12} with $n = 10$ is displayed in Table 7. Here, the proposed CS–GSA algorithm performs well for all functions except F_{12}.

The performance evaluation of GSA, CS, and CS–GSA for the multimodal benchmark functions F_8 to F_{12} with $n = 50$ is displayed in Table 8. Here, the new

Table 5 Performance evaluation of GSA, CS, and CS–GSA for the unimodal benchmark functions (displayed in Table 1) with $n = 10$

		GSA	CS	CS–GSA
F_1	Best	1.4442e-041	1.6724e-018	**3.2886e-042**
	Median	**1.5288e-040**	2.2921e-016	5.4848e-039
	Average	**1.5737e-039**	2.2606e-016	8.6919e-027
F_2	Best	**4.2795e-020**	7.2314e-005	9.7964e-020
	Median	**1.1788e-019**	2.2356e-004	4.8926e-019
	Average	**1.1491e-019**	1.1321e-003	7.1335e-016
F_3	Best	3.6769	4.4477	**2.2928**
	Median	29.0889	33.4532	**7.3735**
	Average	25.7047	28.9740	**18.1172**
F_4	Best	**1.3869e-020**	1.7780e-003	6.2616e-019
	Median	**9.5122e-020**	2.6564e-002	1.4920e-018
	Average	**1.5330e-019**	0.2551	1.2400e-012
F_5	Best	7.7969	12.4932	**7.7306**
	Median	**7.9811**	15.1342	8.2032
	Average	36.4058	39.2132	**35.7741**
F_6	Best	**0**	0.0121	**0**
	Median	**0**	0.0316	**0**
	Average	**0**	0.1276	**0**
F_7	Best	4.9725e-004	0.0217	**4.7790e-004**
	Median	0.0042	0.0319	**9.8222e-004**
	Average	0.0040	0.0332	**0.0011**

hybrid CS–GSA algorithm performs well for all functions except F_9. For the function F_9, GSA performs better.

A comparison of these algorithms for the multimodal benchmark functions F_8 and F_{12} with $n = 50$ is shown in Fig. 5. It is observed that the performance of the CS–GSA algorithm is better compared to other algorithms. Here, the maximum number of iterations is 1000. From Fig. 5, it is observed that the GSA is the second contestant.

The performance evaluation of GSA, CS, and CS–GSA for the multimodal benchmark functions F_{13} to F_{23} with *fixed dimension* is displayed in Table 9. Here, the new hybrid CS–GSA algorithm performs well for all functions except F_{14} and F_{15}. For the function F_{14}, CS performs better than GSA and CS–GSA algorithms. For the function F_{15}, GSA performs better than CS and CS–GSA algorithms. From the knowledge of the 'Best,' 'Median,' and 'Average' values, one can claim that CS–GSA can be used for optimization of such type of functions.

A comparison of these algorithms for the multimodal benchmark functions F_{15} and F_{17} with *fixed dimension* is shown in Fig. 6. It is observed that the performance of the CS–GSA algorithm is better compared to other algorithms. In this study, the

Table 6 Performance evaluation of GSA, CS, and CS–GSA for the unimodal benchmark functions (displayed in Table 1) with $n = 50$

		GSA	CS	CS–GSA
F_1	Best	17.9649	32.9181	**17.7891**
	Median	**204.0765**	449.6065	210.6723
	Average	235.1073	438.9923	**235.0924**
F_2	Best	**0.0055**	0.2223	0.0063
	Median	0.3428	0.4013	**0.1391**
	Average	0.4440	0.6051	**0.3131**
F_3	Best	**1.1520e+003**	5.4531e+003	1.8343e+003
	Median	2.3741e+003	8.0569e+003	**2.0492e+003**
	Average	**2.5979e+003**	7.9411e+003	2.7163e+003
F_4	Best	6.9320	6.0721	**4.6871**
	Median	9.6256	9.7903	**6.1135**
	Average	9.9463	8.9143	**5.9281**
F_5	Best	486.0379	473.9232	**290.1693**
	Median	1.4569e+003	7.8442e+003	**1.0498e+003**
	Average	1.8367e+003	7.8224e+003	**1.6808e+003**
F_6	Best	**84**	221	231
	Median	447	463	**422**
	Average	492.3667	498.4113	**460.2314**
F_7	Best	0.0704	0.1822	**0.0150**
	Median	0.1502	0.1890	**0.0347**
	Average	0.1659	0.3907	**0.0361**

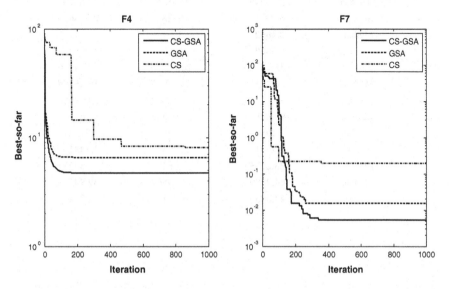

Fig. 4 Performance comparison of GSA, CS, and CS–GSA for the unimodal functions F_4 and F_7 with n = 50

Table 7 Performance evaluation of GSA, CS, and CS–GSA for the multimodal benchmark functions given in Table 2 with $n = 10$

		GSA	CS	CS–GSA
F_8	Best	0.9950	2.2075	**0**
	Median	2.9849	5.0666	**0.9950**
	Average	3.5276	5.5697	**0.8203**
F_9	Best	4.4409e-015	1.1146e-008	**8.8818e-016**
	Median	4.4409e-015	1.5178e-006	**4.4409e-015**
	Average	4.4409e-015	1.4342e-004	**4.1179e-015**
F_{10}	Best	1.9788	1.9305	**1.5922**
	Median	4.3241	4.8115	**2.0312**
	Average	4.4531	5.1299	**2.0954**
F_{11}	Best	4.7116e-032	2.4777	**2.8302e-004**
	Median	4.1906e-005	3.2834	**0.0012**
	Average	0.0585	3.0768	**0.0028**
F_{12}	Best	**1.3498e-032**	1.7255e-005	5.630×10^{-19}
	Median	**1.3498e-032**	1.0319e-004	1.106×10^{-18}
	Average	**1.3498e-032**	1.0445e-002	3.666×10^{-4}

Table 8 Performance evaluation of GSA, CS, and CS–GSA for the multimodal benchmark functions given in Table 2 with $n = 50$

		GSA	CS	CS–GSA
F_8	Best	19.9081	48.6626	**6.0071**
	Median	36.9289	53.0276	**12.2704**
	Average	35.7943	52.6585	**13.6570**
F_9	Best	0.3921	1.8128	**0.2686**
	Median	**0.4660**	2.3657	1.9741
	Average	**0.5669**	2.6760	1.9784
F_{10}	Best	**155.8180**	206.0770	185.4950
	Median	204.1554	281.4023	**199.5789**
	Average	203.1253	286.6942	**201.3960**
F_{11}	Best	**0.7625**	3.7943	0.9637
	Median	2.4986	5.0298	**1.7126**
	Average	2.7001	6.0326	**2.3431**
F_{12}	Best	**23.2165**	26.3823	24.5563
	Median	44.2780	41.5956	**39.8451**
	Average	**44.8839**	48.2117	47.2341

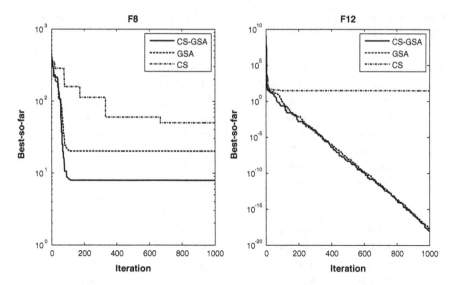

Fig. 5 Performance comparison of CS–GSA, CS, and GSA for the multimodal functions F_8 and F_{12} with n = 50

maximum number of iterations is 1000. From Fig. 6, it is seen that the GSA is the second contestant.

The performances of the proposed algorithm are summarized as follows:

- *For the unimodal test functions* (F_1–F_7): When the best results are concerned, CS–GSA outperforms GSA and CS. When the median and the average values are concerned, GSA outperforms CS–GSA and CS. However, CS–GSA has significant improvements over the CS.
- *For the multimodal test functions* (F_8–F_{12}): For functions F_8 to F_{12} (except F_{11}), the results are dominated by the CS–GSA over GSA and CS.
- *For the multimodal test functions with fixed dimensions* (F_{13}–F_{23}): The result in these functions is not varying so much, but still CS–GSA outperforms the other two algorithms GSA and CS.

The convergence of four benchmark functions, out of 23 such functions, is shown in Figs. 4, 5 and 6 using CS–GSA, GSA, and CS. Here, we consider 1000 iterations. In most of the cases, CS–GSA has shown a better convergence rate than GSA and CS. Reason is that CS has the ability to abandon the worst solutions, while searching for the best solutions quickly. From Figs. 4, 5 and 6, it is observed that CS–GSA provides best fitness function values compared to GSA and CS algorithms, because of the fact that the GSA has the ability to provide the best local search mechanism. Hence, by combining these features of CS and GSA in the hybridized CS–GSA, we get the best results.

Table 9 Performance evaluation of GSA, CS, and CS–GSA for the multimodal benchmark functions given in Table 3 with fixed dimension

		GSA	CS	CS–GSA
F_{13}	Best	−1.7928e+003	−2.5756e+003	**−1.7211e+003**
	Median	**−1.6104e+003**	−2.3242e+003	−1.8677e+003
	Average	**−1.5480e+003**	−2.3359e+003	−1.9888e+003
F_{14}	Best	**0.9980**	**0.9980**	**0.9980**
	Median	3.9711	**0.9985**	0.9984
	Average	5.1573	**1.0009**	1.0351
F_{15}	Best	**6.4199e-004**	9.4845e-004	7.4790e-004
	Median	**0.0038**	0.0016	0.0012
	Average	**0.0042**	0.0018	0.0014
F_{16}	Best	**−1.0316**	−1.0314	**−1.0316**
	Median	**−1.0316**	−1.0305	**−1.0316**
	Average	**−1.0316**	−1.0302	**−1.0316**
F_{17}	Best	**0.3980**	0.3979	**0.3980**
	Median	**0.3995**	0.3997	0.3994
	Average	**0.3999**	0.4007	0.4001
F_{18}	Best	**3.0000**	3.0014	**3.0000**
	Median	**3.0000**	3.0169	**3.0000**
	Average	**3.0000**	3.0235	**3.0000**
F_{19}	Best	**−3.8628**	−3.8623	**−3.8628**
	Median	−3.8596	−3.8593	**−3.8628**
	Average	−3.8593	−3.8590	**−3.8624**
F_{20}	Best	**−3.3220**	−3.2779	**−3.3220**
	Median	**−3.3220**	−3.0968	**−3.3220**
	Average	**−3.3220**	−3.1105	**−3.3220**
F_{21}	Best	**−10.1532**	−8.8466	**−10.1532**
	Median	−2.9417	−4.2211	**−10.1531**
	Average	−5.4547	−4.6898	**−7.3326**
F_{22}	Best	**−10.4029**	−9.3378	**−10.4029**
	Median	**−10.4029**	−5.1803	**−10.4029**
	Average	**−10.4029**	−5.4778	**−10.4029**
F_{23}	Best	**−10.5364**	−8.7847	**−10.5364**
	Median	**−10.5364**	−5.1446	**−10.5364**
	Average	−10.2659	−5.4009	**−10.5364**

4.2 Solving the Constrained Optimization Problems

In this Section, we discuss the use of CS–GSA algorithm for solving the constrained optimization problems. Here, we consider two different constrained optimization issues. These examples are very interesting and may create interest among the readers to explore the idea further.

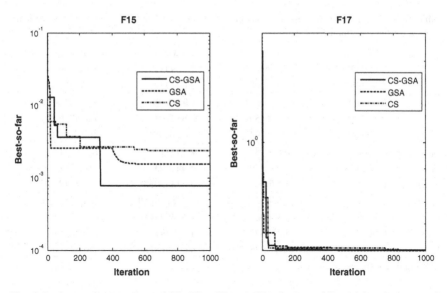

Fig. 6 Performance comparison of CS–GSA, CS, and GSA for the multimodal functions F_{15} and F_{17} with fixed dimension

4.2.1 Minimizing the Function

Here, we present an application of the proposed CS–GSA algorithm for function minimization. This is a constrained optimization problem. We like to minimize the function given in Eq. (21) [39]

$$f(x) = (x_1 - 10)^2 + 5(x_2 - 12)^2 + x_3^4 + 3(x_4 - 11)^2 + 10x_5^6 + 7x_6^2 + x_7^4 - 4x_6 x_7$$
$$- 10x_6 - 8x_7$$

$$(21)$$

subject to the following constraints [39]:

$$g_1(x) = 127 - 2x_1^2 - 3x_2^4 - x_3 - 4x_4^2 - 5x_5 \geq 0$$
$$g_2(x) = 282 - 7x_1 - 3x_2 - 10x_3^2 - x_4 + x_5 \geq 0$$
$$g_3(x) = 196 - 23x_1 - x_2^2 - 6x_6^2 + 8x_7 \geq 0$$
$$g_4(x) = -4x_1^2 - x_2^2 + 3x_1 x_2 - 2x_3^2 - 5x_6 + 11x_7 \geq 0$$
$$-10 \leq x_i \leq 10, \quad i = 1, 2, 3, 4, 5, 6, 7.$$

For the evaluation of constrained problem described in this section, we have taken parameter for various algorithms given in Table 4. From Table 10, it is seen that the proposed CS–GSA scheme is well suited for this constrained optimization problem. The attribute values obtained by CS–GSA (marked as bold face letters)

Table 10 Comparison of the best solutions given by GSA, CS, and CS–GSA with the optimal solution for the constrained problem

EV	GSA	CS	CS–GSA	Optimal
x_1	1.619429	−0.658157	**2.169951**	2.330499
x_2	2.343357	1.408300	**2.041937**	1.951372
x_3	0.965166	−1.128276	**−0.936082**	−0.477541
x_4	2.737996	5.261455	**4.052961**	4.365726
x_5	−0.119048	0.671794	**−0.542999**	−0.624487
x_6	0.406691	0.824787	**0.594817**	1.038131
x_7	0.854488	1.690929	**1.462593**	1.594227
$g_1(x)$	0.934544	1.370761	**3.37333058**	4.464147e-05
$g_2(x)$	2.514614e+02	2.650624e+02	**2.473260e+02**	2.525617e+02
$g_3(x)$	1.591053e+02	2.186001e+02	**1.514995e+02**	1.448781e+02
$g_4(x)$	0.905996	5.433623	**1.650391**	7.632134e-06
$f(x)$	731.535628	761.432542	**688.856815**	680.630111

'EV' stands for the estimated value

Table 11 Comparisons of the statistical results of GSA, CS, and CS–GSA for the constrained problem

Algorithm	Best	Median	Average
GSA	731.535628	736.213432	742.234546
CS	761.432542	776.324511	780.753411
CS–GSA	**688.856815**	**696.062212**	**699.768549**

seem to be very close to the optimal values, presented in the table for a ready reference. From Table 11, it is seen that the 'Best,' the 'Median,' and the 'Average' values obtained by CS–GSA (marked as bold face letters) seem to be better than the other two algorithms.

4.2.2 One-Dimensional (1-D) Recursive Filter Design

Newly, an increasing interest is seen in the application of the EC algorithms for solving the problems of traditional filter design methods. There is a merit of not necessitating virtuous first estimate of the filter coefficients. Here, we present the design of 1-D recursive filters using GSA, CS, and CS–GSA algorithms. Note that the design of the IIR digital filter is well-thought-out here as a constrained optimization tricky. Inbuilt constraint handling is found to guarantee stability. Here, the best results are obtained through the convergence of the proposed CS–GSA method in order to confirm the quality. Interestingly, a faster result is achieved through the convergence of a meta-heuristic hybrid algorithm coined as CS–GSA algorithm. To be precise, the proposed constraint management competence makes the proposed method very eye-catching in the design of 1-D IIR digital filters. Results are compared to GSA and CS techniques.

Fig. 7 One-dimensional
(1-D) recursive filter
optimization using EC
methods

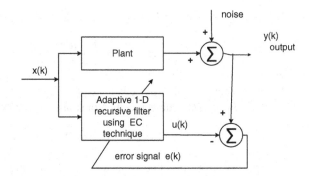

The block diagram showing IIR filter optimization is displayed in Fig. 7. Evolutionary computational technique is very useful for synthesis of digital IIR filters. The filter coefficients are chosen very accurately using evolutionary algorithms. To reduce the error, the filter coefficients are optimized. Then it is used for different applications. For this reason, EAs proved to be beneficial for the design of 1-D digital IIR filters. In this section, we discuss the design of 1-D IIR digital filter design by three fairly new EC methods.

The GSA was used for the design of 1-D recursive filters [27] and IIR filter design [40]. However, they are silent regarding the constraint treatment and diminishing tool, which is required to guarantee stability of the 1-D recursive filters. The design was not considered as the constrained optimization problem. Recently, the authors in [40] considered this as a constrained optimization work and then resolved this in designing the 1-D digital IIR filters. In this section, three optimization algorithms GSA, CS, and CS–GSA are deployed to design 1-D IIR filters.

Note that the system is called as recursive provided the preset output depends on the present input, the past input, and the past output of the system. So a 1-D system can be represented as

$$y(m) = \{x(m), x(m-1), \ldots, x(m-M), y(m-1), \ldots, y(m-M)\}. \qquad (22)$$

For the evaluation of our proposed algorithm, let us consider a 1-D recursive filter transfer function. The transfer function can be represented as

$$H(z) = H_0 \frac{\sum_{i=0}^{S} a_i z^i}{\sum_{i=0}^{S} b_i z^i}, \ a_0 = 1, \ \text{and} \ b_0 = 1. \qquad (23)$$

The stability conditions are described as

$$H(z) = \frac{A(z)}{B(z)}, \ \text{with} \ B(z) \neq 0, \ \text{and} \ |z| \geq 1. \qquad (24)$$

For $S = 2$, one can write

$$H(z) = H_0 \frac{1 + a_1 z + a_2 z^2}{1 + b_1 z + b_2 z^2}. \tag{25}$$

In this section, the objective is to optimize the vector (X), where $X = [a_1, a_2, b_1, b_2, H_0]$, subject to the constraints given below:

$$(1 + b_i) > 0, \text{ and } (1 - b_i) > 0. \tag{26}$$

Hence, this problem is known as a constrained optimization problem. Here, an attempt is made to solve this problem using three different EC algorithms GSA, CS, and CS–GSA. To evaluate the proposed 1-D recursive function, let us consider M_d as the desired magnitude response of the digital IIR filter expressed as a function of frequency $\omega \in [0, \pi]$:

$$M_d(\omega) = \begin{bmatrix} 1 & \text{if } \omega \leq 0.05\pi \\ \frac{1}{\sqrt{2}} & \text{if } 0.05\pi < \omega \leq 0.08\pi \\ \frac{1}{2\sqrt{2}} & \text{if } 0.08\pi < \varpi \leq 0.1\pi \\ 0 & \text{otherwise} \end{bmatrix}. \tag{27}$$

Here, our objective is to find $H(z)$, such that it will closely approximate the desired magnitude response. The approximation can be attained by a fitness function J such as

$$J = \sum_{n=0}^{P} [|M(\omega)| - M_d(\omega)]^2, \tag{28}$$

where $M(\omega)$ is the Fourier transform of the $H(z)$, i.e., $M(\omega) = H(Z)|_{z=e^{-j\omega}}$ and $\omega = (\pi/P)n$.

The results (the filter parameters) are shown in Table 12.

Table 12 Comparisons of the best solution given by GSA, CS, and CS–GSA for the 1-D recursive filter design. "EP" refers to the estimated parameters

EP	GSA	CS	CS–GSA
a_1	0.732505	0.746526	**0.883583**
a_2	0.580788	1.321402	**0.741019**
b_1	−0.852237	−0.928369	**−0.982292**
b_2	−0.274744	−0.197190	**−0.134745**
H_0	0.061419	0.041052	**0.050245**
$1 + b_1$	0.147762	0.071630	**0.017707**
$1 - b_1$	1.852237	1.928369	**1.982292**
$1 + b_2$	0.725255	0.802809	**0.865254**
$1 - b_2$	1.274744	1.197190	**1.134745**
J	0.596505	0.617075	**0.578251**

Table 13 Comparisons of statistical results of GSA, CS, and CS–GSA for 1-D recursive filter design

Algorithm	Best	Median	Average
GSA	0.596505	0.796212	0.857275
CS	0.617075	0.776405	0.785817
CS–GSA	**0.578251**	**0.669519**	**0.673789**

From Table 12, it is observed that the filter parameter obtained by CS–GSA algorithm is better than CS and GSA algorithms. They are optimized using CS–GSA algorithm to reduce the error. A comparison of the statistical results of GSA, CS, and CS–GSA for 1-D recursive filter design is presented in Table 13.

From Table 13, it is seen that the 'Best,' 'Median,' and 'Average' values of the filter parameters for 50 independent runs obtained by CS–GSA algorithm are better than CS and GSA algorithms. These statistical parameters are obtained using CS–GSA algorithm for a better performance.

Figure 8 displays the frequency responses of the 1-D recursive filter designed using CS–GSA, GSA, and CS as the three competitive methods. The above

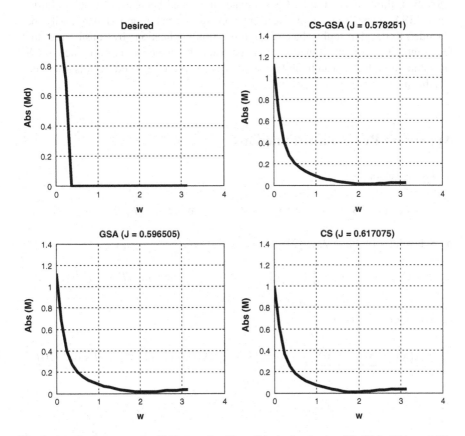

Fig. 8 Amplitude responses of 1-D recursive filter of desired and using CS–GSA, GSA, and CS

amplitude responses are plotted exhausting the solutions achieved by minimizing J for 50,000 function evaluations, and the parameter is given in Table 4. Note that in the cited numerical example, the solution attained by utilizing CS–GSA method offers us a better approximation of the proposed transfer function than latter methods. It seems closer to that of the desired frequency response. The GSA method is the second contestant in this work. The performance of the CS–GSA is quite better than the latter methods. From Fig. 8, we see that the frequency response attained by utilizing CS–GSA method is better than the frequency responses exhibited by other two algorithms CS and GSA.

5 Conclusions

In this chapter, the proposed hybrid algorithm CS–GSA outperforms both CS and GSA algorithms in terms of obtaining best solutions. In fact, the GSA is used to explore the local search ability, whereas CS is used to speed up the convergence rate. The convergence speed of the proposed hybrid algorithm is faster than CS and GSA algorithms. Interestingly, CS simulates the social behavior of Cuckoo birds, while GSA inspires by a physical phenomenon. This proposal can easily be extended to develop multi-objective optimization applications by considering different inbuilt constraints. Finally, it may be noted that the better convergence of CS algorithm and local search ability of the GSA produce good results that are beneficial.

Appendix: Benchmark Functions

a. Sphere Model

$$F_1(X) = \sum_{i=1}^{n} x_i^2, \ -95 \leq x_i \leq 95, \ \text{and} \ \min(F_1) = F_1(0,\ldots,0) = 0$$

b. Schwefel's Problem 2.22 [21, 42]

$$F_2(X) = \sum_{i=1}^{n} |x_i| + \prod_{i=1}^{n} |x_i|, \ -12 \leq x_i \leq 12, \ \text{and} \ \min(F_2) = F_2(0,\ldots,0)$$
$$= 0$$

c. Schwefel's Problem 1.2

$$F_3(X) = \sum_{i=1}^{n} \left(\sum_{j=1}^{i} x_j \right)^2, \ -90 \leq x_i \leq 90, \ \text{and} \ \min(F_3) = F_3(0,\ldots,0)$$
$$= 0$$

d. Schwefel's Problem 2.21

$$F_4(X) = \max_i\{|x_i|, 1 \le i \le n\}, \quad -90 \le x_i \le 90, \quad \text{and} \quad \min(F_4) = F_4(0,\ldots,0)$$
$$= 0$$

e. Generalized Rosenbrock's Function

$$F_5(X) = \sum_{i=1}^{n-1} \left[100\left(x_{i+1} - x_i^2\right)^2 + (x_i - 1)^2\right], \quad -30 \le x_i \le 30$$
$$\min(F_5) = F_5(0,\ldots,0) = 0.$$

f. Step Function

$$F_6(X) = \sum_{i=1}^{n} \left(\lfloor x_i + 0.5 \rfloor\right)^2, \quad -100 \le x_i \le 100, \quad \text{and} \quad \min(F_6) = F_6(0,\ldots,0)$$
$$= 0.$$

g. Quartic Function i.e. Noise

$$F_7(X) = \sum_{i=1}^{n} i x_i^4 + \text{random}[0,1), \quad -1.28 \le x_i \le 1.28$$
$$\min(F_7) = F_7(0,\ldots,0) = 0$$

h. Generalized Rastrigin's Function

$$F_8(X) = \sum_{i=1}^{n} \left[x_i^2 - 10\cos(2\pi x_i) + 10\right], \quad -5.12 \le x_i \le 5.12$$
$$\min(F_8) = F_8(0,\ldots,0) = 0.$$

i. Ackley's Function

$$F_9(X) = -20\exp\left(-0.2\sqrt{\frac{1}{n}\sum_{i=1}^{n} x_i^2}\right) - \exp\left(\frac{1}{n}\sum_{i=1}^{n}\cos(2\pi x_i)\right) + 20 + e$$
$$-32 \le x_i \le 32, \quad \text{and} \quad \min(F_9) = F(0,\ldots,0) = 0.$$

j. Generalized Griewank Function

$$F_{10}(X) = \frac{1}{4000}\sum_{i=1}^{n} x_i^2 - \prod_{i=1}^{n}\cos\left(\frac{x_i}{\sqrt{i}}\right) + 1, \quad -600 \le x_i \le 600$$
$$\min(F_{10}) = F_{10}(0,\ldots,0) = 0.$$

k. Generalized Penalized Function 1

$$F_{11}(X) = \frac{\pi}{n}\left\{10\sin(\pi y_i) + \sum_{i=1}^{n} (y_i - 1)^2\left[1 + 10\sin^2(\pi y_{i+1})\right] + (y_n - 1)^2\right\}$$
$$+ \sum_{i=1}^{n} u(x_i, 10, 100, 4),$$

where

$$u(x_i, a, k, m) = \begin{cases} k(x_i - a)^m, & x_i > a \\ 0, & -a < x_i < a, \\ k(-x_i - a)^m, & x_i < -a \end{cases} \quad \text{and} \quad y_i = 1 + \frac{1}{4}(x_i + 1)$$

$-50 \le x_i \le 50, \quad \text{and} \quad \min(F_{11}) = F_{11}(1, \ldots, 1) = 0.$

l. Generalized Penalized Function 2

$$F_{12}(X) = 0.1 \Big\{ \sin^2(3\pi x_1) + \sum_{i=1}^{n} (x_i - 1)^2 \big[1 + \sin^2(3\pi x_i + 1) \big] + (x_n - 1)^2 \cdot \\ \big[1 + \sin^2(2\pi x_n) \big] \Big\} + \sum_{i=1}^{n} u(x_i, 5, 100, 4),$$

where

$$u(x_i, a, k, m) = \begin{cases} k(x_i - a)^m, & x_i > a \\ 0, & -a < x_i < a \\ k(-x_i - a)^m, & x_i < -a \end{cases}, \quad \text{and} \quad y_i = 1 + \frac{1}{4}(x_i + 1)$$

$-50 \le x_i \le 50, \quad \text{and} \quad \min(F_{12}) = F_{12}(1, \ldots, 1) = 0.$

m. Generalized Schwefel's Problem 2.26

$$F_{13}(X) = \sum_{i=1}^{30} -x_i \sin\left(\sqrt{|x_i|}\right), \quad -500 \le x_i \le 500$$

$$\min(F_{13}) = F_{13}(420.9687, \ldots, 420.9687) = -12569.5.$$

n. Shekel's Foxholes Function

$$F_{14}(X) = \left(\frac{1}{500} + \sum_{j=1}^{25} \frac{1}{j + \sum_{i=1}^{2} (x_i - a_{ij})^6} \right)^{-1}, \quad -65.536 \le x_i \le 65.536$$

$$\min(F_{14}) = F_{14}(-32, -32) \approx 1,$$

where

$$a_{ij} = \begin{pmatrix} -32 & -16 & 0 & 16 & 32 & -32 & \cdots & 0 & 16 & 32 \\ -32 & -32 & -32 & -32 & -32 & -16 & \cdots & 32 & 32 & 32 \end{pmatrix}.$$

o. Kowalik's Function

$$F_{15}(X) = \sum_{i=1}^{11} \left[a_i - \frac{x_1(b_i^2 + b_i x_2)}{b_i^2 + b_i x_3 + x_4} \right]^2, \quad -5 \le x_i \le 5$$

$\min(F_{15}) \approx F_{15}(0.1928, 0.1908, 0.1231, 0.1358) \approx 0.0003075$. The coefficients are displayed in Table 14 [21, 41, 42].

p. Six-Hump Camel-Back Function

$$F_{16}(X) = 4x_1^2 - 2.1x_1^4 + \frac{1}{3}x_1^6 + x_1x_2 - 4x_2^2 + 4x_2^4, \quad -5 \leq x_i \leq 5$$

$$X_{min} = (0.08983, -0.7126), (-0.08983, 0.7126)$$
$$\min(F_{16}) = -1.0316285.$$

q. Branin Function

$$F_{17}(X) = \left(x_2 - \frac{5.1}{4\pi^2}x_1^2 + \frac{5}{\pi}x_1 - 6\right) + 10\left(1 - \frac{1}{8\pi}\right)\cos x_1 + 10$$
$$-5 \leq x_1 \leq 10, \ 0 \leq x_2 \leq 15$$

$$X_{min} = (-3.142, 12.275), (3.142, 2.275), (9.425, 2.425)$$
$$\min(F_{17}) = 0.398.$$

r. Goldstein-Price Function

$$F_{18}(X) = \left[1 + (x_1 + x_2 + 1)^2(19 - 14x_1 + 3x_1^2 - 14x_2 + 6x_1x_2)\right] \times \left[(2x_1 - 3x_2)^2\right.$$
$$\times \left. \left(18 - 32x_1 + 12x_1^2 + 48x_2 - 36x_1x_2 + 27x_2^2\right) + 30\right]$$
$$-2 \leq x_i \leq 2, \ \text{and} \ \min(F_{18}) = F_{18}(0, -1) = 3.$$

Table 14 Kowalik's function F_{15}	i	a_i	b_i^{-1}
	1	0.1957	0.25
	2	0.1947	0.5
	3	0.1735	1
	4	0.1600	2
	5	0.0844	4
	6	0.0627	6
	7	0.0456	8
	8	0.0342	10
	9	0.0323	12
	10	0.0235	14
	11	0.0246	16

s. Hartman's Family

$$F(X) = - \sum_{i=1}^{4} c_i \exp \left(- \sum_{j=1}^{n} a_{ij} \left(x_j - p_{ij} \right)^2 \right), \, 0 \le x_j \le 1, \, n = 3, 6$$

for $F_{19}(X)$ and $F_{20}(X)$, respectively. X_{min} of $F_{19} = (0.114, 0.556, 0.852)$, and $\min(F_{19}) = -3.86$. X_{min} of $F_{20} = (0.201, 0.150, 0.477, 0.275, 0.311, 0.657)$, and $\min(F_{20}) = -3.32$.

The coefficients are shown in Tables 15 and 16, respectively.

t. Shekel's

Family$F(X) = - \sum_{i=1}^{m} \left[(X - a_i)(X - a_i)^T + c_i \right]^{-1}$, $m = 5, 7$, and 10, for F_{21}, F_{22}, and F_{23} $0 \le x_j \le 10$, $x_{local\_optima} \approx a_i$, and $F(x_{local\_optima}) \approx 1/c_i$ for $1 \le i \le m$.

These functions have five, seven, and ten local minima for F_{21}, F_{22}, and F_{23}, respectively. The coefficients are shown in Table 17.

Table 15 Hartman function F_{19}

i	a_{i1}	a_{i2}	a_{i3}	c_i	p_{i1}	p_{i2}	p_{i3}
1	3	10	30	1	0.3689	0.1170	0.2673
2	0.1	10	35	2	0.4699	0.4387	0.7470
3	3	10	30	3	0.1091	0.8732	0.5547
4	0.1	10	35	4	0.038150	0.5743	0.8828

Table 16 Hartman function F_{20}

i	a_{i1}	a_{i2}	a_{i3}	a_{i4}	a_{i5}	a_{i6}	c_i
1	10	3	17	3.5	1.7	8	1
2	0.05	10	17	0.1	8	14	1.2
3	3	3.5	1.7	10	17	8	3
4	17	8	0.05	10	0.1	14	3.2
i	p_{i1}	p_{i2}	p_{i3}	p_{i4}	p_{i5}	p_{i6}	
1	0.1312	0.1696	0.5569	0.0124	0.8283	0.5886	
2	0.2329	0.4135	0.8307	0.3736	0.1004	0.9991	
3	0.2348	0.1415	0.3522	0.2883	0.3047	0.6650	
4	0.4047	0.8828	0.8732	0.5743	0.1091	0.0381	

Table 17 Shekel function F_{21}, F_{22}, and F_{23}

i	a_{i1}	a_{i2}	a_{i3}	a_{i4}	c_i
1	4	4	4	4	0.1
2	1	1	1	1	0.2
3	8	8	8	8	0.2
4	6	6	6	6	0.4
5	3	7	3	7	0.4
6	2	9	2	9	0.6
7	5	5	3	3	0.3
8	8	1	8	1	0.7
9	6	2	6	2	0.5
10	7	3.6	7	3.6	0.5

References

1. Du W, Li B (2008) Multi-strategy ensemble particle swarm optimization for dynamic optimization. Inf Sci 178:3096–3109
2. Panda R, Naik MK (2012) A crossover bacterial foraging optimization algorithm. Appl Comput Intell Soft Comput, 1–7. Hindawi Publication
3. Mastorakis NE, Gonos IF, Swamy MNS (2003) Design of two-dimensional recursive filters using genetic algorithm. IEEE Trans Circuits Syst-I Fundam Theory Appl 50:634–639
4. Panda R, Naik MK (2013) Design of two-dimensional recursive filters using bacterial foraging optimization. In: Proceedings of the 2013 IEEE Symposium on Swarm Intelligence (SIS), pp 188–193
5. Cordon O, Damas S, Santamari J (2006) A fast and accurate approach for 3D image registration using the scatter search evolutionary algorithm. Pattern Recogn Lett 26:1191–1200
6. Panda R, Agrawal S, Bhuyan S (2013) Edge magnitude based multilevel thresholding using cuckoo search technique. Expert Syst Appl 40:7617–7628
7. Panda R, Naik MK, Panigrahi BK (2011) Face recognition using bacterial foraging strategy. Swarm Evol Comput 1:138–146
8. Liu C, Wechsler H (2000) Evolutionary pursuit and its application to face recognition. IEEE Trans Pattern Anal Mach Intell 22:570–582
9. Zheng WS, Lai JH, Yuen PC (2005) GA-Fisher: a new LDA-based face recognition algorithm with selection of principal components. IEEE Trans Syst Man Cybern Part B 35:1065–1078
10. Mitchell M (1998) An introduction to genetic algorithms. MIT Press, Cambridge
11. Dorigo M, Maniezzo V, Colorni A (1996) The ant system: optimization by a colony of cooperating agents. IEEE Trans Syst Man Cybern Part B 26:29–41
12. Kennedy J, Eberhart RC (1995) Particle swarm optimization. In: Proceedings of the IEEE international conference on neural networks vol 4, pp 1942–1948
13. Gazi V, Passino KM (2004) Stability analysis of social foraging swarms. IEEE Trans Syst Man Cybern Part B 34:539–557
14. Yang XS, Deb S (2009) Cuckoo search via Lévy flights. In: Proceedings of the world congress on nature and biologically inspired computing, (NaBIC 2009), pp 210–214
15. Yang XS, Deb S (2013) Cuckoo search: recent advances and applications. Neural Comput Appl 24(1):169–174
16. Cuckoo Search and Firefly Algorithm. http://link.springer.com/book/10.1007%2F978-3-319-02141-6

17. Pinar C, Erkan B (2011) A conceptual comparison of the Cuckoo-search, particle swarm optimization, differential evolution and artificial bee colony algorithms. Artif Intell Rev, Springer. doi:10.1007/s10462-011-9276-0
18. Chakraverty S, Kumar A (2011) Design optimization for reliable embedded system using cuckoo search. In: Proceedings of the international conference on electronics, computer technology, pp 164–268
19. Barthelemy P, Bertolotti J, Wiersma DS (2008) A Lévy flight for light. Nature 453:495–498
20. Rashedi E, Nezamabadi S, Saryazdi S (2009) GSA: a gravitational search algorithm. Inf Sci 179:2232–2248
21. Yao X, Liu Y, Lin G (1999) Evolutionary programming made faster. IEEE Trans Evol Comput 3:82–102
22. Chetty S, Adewumi AO (2014) Comparison study of swarm intelligence techniques for annual crop planning problem. IEEE Trans Evol Comput 18:258–268
23. Chen J-F, Do QH (2014) Training neural networks to predict student academic performance: a comparison of cuckoo search and gravitational search algorithms. Int J Comput Intell Appl 13 (1):1450005
24. Swain KB, Solanki SS, Mahakula AK (2014) Bio inspired cuckoo search algorithm based neural network and its application to noise cancellation. In: Proceedings of the international conference on signal processing and integrated networks (SPIN), pp 632–635
25. Khodier M (2013) Optimisation of antenna arrays using the cuckoo search algorithm. IET Microwaves Antennas Propag 7(6):458–464
26. Zhao P, Li H (2012) Opposition based Cuckoo search algorithm for optimization problems. In: Proceedings of the 2012 fifth international symposium on computational intelligence and design, pp 344–347
27. Saha SK, Kar R, Mandal D, Ghosal SP (2013) Gravitational search algorithm: application to the optimal IIR filter design. Journal of King South University, 1–13
28. Rashedi E, Nezamabadi-pour H, Saryazdi S (2011) Filter modeling using gravitational search algorithm. Eng Appl Artif Intell 24:117–122
29. Rashedi E, Nezamabadi-pour H, Saryazdi S (2011) Disruption: A new operator in gravitational search algorithm. Sci Iranica D 18:539–548
30. Doraghinejad M, Nezamabadi-pour H, Sadeghian AH, Maghfoori M (2012) A hybrid algorithm based on gravitational search algorithm for unimodal optimization. In: Proceedings of the 2nd international conference on computer and knowledge engineering (ICCKE), pp 129–132
31. Yazdani S, Nezamabadi-pour H, Kamyab S (2013) A gravitational search algorithm for multimodal optimization. Swarm Evol Comput 1–14
32. Rashedi E, Nezamabadi-pour H, Saryazdi S (2009) BGSA: binary gravitational search algorithm. Nat Comput 9(3):727–745
33. Mirjalili S, Hashim SZM (2010) A new hybrid PSOGSA algorithm for function optimization. In: 2010 international conference on computer and information application, pp 374–377
34. Jiang S, Ji Z, Shen Y (2014) A novel hybrid particle swarm optimization and gravitational search algorithm for solving economic emission load dispatch problems with various practical constraints. Electr Power Energy Syst 55:628–644
35. Ghodrati A, Lotfi S (2012) A hybrid CS/PSO algorithm for global optimization. Lect Notes Comput Sci 7198:89–98
36. Guo Z (2012) A hybrid optimization algorithm based on artificial bee colony and gravitational search algorithm. Int J Digit Content Technol Appl 6(17):620–626
37. Sun G, Zhang A (2013) A hybrid genetic algorithm and gravitational using multilevel thresholding. Pattern Recognit Image Anal 7887:707–714
38. Yin M, Hu Y, Yang F, Li X, Gu W (2011) A novel hybrid K-harmonic means and gravitational search algorithm approach for clustering. Expert Syst Appl 38:9319–9324
39. Liu H, Cai Z, Wang Y (2010) Hybridizing particle swarm optimization with differential evolution for constrained numerical and engineering optimization. Appl Soft Comput 10:629–640

40. Sarangi SK, Panda R, Dash M (2014) Design of 1-D and 2-D recursive filters using crossover bacterial foraging and cuckoo search techniques. Eng Appl Artif Intell 34:109–121
41. He J (2008) An experimental study on the self-adaption mechanism used by evolutionary programing. Prog Nat Sci 10:167–175
42. Ji M (2004) A single point mutation evolutionary programing. Inf Process Lett 90:293–299

Study of Economic Load Dispatch by Various Hybrid Optimization Techniques

Dipankar Santra, Arindam Mondal and Anirban Mukherjee

Abstract The economic load dispatch (ELD) is one of the most complex optimization problems of electrical power system. Classically, it is to identify the optimal combination of generation level of all power generating units in order to minimize the total fuel cost while satisfying the loads and losses in power transmission system. In view of the sharply increasing nature of cost of fossil fuel, energy management has gained lot of significance nowadays. Herein lies the relevance of continued research on improving the solution of ELD problem. A lot of research work have been carried out on this problem using several optimization techniques including classical, linear, quadratic, and nonlinear programming methods. The objective function of the ELD problem being of highly nonlinear and non-convex nature, the classical optimization methods cannot guarantee convergence to the global optimal solution. Some soft computing techniques like *Artificial Bee Colony (ABC)*, *Particle Swarm Optimization (PSO)*, *Clonal Selection Algorithm (CSA)*, *Ant Colony Optimization (ACO)*, *Simulated Annealing (SA)*, *Genetic Algorithm (GA)*, etc. are now being applied to find even better solution to the ELD problem. An interesting trend in this area is application of hybrid approaches like *GA-PSO, ABC-PSO, CSA-SA,* etc. and the results are found to be highly competitive. In this book chapter, we focus on the hybrid soft computing approaches in solving ELD problem and present a concise and updated technical review of systems and approaches proposed by different research groups. To depict the differences in technique of the hybrid approaches over the basic soft computing methods, the individual methods are introduced first. While the basic working principle and case studies of each hybrid approach are described briefly, the achievements of the approaches are discussed separately. Finally, the challenges in the present problem and some of the most promising approaches are highlighted and the possible future direction of research is hinted.

D. Santra · A. Mondal · A. Mukherjee (✉)
RCC Institute of Information Technology, Kolkata, India
e-mail: anirbanm.rcciit@gmail.com

© Springer India 2016
S. Bhattacharyya et al. (eds.), *Hybrid Soft Computing Approaches*,
Studies in Computational Intelligence 611,
DOI 10.1007/978-81-322-2544-7_2

Keywords Economic load dispatch · Artificial bee colony · Particle swarm optimization · Clonal selection algorithm · Ant colony optimization · Simulated annealing · Genetic algorithm · Firefly algorithm · Gravitational search algorithm

1 Introduction

The effective and economic operation and management of electrical power generating system has always been an important concern in the electrical power industry. The growing size of power grids, huge demand and crisis of energy across the world, continuous rise in price of fossil fuel necessitate the optimal combination of generation level of power generating units. The classic problem of Economic Load Dispatch (ELD) is to minimize the total cost of power generation (including fuel consumption and operational cost) from differently located power plants while satisfying the loads and losses in the power transmission system. The objective is to distribute the total load demand and total loss among the generating plants while simultaneously minimizing generation costs and satisfying the operational constraints.

The ELD problem concerns two different problems—one is the pre-dispatch problem requiring optimal selection of the generating units out of the available ones to meet the demand and produce an expected margin of operating reserve over specified time-slots. The other problem is the online dispatch in such an economic manner that the total cost of supplying the dynamic requirements of the system is minimized. Since the power generation cost in fossil fuel fired plants is very high, an optimum load dispatch saves a considerable amount of fuel and expenditure therein.

The ELD problem can be conceived as an optimization problem of minimizing the total fuel cost of all generating units while satisfying the demand and losses.

Consider a system with n power generating units. The objective function is to minimize the total fuel cost (F) given by the following expression:

$$F = \sum_{i=1}^{n} C_i(P_i) = \sum_{i=1}^{n} a_i + b_i P_i + c_i P_i^2 \tag{1}$$

Here n is the total number of generation units, a_i, b_i, c_i are the cost coefficients of ith power generation unit, P_i is the output of ith power generation unit, and C_i is the cost function of ith generating unit. $i = 1, 2 \ldots n$. The operational constraints are given by:

- *Power Balance Equation* In ELD of power, the total power generated should exactly match with the load demand and losses which is represented by the following equation. It is a kind of equality constraint.

$$\sum_{i=1}^{n} P_i = P_D + P_L \tag{2}$$

Here P_i is the power output from ith generating unit, n is the number of generating units, P_L is the Transmission Loss, and P_D is the Load Demand. P_L is calculated using B-coefficient as:

$$P_L = \sum_{i=1}^{n}\sum_{j=1}^{n} P_i B_{ij} P_j \tag{3}$$

- *Generator Constraints* The output power of each generating unit is restricted by its upper (P^{max}) and lower (P^{min}) limits of actual power generation and is given by:

$$P_i^{min} \le P_i \le P_i^{max} \tag{4}$$

- *Ramp Rate Limits* In practice, the power output of a generator is not instantaneously adjustable. The operating range of all such units is restricted by their ramp rate limits during each dispatch period. So, the dispatch output of a generator should be restricted between the upper (UR_i) and down (DR_i) ramp rate constraints as expressed in Eq. (5).

$$\max\left(P_i^{min}, UR_i - P_i\right) \le P_i \le \min(P_i^{max}, P_i^o - DR_i) \tag{5}$$

Here P_i is the current power output of ith unit and P_i^o is the power generated by the ith unit at previous hour.
- *Prohibited Operating Zone* In prohibited operating zone, if any, a unit has discontinuous cost characteristics. So operation of the unit is not desirable in prohibited zones. The following constraints may be considered for such cases:

$$\begin{aligned}
P_i^{min} &\le P_i \le P_{i,1}^L (i = 1, 2, \ldots n) \\
P_{i,j-1}^U &\le P_i \le P_{i,1}^L (j = 2, 3, \ldots n_z)\,(i = 1, 2, \ldots n) \\
P_{i,nz}^U &\le P_i \le P_i^{max} (i = 1, 2, \ldots n)
\end{aligned} \tag{6}$$

Here, $P_{i,j-1}^U$ and $P_{i,1}^L$ are the upper and lower boundaries of jth prohibited zone of ith unit and n_z is the number of prohibited zones of ith unit.

Some of the other inequality constraints are *Reserve Contribution* [1, 2] and *Transmission Line Limits* [1, 2].

Thus, characteristically the ELD problem is a nonlinear and complex problem having heavy equality and inequality constraints like Ramp Rate Limits, Prohibited Operating Zone, etc. Therein lies the difficulty of the problem of finding the optimal solution. Classical methods for optimization such as Lambda Iteration [3], Newton's method [4], and Lagrange Multiplier method [5] can solve ELD problem assuming that the incremental cost curves corresponding to the generating units are monotonically increasing linear piecewise functions. However, in reality, there is distinct nonconvexity in the fuel cost function of the generating units. Classical calculus-based methods cannot address this type of problem adequately and lead to suboptimal solutions. Dynamic programming [6] can be used to solve ELD problem with cost curves discontinuous and nonlinear in nature, but it is computationally extensive and suffers from finding only local optima owing to premature convergence.

To overcome the problems of—restriction in shape of cost curves, unidirectional search, premature and slow convergence, sub-optimal solution and significant computational overhead, non-conventional stochastic and intelligent techniques are now used to resolve the complexities of ELD problem reasonably well. These techniques include *Genetic Algorithm (GA)* [7–16], *Particle Swarm Optimization (PSO)* [17–54], *Evolutionary Programming (EP)* [55], *Differential Evolution (DE)* [56–61], *Artificial Bee Colony (ABC)* optimization [62–64], *Ant Colony Optimization (ACO)* [65–69], *Artificial Immune System (AIS)-Clonal Selection Algorithm (CSA)* [70], *Simulated Annealing (SA)* [71], *Gravitational Search (GS)* algorithm [72], *Cuckoo Search Algorithm (CSA)* [73] besides others. Unlike classical optimization methods, the intelligent stochastic techniques work on a population of possible solutions in the search space and create an advantage of getting multiple suitable solutions in a single run; they are easy to implement, robust, and computationally less expensive. When applied to complex optimization problems like the ELD problem, these techniques have high probability of finding the optima quickly through competition and collaboration among the possible solutions. Basic working principles of different intelligent techniques which are used to find optimal and near optimal solution for ELD problem are briefly presented in the following sections.

1.1 Genetic Algorithm (GA)

The GA is basically an evolutionary algorithm, some of the other of its kind being evolution strategies, genetic programming, and EP. An evolutionary algorithm sustains a population of candidate solutions to an optimization problem. The population changes through repeated application of stochastic operators. Using

Tomassini's [74] terms, GA consider the ELD problem as the environment where the living individuals are the feasible solutions. Finding globally acceptable solutions to the problem is analogous to adjusting to the surrounding by a natural habitat. Just as a new generation of a population is promoted by elimination of useless traits and by developing useful features, a new and better solution is found by iterative fine-tuning of fitness function. Generally, GA enters a loop with an initial population created with a set of individuals generated randomly. In each iteration (called "generation"), a fresh population is created applying a number of stochastic operators to the earlier population (causing effect equivalent to genetic crossover and mutation).

1.2 Particle Swarm Optimization (PSO)

Particle Swarm Optimization is a population-based stochastic optimization, inspired by social behavior of bird flocking or fish schooling. In PSO, each single solution is a "bird" (particle) in the search space of food (the best solution). All particles have fitness values evaluated by the fitness function (the cost function for ELD problem), and have velocities that direct the "flying" (or evaluation) of the particles. Initialized with a set of random particles (solutions), PSO searches for the optimal solution by updating generations in each iteration. All particles are updated by two "best" values—one called the *pbest* or personal best implying the best solution or fitness a particle has achieved so far while the other one is the *gbest* or global best implying the best value obtained by any particle in the population. The best value obtained in the topological neighbour or on part of a population is a local best and is called *pbest*. Upon finding the two best values, the velocity and position of the particle are updated using following equations:

$$V_i^{t+1} = wV_i^t + c_1\, rand_1()\left(pbest_i - X_i^t\right) + c_2\, rand_2()\left(gbest_i - X_i^t\right)$$
$$X_i^{t+1} = X_i^{(t)} + V_i^{(t+1)} \tag{7}$$

Here, i is the index of each particle, t is the current iteration number, $rand_1()$ and $rand_2()$ are random numbers between 0 and 1. $pbest_i$ is the best previous experience of the ith particle while $gbest_i$ is the best particle among the entire population. Constants c_1 and c_2 are the weightage factors of the stochastic acceleration terms, which pull each particle toward the $pbest_i$ and $gbest_i$, w being the inertia weight controlling the exploration properties of the algorithm. If $c_1 > c_2$, the particle tends to reach $pbest_i$, the best position identified by the particle, rather than converge to $gbest_i$ found by the population and vice versa.

PSO is very effective in finding global best of ELD problem and that is the reason why most of the hybrid techniques of solving ELD problem are found to hybridize PSO with other intelligent optimization techniques. PSO shares many similarities with GA though unlike GA, PSO has no evolution operators such as crossover and mutation. Compared to GA, the advantages of PSO are that it is easy to implement and there are few parameters to adjust. Here the particles update themselves with the internal velocity and position parameters only.

1.3 Ant Colony Optimization (ACO)

Ant Colony Optimization is another powerful swarm-based optimization technique often used to solve the ELD problem. The algorithm follows ant's movement in search of food. The ant that reaches the food in shorter path returns to the nest earlier. Other ants in the nest have high probability of following the shorter route because pheromone deposited in shorter path is more than that deposited by ants traversing longer paths.

In ACO algorithm, a number of search procedure, analogous to "ants", work parallel to find the best solutions of the ELD problem. An ant develops a solution and shares its information ("pheromone") with other ants [75]. Though each ant can build a solution, better solutions are found through this information exchange [76] within a structural neighbourhood. While developing a solution, each ant uses two information sources—one is the personal information (ant's local memory storing previously visited nodes) and the other one is an ant-decision table defined by functional combination of the publicly available (pheromone trail) and problem-specific heuristic information [77]. The publicly available information is the set of ant's decisions from the beginning of the search process. The concept of pheromone evaporation is used to prevent stagnation owing to large accumulations.

1.4 Artificial Bee Colony (ABC)

Like ACO, ABC algorithm [78] is a swarm-based metaheuristic algorithm simulating the behavior of honeybees. When applied to ELD problem, the solution produced by the algorithm is represented by the location of source of nectar while the amount of nectar represents the quality (fitness) of the solution. Employee bees fly around in search of source of nectar (representing trial in a search space) and select their preferred source of nectar based on their experience. Once search is completed, they share their findings (source) with the onlooker bees waiting in the hive. The onlooker bees then make a probabilistic selection of new source of nectar based on the information received from the employee bees. Only if the amount of nectar of the new source is higher than that of the old one, the onlookers choose the new position. If the quality of solution is not improved by a predetermined number of trials (finding sources with higher nectar), then the scout bees fly to choose new source randomly abandoning the old source. If the abandoned source is $x_{pq}, q \in$

$\{1, 2 \ldots, D\}$, where p *is* a solution and q is a randomly chosen index, then a new nectar source chosen by a bee is given by:

$$x_{pq} = x_{q\,min} + \text{rand}(0, 1) * (x_{q\,max} - x_{q\,min}) \tag{8}$$

Here, $x_{q\,max}$ and $x_{q\,min}$ are the maximum and minimum limits of the ELD parameter to be optimized. Each solution in the ELD solution space is a vector, D being the number of optimization parameters [79].

1.5 Firefly Algorithm (FA)

Firefly algorithm is a novel metaheuristic optimization algorithm and it has been applied successfully for ELD problem by Sudhakara et al. [80] and for Economic Emission Dispatch (EED) problem by Apostolopoulos et al. [81]. The algorithm is based on the social flashing behavior of fireflies; two important issues are the variation of light intensity (associated with the objective function) and the formulation of the attractiveness. Since a firefly's attractiveness (β) is proportional to the light intensity (I) seen by adjacent fireflies and I varies as: $I(r) = I_0 e^{-\gamma r^2}$, the attractiveness $\beta(r)$ at a distance r is determined by: $\beta(r) = \beta_0 e^{-\gamma r^2}$. The movement of a firefly i attracted to another firefly j is determined by:

$$x_{i+1} = x_i + \beta_0 e^{-\gamma r^2} (x_j - x_i) + \alpha \left(\text{rand} - \frac{1}{2} \right) \tag{9}$$

Here, *rand* is a random number generator distributed uniformly in [0, 1] space and α is a randomization parameter. The third term in Eq. (9) is used to represent random movement of fireflies in case there are no brighter fireflies to attract the movement.

The FA has many similarities with other swarm intelligence-based algorithms, e.g., PSO, ACO, and ABC but from implementation point of view it is simpler than the others.

1.6 Direct Search (DS)

The Direct Search (DS) [82] optimization methods including Pattern Search (PS) algorithm, Simplex Method (SM), Powell Optimization (PO), etc. are appropriate for solving non-continuous, non-differentiable, and multimodal optimization problems such as the Economic Dispatch. The generic method is based on direct search and gradual reduction in search space. It is designed to explore a set of points, in the neighbourhood of the current position, targeting a smaller value of objective function. Since the variable's values are selected around the best position found in the previous iteration, there is better chance of convergence to the local optima. The starting point of the algorithm is a solution vector $Q(c)$ and n trial solution vectors Q_i are generated around $Q(c)$ as:

$$Q_i = Q(c) + R(c) \; rand(-0.5, 0.5)$$
$$R(c) = \psi(P_{i \; max} - P_{i \; min}) \tag{10}$$

Here, $R(c)$ is the initial range vector; ψ is the multiplication factor from 0 to 1. The best solution $(Q(c) = Q_{best})$ is found which minimizes the objective function. With a reduced range vector $R(c+1) = R(c)(1 - \beta)$, β being the reduction factor, the algorithm is repeated unless the best solution does not change for a pre-specified interval of generations. The main advantages of DS are the ease of framing the problem in computing language, speed in obtaining the optimum point and consistency of the results.

1.7   Biogeography-Based Optimization (BBO)

Biogeography concerns migration of species from one area to another, evolution of new species, and extinction of existing species. A habitat is any area that is geographically isolated from another area. Habitats with a high habitat suitability index (*HSI*) tend to have a large number of species, while those with low *HSI* have a small number of species. Habitats with a high *HSI* have a low immigration rate and a high emigration rate of species because they are overpopulated. On the contrary, habitats with a low *HSI* have a high immigration rate and low emigration rate because of their sparse population of species. Following this natural phenomena, in the ELD optimization scenario, the best solution is assumed to have the best possible features. As higher *HSI* implies lesser chance of sharing its features, a solution having better feature also has greater probability of sharing those features. This Biogeography-based Optimization (BBO) approach has yielded fairly good solution values for ELD problem [83].

1.8   Gravitational Search Algorithm (GSA)

The basis of Gravitational Search Algorithm (GSA) [84] is the law of gravity. In GSA, performances of the objects (agents) are measured with a fitness function which is expressed in terms of the masses of the objects. Assuming a system with n masses X_i ($i = 1, 2, 3, \ldots, n$), the force acting on ith mass (M_i) from jth mass (M_j), at a given iteration (k), is defined as follows:

$$F_{ij}^d(k) = G(t)\frac{M_{pi}(k) \times M_{aj}(k)}{R_{ij}(k) + \varepsilon}\left(X_j^d(k) - X_i^d(k)\right) \tag{11}$$

Here $G(k)$ is the gravitational constant, ε is a small constant, and $R_{ij}(k)$ is the Euclidian distance between ith and jth agents. New positions imply new masses. In each iteration, the masses (m and M) and acceleration (a) are updated by the following equations:

$$m_i(k) = \frac{fit_i(k) - worst(k)}{best(k) - worst(k)}; \; M_i(k) = \frac{m_i(k)}{\sum_{j=1}^{N} m_j(k)}; \; a_i^d(k) = \frac{F_i^d(k)}{M_i(k)} \tag{12}$$

Here $fit_i(k)$ represents the fitness value of the ith agent at iteration k. For a minimization problem like ELD, the worst and best values at iteration k, i.e., *worst* (k) and *best*(k), respectively, are defined as follows:

$$best(k) = min\{fit_i(k)\}; \; worst(k) = max\{fit_i(k)\} \tag{13}$$

1.9   Clonal Selection Algorithm (CSA)

The Artificial Immune System (AIS) is a powerful computational intelligence method based on the natural immune system of human body. CSA is a class of AIS algorithm inspired by the clonal selection theory of response to infections by the immune system. In CSA, a candidate solution is called an antigen—an agent that invades the body, which is recognized by the antibody—the defense agent that destroys antigen. An antibody represents a possible improved solution to the problem. CSA when applied to find optimal solution of ELD [70], first an initial population of N antibodies is randomly produced in the problem space and the affinity of each antibody is determined (by evaluating objective function). Then antibodies which have the highest affinity are selected and copied to generate an improved new population of antibodies as per Eq. (14). Higher the affinity of an antibody, more copies will be generated.

$$nc = round\left(\frac{\beta.N}{i}\right), \quad i = 1, \ldots, n \tag{14}$$

where nc is the number of copies of antibodies from ith antibody (parent) and β is a constant which indicates the rate of copy. Finally, the number of antibodies (Nc) in the regenerated population would be

$$Nc = round \sum_{i=1}^{n}\left(\frac{\beta.N}{i}\right) \tag{15}$$

Nc antibodies are mutated in proportion to their affinities and after determining the affinity of each mutated antibodies, *m* antibodies with higher affinity are selected which enter the next generation directly. *p* new antibodies are generated randomly and increase the population. These new antibodies add to the diversity of the solution and subsequent convergence to local optima is avoided. This cycle is repeated until the termination criterion is met.

1.10 Simulated Annealing (SA)

Simulated Annealing [71] is a powerful algorithm for many optimization problems and it has been successfully applied for ELD problem as well. The idea behind this algorithm is the annealing process of metals. In this process a metal is heated up to a high temperature and then cooled down step-by-step till the metal reaches its lowest energy state. At each step of cooling, the temperature is fixed for a certain period of time allowing thermal equilibrium of the system. In SA, the objective function corresponds to the energy state of the metal and iterations emulate the temperature levels in the annealing process. When SA is applied in an ELD problem, it starts from a random point (starting temperature) and during processing new points (temperature levels) are generated resulting in convergence to the global optima.

The objective function determines the strength of every new point and calculates the change in energy (ΔF). If $\Delta F < 0$, the new point replaces the old point. Else (i.e., if $\Delta F \geq 0$) the new point is retained with some probability (following Boltzmann probability distribution). The efficiency of SA in solving ELD mostly depends on selection of starting point. Linear or exponential decrement of temperature (in each iteration) reduces the probability of acceptance of the worse point which in turn helps avoiding local minima.

1.11 Differential Evolution (DE)

Differential Evolution is a population-based stochastic parallel search technique (similar to GA) that has been proved to be effective for solving ELD problems [56–61]. DE starts with an initial population of feasible solutions (parents) and generates new solutions (child) using three genetic operators—mutation, crossover, and selection until the optimal solution is reached. The mutation operation involves three solution vectors (X_{ra}^G, X_{rb}^G, and X_{rc}^G), which are randomly selected from a population of N_P solution vectors. A mutant vector (V_i^{G+1}) is created by combining one vector with the difference of two other vectors as per following equation:

$$V_i^{G+1} = X_{ra}^G + F\left[X_{rb}^G - X_{rc}^G\right], \; i = 1, 2\ldots, N_P \tag{16}$$

where F is the scaling factor and $ra \neq rb \neq rc \neq i$.

In the crossover operation, certain parameter(s) of the targeted vector is replaced by the corresponding parameter of the mutant vector based on a probability distribution to create a new trial vector (child). Thus, the crossover operator efficiently extracts information from successful combinations thereby triggering search in a better solution space. DE also uses nonuniform crossover where child vector is often taken from only one parent. So the parent competes with the child. The fittest individual survives until the next generation.

1.12   Bacteria Foraging (BF)

Bacteria Foraging is a swarm optimization method that provides certain advantages in handling complex dynamic ED problem in area of power system optimization [85]. Chemotaxis is a foraging behavior whereby a bacterium tries to reach toward more nutrient concentration. If θ be the initial position of bacterium then J $(\theta) < 0$, $= 0$, and > 0 represent nutrient rich, neutral, and noxious environment of the bacteria, respectively. By chemotaxis, lower values of $J(\theta)$ are searched and positions corresponding to $J(\theta) \geq 0$ are avoided. In the context of ELD problem, the objective function (cost function) represented by $J(\theta)$ is calculated at each chemotactic step j, the step size being $C(i)$. If at position $\theta(j + 1)$, the value J is greater than that at position $\theta(j)$, the process will be repeated for the subsequent steps (subject to a maximum N_s number of steps) until a minimum value of J is reached. After N_c chemotactic steps, a reproduction step (N_{re}) is taken in which the population is stored in ascending order of J value. The least healthy bacteria are thereby replaced by copies of the healthiest bacteria. This is followed by the process wherein each bacterium in the population undergoes elimination (dispersal to a random location) with probability p_{ed} [86].

2   Hybrid Approaches for Solving ELD Problem

Though different soft computing methods like those discussed above were applied in solving ELD problem to overcome the limitations of the classical optimization methods, they too have their own limitations. This includes stagnation of fitness function at a local best value for a long time, dead loop of idle individual, loss of fitness quality with iteration, and slow exploration of search space. To further eliminate these limitations and to improve the quality of solutions, various hybrid approaches have evolved by combining the individual soft computing techniques in pairs. Fifteen such selected (including one new) hybrid methods are presented in this section to represent the state-of-the-art in hybrid soft computing application in ELD problem. The hybrid combinations are found to generate high quality solution with sure, fast, and stable convergence, modeling flexibility and robustness, greater consistency, and less computational time compared to the individual soft computing techniques.

Most hybrid methods, found in the literature, for solving ELD problems use PSO technique in combination with other soft computing techniques described in Sect. 1. Hence, approaches that use PSO as a common hybrid component are discussed first followed by other hybrid approaches not using PSO.

2.1 Particle Swarm Optimization—Direct Search Method (PSO-DS)

Application of hybrid PSO-DS technique in solving ELD problem was done by Victoire and Jeyakumar [87]. The steps of the method are given below.

Step 1 Input data
Step 2 Random initialization of search points and velocities of PSO agents
Step 3 Do while (Termination criterion not met)
Step 4 Evaluate the objective function and update the weights
Step 5 Modify the searching points and velocities
Step 6 If solution improves, then
Step 7 Fine-tune the search region using DS
Step 8 End while

Different test cases of Economic Dispatch (ED) problem (based on fuel cost functions) were studied by Victoire et al. [87] to illustrate the flexibility and effectiveness of the PSO-DS approach using MATLAB 6.1.

Case 1 This test was done with 13 generating units. To simulate the valve-point loading effects of generating units, a sinusoid component is added to the quadratic fuel cost function (Eq. 17) thereby making the model more realistic. This also increases the non-linearity and number of potential local optima in the solution space.

$$F_i(P_i) = a_i P_i^2 + b_i P_i + c_i + |e_i(\sin(f_i(P_{i\,min} - P_i))| \tag{17}$$

Here e_i and f_i are cost coefficient of the ith generating unit.

A comparison of PSO-DS performance (with GA, SA, and hybrid GA-SA methods) for a load demand of 2520 MW is shown in Table 1. The overall result of PSO-DS method is found to be better than the other three methods. [87] can be referred for detailed data for this test system.

Case 2 This test was done with 3 subsystems and 10 generating units with multiple-fuel options. Unlike conventional ED problem which has a quadratic cost function, this system has piecewise quadratic fuel cost function (Eq. 18) as multiple fuels are used for generation.

Table 1 Comparison of results for Case 1 (13 units, 2520 MW load)

Generator	Cost per unit generation (MW)			
	GA	SA	GA-SA	PSO-DS
z1	628.32	668.40	628.23	628.3094
z2	356.49	359.78	299.22	298.9996
z3	359.43	358.20	299.17	298.8181
a1	159.73	104.28	159.12	159.7441
a2	109.86	60.36	159.95	159.5509
a3	159.73	110.64	158.85	159.1718
a4	159.63	162.12	157.26	159.5712
b1	159.73	163.03	159.93	159.5940
b2	159.73	161.52	159.86	159.4003
b3	77.31	117.09	110.78	113.6156
c1	75.00	75.00	75.00	113.2250
c2	60.00	60.00	60.00	55.0000
c3	55.00	119.58	92.62	55.0000
Total cost ($/h)	24398.23	24970.91	24275.71	24182.55

$$F_i(P_i) = \begin{cases} a_{i1} + b_{i1}P_i + c_{i1}P_i^2, & P_{i\,min} \leq P_i \leq P_1 \\ a_{i2} + b_{i2}P_i + c_{i2}P_i^2, & P_1 \leq P_i \leq P_2 \\ a_{i3} + b_{i3}P_i + c_{i3}P_i^2, & P_2 \leq P_i \leq P_{i\,max} \end{cases} \tag{18}$$

Here, a_{ij}, b_{ij}, c_{ij} are cost coefficients of the ith generating unit for fuel type j, ($j = 1, 2, 3$).

The generating units were expected to supply load demands of 2400 MW, 2500 MW, 2600 MW, and 2700 MW, [88] can be referred for detailed data for this system. In Table 2, the performance of the PSO-DS method for Case 2 is compared with those of Numerical Method (NM), Enhanced Lagrangian Neural Network (ELNN) method [88], GA [89], and EP [90].

The authors observed that PSO gets stagnated after the 15th iteration and generated a solution which is local optimum. Whereas PSO-DS further explores to find much optimized solution than the one generated by PSO. Also, the PSO-DS is applied in Case 2 experiment with various agents and the experiment proved that, the number of agents above 200 does not have considerable influence on the convergence characteristics and quality of solution.

2.2 Chaotic Particle Swarm Optimization—Quasi-Newton Implicit Filtering Algorithms (PSO or Chaotic PSO–IF)

An application of hybrid chaotic PSO-IF method in ELD problem was examined by Coelho and Mariani in the year 2007 [91]. IF is an applied Quasi-Newton direct

Table 2 Comparison of
results for Case 2 (10 units,
varying load)

P_D (MW)	Fuel cost ($/h)				
	NM	ELNN	GA	EP	PSO-DS
2400	488.50	481.74	482.00	481.79	481.72
2500	526.70	526.27	526.24	526.24	526.24
2600	574.03	574.41	574.40	574.39	574.38
2700	625.18	623.88	623.81	623.81	623.81

search method; it is a generalization of the gradient projection algorithm [92] that calculates derivatives with difference quotients. The step sizes in the difference quotients are iteratively changed to avoid local minima attributed to high-frequency, low-amplitude oscillations. Chaotic PSO is different from traditional PSO in the sense instead of random processing, chaotic mapping is done with the stochastic properties of PSO to improve convergence to the global optima. In chaotic motion every possible state in a search space is visited only once thereby having no specific periodicity.

The PSO-IF approach proposed in [91] is based on Hénon map [93] which is a simplified version of the Poincaré map of the Lorenz system [93]. The Hénon equations are as follows:

$$y(t) = 1 - a\{y(t-1) + z(t-1)\}$$
$$z(t) = b\{y(t-1)\}$$
(19)

$$V_i^{t+1} = wV_i^t + c_1 \, rand_1()\left(pbest_i - X_i^t\right) + c_2 \, rand_2()\left(gbest - X_i^t\right)$$
$$+ (c_1 + c_2)Z_{i,j}(t)d_j[P_j - x_i^t]$$

Here $Z_{i,j}(t)$ values between 0 and 1 are found from Hénon map; P_j is the mean value of previous best positions of the jth dimension; and d_j is a distance factor of the jth dimension based on historical knowledge sources [94]. d_j is calculated as follows:

$$d_j = sgn\left(\sum_{i=0}^{n} sgn(v_i(t))\right)$$
(20)

where the function $sgn(x)$ returns the sign of x.

The authors Coelho and Mariani [91] presented a report on three case studies involving 13 thermal units of generation with the effects of valve-point loading; one of the case studies is given in Table 3. In the test cases the load demand (P_D) was 1800 MW. The result establishes the supremacy of the hybrid PSOs (particularly Chaotic PSO-IF) over PSO, Chaotic PSO, and IF.

Table 3 Convergence results for 50 runs (13 units, 1800 MW load)

Optimization method	Mean time (s)	Minimum cost ($/h)	Mean cost ($/h)	Maximum cost ($/h)
IF	1.4	18812.3852	18962.0139	19111.6426
PSO	2.6	18874.7634	19159.3967	19640.4168
Chaotic PSO	3.3	18161.1013	18809.8275	19640.7556
PSO-IF	14.8	18605.1257	18854.1601	19111.6426
Chaotic PSO-IF	15.3	17963.9571	18725.2356	19057.2663

2.3 Evolutionary Programming—Efficient Particle Swarm Optimization (EP-EPSO)

In 2010, Pandian and Thanushkodi [95] proposed a new hybrid method combining EP and Efficient Particle Swarm Optimization (EPSO) techniques to solve ELD problems with transmission losses. In the EP sub-problem, after random initialization of population (of solutions), a new population is generated by adding a Gaussian random number with zero mean and fixed standard deviation. Then a stochastic tournament method is used for selection. The proposed method extends the basic PSO by modifying the formulation for updating the position, velocity, *pbest*, and *gbest* while satisfying the constraints in a different way.

The effectiveness of the proposed hybrid method in solving ELD problem is tested on the data of a 40-unit system; two ELD problems are considered—one with smooth and another with non-smooth cost functions, the later considering valve-point loading effects. Simulation in MATLAB shows that the EP-EPSO method yields a lower production cost compared to the cost for Neural Network (NN), EP, EPSO, and hybrid NN-EPSO as shown in Table 4.

2.4 Genetic Algorithm—Particle Swarm Optimization (GA-PSO)

A hybrid GA-PSO method was proposed by Younes and Benhamida in the year 2011 [96]. In this approach both algorithms are executed simultaneously. After N iterations, values of P individuals selected by each algorithm are interchanged.

Table 4 Comparison of results (40 units)

Optimization method	Cost ($/h)	Simulation time (s)
NN	146069.74	28.07
EPSO	130330.36	7.232
EP	143799.00	9.242
NN-EPSO	130328.32	8.3529
EP-EPSO	130227.33	7.7590

The individual with larger fitness is finally selected. The steps of the hybrid approach are as follows:

Step 1   Initialize GA and PSO

Step 2   Run GA and PSO simultaneously

Step 3   Stop once the best individual in any process satisfies the termination criterion. Remember the best solution

Step 4   Select P individuals from both processes according to their fitness and exchange values. Go to step 3.

The GA-PSO approach has been tested with five demand load situations for IEEE 25-bus system and the results have been compared with those obtained for classical optimization methods such as Broyden–Fletcher–Goldfarb–Shanno (BFGS) and intelligent methods such as Binary-Coded Genetic Algorithm (BCGA) [97], Real-Coded Genetic Algorithms (RCGA) [98], and PSO separately. Table 5 depicts the comparative where GA-PSO approach shows the most economic value.

2.5   Fuzzy Adaptive Particle Swarm Optimization (FA-PSO)

In 2012, Soni and Pandit [99] proposed two hybrid approaches based on PSO—one is Self-Organizing Hierarchical PSO or SOH-PSO and the other is Fuzzy Adaptive PSO or FA-PSO. The FA-PSO being more useful (of the two) with respect to ELD, is briefly described here.

Observing the long-time stagnation (at one fitness value) often caused by PSO, fuzzy technique is introduced in PSO. The output variables, i.e., the inertia weight (w) and learning factors (c_1 and c_2) are adjusted with input variables—the best unchanged fitness (NU) and the number of generations. The best fitness (BF) represents the best candidate solution. The ELD problem has different ranges of the best fitness values which are normalized into [0, 1] using the following formula:

$$NBF = \frac{BF - BF_{min}}{BF_{max} - BF_{min}} \tag{21}$$

Table 5 Comparison of results (IEEE 25-bus system, constant loss 414,487 MW)

P_D (MW)	BFGS	BCGAs	RCGAs	PSO	GA-PSO
P1	211.30	206.72	213.68	197.45	211.54
P2	126.30	121.64	127.46	114.93	122.46
P3	151.29	151.82	141.93	168.29	140.117
P4	71.24	33.21	29.53	29.08	27.358
P5	211.31	358.05	258.86	259.29	267.514
Total cost ($/h)	2029.3	2011.0	2010.8	2099.00	2007.44

Here, BF_{max} and BF_{min} are the maximum and minimum BF values, respectively. NU values are normalized in a similar way. The bound values for w, c_1, and c_2 are: $0.2 \leq w \leq 1.2$, $1 \leq c_1$ and $c_2 \leq 2$. Following are the steps of the FA–PSO algorithm for solving ELD problem:

Step 1 Randomly generate initial population and velocity of each particle

Step 2 Evaluate the objective function for each particle

Step 3 Select global best position for the ith particle having least value of objective function

Step 4 Select best local position for the ith particle

Step 5 Update the population and velocity parameters

Step 6 Find the next position for each particle based on the updated parameters and Eq. (21)

Step 7 If all particles are selected, Go to Step 8,
Else $i = i + 1$
Go to Step 4

Step 8 Stop search if number of iteration reaches terminal value,
Else go to Step 2

The method has been tested with a system of 6 generators, 46 transmission lines, and 26 buses for a demand of 1263 MW and a comparison of performance of FA-PSO with SOH-PSO and traditional PSO is shown in Table 6.

2.6 Artificial Bee Colony—Particle Swarm Optimization (ABC-PSO)

Manteaw and Odero [100] proposed in 2012 a hybrid optimization approach for solving combined ELD and ED (Emission Dispatch) problem involving ABC algorithm and PSO technique. The objectives are combined using weighting function and the best combined objectives are determined by cardinal priority ranking method through normalized weights.

In this hybridization, ABC executes till the terminal criterion is reached. After this, the best values of individuals generated by the ABC are given as input to the PSO. Generally, the PSO randomly generates its first individual sets but in this hybridized approach, the output of ABC is the input of PSO. The logical steps of the ABC-PSO approach are:

Table 6 Comparison of results (6 units, load 1263 MW)

Method	Best cost ($)	Worst cost ($)	Average cost ($)
FA–PSO	15,445.24	15,451.60	15,448.05
SOH–PSO	15,446.00	15,451.7	15,450.3
Simple PSO	15,466.61	15,451.7	15,450.3

Step 1 Execute ABC
Step 2 Generate best values for all individuals
Step 3 Take these values as input to PSO
Step 4 Execute PSO until stopping criterion is reached.

The authors have reported testing of the ABC-PSO method in a 10-generator system with 2000 MW demand load simulated in MATLAB 2009. Comparison (of performance) with some other methods such as DE, Non-Sorting Genetic Algorithm (NSGA), and Strength Pareto Evolutionary Algorithm (SPEA) is shown in Table 7.

2.7 Particle Swam Optimization—Gravitational Search Algorithm (PSO-GSA)

Very recently, Dubey et al. [103] have proposed a hybrid PSO-GSA method to solve ELD problem.

The main idea of PSO-GSA is to combine the social behavior (*gbest*) in PSO with the localized search capability of GSA. In PSO-GSA, all agents are randomly initialized first. After initialization, the force $F_{ij}^d(k)$ acting on agent i, mass $M_i(k)$ and acceleration $a_i^d(k)$ of agent i are calculated using Eqs. (11) and (12) of Sect. 1.8. The best solution (fitness for each agent) should be updated after each iteration. The velocities of all agents are then updated using the following modified PSO expression (refer Eq. (7) in Sect. 1.2):

$$V_i^{k+1} = wV_i^k + c_1 \, rand_1 \, a_i^d(k) + c_2 \, rand_2 \left(gbest_i - X_i^k \right) \qquad (22)$$

The agent positions are also updated following Eq. (7) and the process is repeated until the stopping criterion is met.

In one of the four case studies done by the authors, a 6-unit generator system with 1263 MW total demand has been used to validate the effectiveness of the PSO-GSA method in optimizing ED considering non-equality constraints like ramp rate limits and generator prohibited zones. The algorithm is implemented in MATLAB 7.8. A performance comparison with nine other optimization methods is depicted in Table 8.

The same hybridization approach involving PSO and GSA has been used by Ashouri and Hosseini in year 2013 [102]. In this approach a more effective method has been used in PSO for the movement of particles, considering the worst solutions of every individual and also the global solution. One notable change is in the

Table 7 Comparison of results (10 units, load 2000 MW)

	ABC-PSO	DE	(NSGA)–II	(SPEA)–II
Fuel cost ($/h)	113,420	113,480	113,540	113,520

formulation of updating velocity of agents as given by Eq. (23). The weight factors (ω) have been modified too.

$$V_i^{k+1} = \omega V_i^k + c_1 \, rand_1 a_i^d(k) + c_2 \, rand_2\left(gbest_i - X_i^k\right)$$
$$+ c_3 \, rand_3\left(X_i^k - P_{worst}^k\right) \tag{23}$$

The result of the experiment and comparison done (with a 6-generator system) with hybrid GA and a special class Ant Colony Optimization (GA-API), Tabu Search Algorithm (TSA), GA, SA, DE, Intelligent PSO (IPSO), and traditional PSO is tabulated in Table 9.

In the following sub-sections some hybrid methods are discussed that involve intelligent techniques other than PSO. The comparative (results) with other methods are also shown in each case.

2.8 Simulated Annealing—Clonal Selection Algorithm (SA-CSA)

In 2010, Amjadi and Sharifzadeh [104] developed a new hybrid approach for power generation optimization. In this approach, the hybridization of SA and CSA is accomplished in the selection step of CSA. The selection of population is done at two levels: first using CSA and then using SA.

When the usual method of CSA is applied, the degree of affinity causes selection of antibodies from the whole population. At the next level, antibodies selected from the previous level and the initial population is compared. In the next iteration selection is based on SA following the criteria given in Eq. (24):

Table 8 Comparison of results (6 units, load 1263 MW)

Methods	Min generation cost ($/h)	Time/iter (s)
PSO	15450.00	14.89
GA	15459.00	41.58
NPSO-LRS	15450.00	–
ABF-NM	15443.82	–
DE	15449.77	0.0335
SOH-PSO	15446.02	0.0633
HHS	15449.00	0.14
BBO	15443.09	0.0325
Hybrid SI-based HS	15442.84	0.9481
PSO-GSA	15442.39	0.0420

NPSO-LRS New Particle Swarm with Local Random Search, *ABF-NM* Adaptive Bacteria Foraging with Nelder–Mead Technique [101], *HS* Harmonic Search

Table 9 Comparison of results (6 units, load 1263 MW)

Methods	P_{loss} (MW)	Total cost ($/h)
PSO-GSA	12.72	15444.00
IPSO	12.55	15444.10
GA-API	12.98	15449.70
DE	12.96	15449.70
GA	13.02	15459.00
PSO	12.95	15450.00
TSA	14.34	15451.63
SA	13.13	15461.10

$$P(t+1) = \begin{cases} P_i'(t), F\left(P_i'(t)\right) < F(P_i(t)) \\ P_i'(t), F\left(P_i'(t)\right) > F(P_i(t))h(P_i(t), P_i'(t) > rand \\ P_i(t), \text{ otherwise} \end{cases} \quad (24)$$

$$h\left(P_i(t), P_i'(t)\right) = \exp\left[\frac{F(P_i(t)) - F(P_i'(t))}{F(P_i(t))}/T\right]$$

$$T(iter + 1) = \alpha T(iter) \text{ and } T(0) = T_0$$

In the above equation, normalized difference between the parent and offspring objective functions $\left[\frac{F(P_i(t)) - F(P_i'(t))}{F(P_i(t))}\right]$ has been considered to eliminate the effect of diversity of objective functions.

A case study done with 10-unit system and load of 2700 MW shows applicability of the method in minimizing the fuel cost function of ELD problem. A comparison of performance of SA-CSA method has been done with some contemporary methods as shown in Table 10.

2.9 Bacterial Foraging-Differential Evolution (BF-DE)

Biswas et al. [105] reported a hybrid approach combining BF and DE algorithms in 2009. The resulting algorithm is also referred as the CDE (Chemotactic Differential Evolution).

The authors have incorporated into DE an adapted form of chemotactic step that characterizes BF. BF is a stochastic application of computational chemotaxis that makes local search based on gradient descent method. One disadvantage of DE is that the global optima may not be approached till the population converges to a local optima or any other point. Moreover, new individuals may add to the population but DE gets stagnated and does not proceed toward a better solution [106]. In this hybrid approach, the convergence characteristics of the classical DE have

Table 10 Comparison of results (10 units, load 2700 MW)

Methods	Total generation cost ($/h)		
	Best	Average	Worst
CGA-MU	624.7193	627.6087	633.8652
IGA-MU	624.5178	625.8692	630.8705
PSO-LRS	624.2297	625.7887	628.3214
NPSO-LRS	624.1273	624.9985	626.9981
CBPSO-RVM	623.9588	624.0816	624.2930
SA-CLONAL	623.8143	623.8356	623.8480

PSO-LRS Particle Swarm Optimization—Local Random Search, *NPSO-LRS* New PSO—Local Random Search, *CBPSO-RVM* Combined PSO-Real Value Mutation

been improved by introducing foraging random walk vector. BF successfully breaks the dead loop of an idle individual and helps to jump from the local minima quickly.

The authors have tested their method with a 6-unit system (with a demand load of 1263 MW) and the comparative result with respect to other conventional and hybrid approaches is tabulated in Table 11.

2.10 Genetic Algorithm—Active Power Optimization (GA-APO)

Malik et al. [107] presented a new hybrid approach involving GA and Active Power Optimization (APO) and have reported its use for the solution of ELD problem with valve-point effect. The proposed approach is able to fine-tune the near optimal results produced by GA.

APO is based on Newton's second-order approach (NSOA). APO is developed and implemented by the authors using some technique of storage optimization and classical linear system solution method. In the proposed hybrid approach, GA works as a global optimizer and produces near optimal generation schedule. APO works on this schedule and replaces the power output at the generation buses. It dispatches the active power of the generating units to minimize the cost and produce optimum generation schedule.

GA-APO is implemented in a computational framework called PED Frame [108] in visual C environment. One can input cost curves and other ELD-specific information through PED Frame and can also get output in standard format.

Table 11 Comparison of results (6 units, load 1263 MW)

	BF-DE	PSO	GA	NPSO-LRS	CPSO1
Minimum cost	15444.1564	15,450	15,459	1540	15,447

Table 12 Different IEEE systems results comparison

System	Cost of GA	Cost of GA-APO
6-Bus 3-machines system $P_D = 210$ MW	3463.37	3205.99
IEEE 14-Bus 5-machine system $P_D = 259$ MW	1012.44	905.54
IEEE 30-Bus 6-machine system $P_D = 283.4$ MW	1117.13	984.94

The authors have investigated three test systems to demonstrate the effectiveness of their hybrid approach. The test systems consists 3-machines 6-bus system [109], IEEE 5-machines 14-bus system [110], and IEEE 6-machines 30-bus system [111]. The outputs are compared in Table 12. The coefficients e_i and f_i reflecting valve-point effects are introduced in the system to convert the quadratic convex cost curve into nonconvex cost curves.

2.11 Differential Evolution—Biogeography-Based Optimization (DE-BBO)

In 2010, Bhattacharya and Chattopadhyay [112] presented application of hybrid DE-BBO method in ELD problem taking into account transmission losses, and constraints such as ramp rate limits, valve-point loading, and prohibited operating zones.

Though DE yields optimum ELD solution satisfying all the constraints, one of its major disadvantages is that DE is unable to map its entire unknown variables together efficiently when the system complexity and size increase. Owing to the crossover operation in DE, solutions having good initial fitness value often suffer loss of quality during further processing. BBO has got no crossover stage and solutions gradually mature as the algorithm proceeds through migration operation. The most striking characteristic of DE-BBO is hybridization of migration operation. In this algorithm, new features are developed in child population from corresponding parents through mutation (caused by DE) and migration (caused by BBO). Here, good solutions would be less destroyed, while poor solutions can take a lot of new features from good solutions.

The performance of hybrid approach (developed in MATLAB 7) has been compared for a system of 10 generating units with few other hybrid approaches like NPSO-LRS, PSO-LRS, IGA-MU, and CGA-MU. The result is presented in Table 13.

Table 13 Comparison of results (10 units)

Methods	Generation cost ($/h)		
	Max.	Min.	Average
DE-BBO	605.62	605.62	605.63
NPSO-LRS [113]	626.99	624.13	624.99
PSO-LRS [113]	628.32	624.23	625.79
IGA-MU [114]	630.87	624.52	625.87
CGA-MU [114]	633.87	624.72	627.61

2.12 Hybrid Immune Genetic Algorithm (HIGA)

Hosseini et al. [115], in the year 2012 proposed application of Hybrid Immune Genetic Algorithm (HIGA) in solving ELD problem. This is a hybridization of Immune Algorithm and GA.

The Immune Algorithm creates an initial solution set and iteratively improves its performance using affinity factor, hyper-mutation operator, and clonal selection [116]. The affinity factor ($AF = (TC_n)^{-1}$) is a measure of strength of solutions in optimizing the antigens (objective functions). The hypermutation operator acts like the mutation operator in GA [117], but unlike in GA, the probability of mutation in Immune Algorithm is inversely proportional to the affinity factor of the solution. Thus, if the affinity factor (of a solution) is low, it will be more mutated to be able to explore the solution space and vice versa. In clonal selection, the crossover operator is used to propagate the attributes of high-quality solutions among others. This sometimes causes reproduction of each solution depending on affinity factor of the solution.

The cost comparisons of HIGA optimization on 6-unit and 40-unit power generation systems are presented in Tables 14 and 15, respectively.

2.13 Fuzzified Artificial Bee Colony Algorithm (FABC)

A recent study has been done by Koodalsamy and Simon [118] to demonstrate the efficiency of a hybrid Fuzzy Artificial Bee Colony (FABC) algorithm for solving multi-objective ED problem, i.e., minimizing (i) probability of energy unavailability,[1] (ii) emission cost, and (iii) fuel cost simultaneously while satisfying load demands and operational constraints. In this approach, ABC works with a fixed number of bees that fly around in a multidimensional search space to locate the food sources. With every generation cycle of ABC, the best compromise is chosen from

---

[1]This is equivalent to Expected Energy Not Supplied (EENS); higher the EENS, lower is the reliability level.

Table 14 Comparison of results (6 units, load 1263 MW)

	BFO	PSO	NPSO-LRS	GA	HIGA
Total cost	15443.85	15450.14	15450.00	15457.96	15443.10

Table 15 Comparison of results (40 units, load 10,500 MW)

	DE-BBO	BBO	QPSO	HIGA
Total cost	121420.90	121426.95	121448.21	121416.94

the Pareto optimal set using fuzzy fitness. The normalized membership function for fuzzy fitness is calculated as:

$$FIT_p = \frac{(\mu_c^p + \mu_e^p + \mu_r^p)}{\sum_{p=1}^{m} (\mu_c^p + \mu_e^p + \mu_r^p)} \tag{25}$$

$$\mu_j^p = \begin{cases} 1, & for\ F_i \leq F_{i\,min} \\ \frac{(F_{j\,max} - F_j)}{(F_{j\,max} - F_{j\,min})}, & for\ F_{j\,min} < F_j < F_{j\,max} \\ 0, & for\ F_j \geq F_{j\,max} \end{cases} \tag{26}$$

Here m is the total number of non-dominated solutions or population of bees and p is the pth position of bees or food sources. The fuzzy membership functions μ_c^p, μ_e^p and μ_r^p are related to cost, emission, and reliability objective functions, respectively. The design of μ_c^p, μ_e^p and μ_r^p, i.e., μ_j^p is shown in Eq. (26) where F_j is the degree of the objective function in the fuzzy domain. The best compromise solution corresponds to the maximum value in the population implying food sources with highest quality of nectar information.

The authors have tested the algorithm using MATLAB 7. Out of several test cases carried out to validate the FABC algorithm, the results of solving EED problem of IEEE 30-bus system for two different load conditions (2.834 MW and 2.8339 MW) and comparisons with other reported methods are shown in Table 16.

2.14 Firefly Algorithm-Ant Colony Optimization (FFA-ACO)

Younes [119] proposed the hybrid approach involving FFA and ACO algorithm in 2013.

According to this algorithm, first an initial population n of fireflies x_i is generated. Light intensity of firefly is determined by the objective function. New solutions or attractiveness of fireflies are evaluated and light intensity updated as firefly i is moved towards j. The fireflies are ranked according to attractiveness and the

Table 16 Comparison of results (6 units, IEEE 30 bus system)

Load	2.834 MW		2.8339 MW			
Method	MOPSO	FABC	NSGA	MOHS	MBFA	FABC
Total cost ($/h)	938.91	938.75	938.46	939.92	938.33	938.24

best (solutions) are passed as initial points of ACO. Following the scheduled activities in ACO, the ant's response functions are compared and communication with best ant response is made to get the best solution.

The FFA-ACO approach has been developed using MATLAB 7. It is tested using the modified IEEE 30-bus system consisting of 6 generators (with power demand of 283.40 MW). A comparison of performance of FFA-ACO has been done with few other approaches (refer Table 17).

2.15 Particle Swarm Optimization—Ant Colony Optimization Algorithm (PSO-ACO)

In the previous subsections, we have discussed some hybrid methods that are used to get optimized solution of ELD problem. The studies conducted by the researchers confirm that the PSO method itself can be used as an effective and powerful technique for optimizing ELD solution. However, one of its prominent weaknesses found is that it may get stuck into local optima if the global best and local best positions become identical to the particle's position repeatedly. To alleviate this drawback, hybrid methods combining PSO with other global optimization algorithms like GA, IF, EP, FA, ABC, GSA have been used. Now, in addition to these, a new hybrid of PSO is suggested by the present authors by combining PSO with ACO to study whether better optimization of ELD solution can be achieved. To the best of our knowledge, study of ELD problem solving using PSO-ACO hybrid approach has not been reported in the literature. In this approach, new generation members can be produced at each iteration using PSO and then ACO algorithm can be applied to create extended opportunity of fine-tuning the members.

In PSO algorithm, if the *gbest* value does not change over few iterations, other particles are drawn closer to the *gbest* position. As the velocity of the *gbest* particle gradually reduces by iteration, exploring the local search space (by the best agent) also diminishes. Here the ACO algorithm comes into play. Taking the *gbest*

Table 17 Comparison of results (6 units, load 283.40 MW)

	MDE-OPF [120]	PSO	ACO	FFA	FFA-ACO
Cost ($/h)	802.62	801.77	801.77	801.01	800.79
Time (s)	23.07	16.26	14.97	13.83	10.73

particles as the input, following the schedule activities of ACO, the ant's response functions are compared and communication with best ant response is made to eventually get the best solution.

The suggested steps of the PSO-ACO method are given below.

Read the input data
Initialize the search points and velocities in PSO
While (Termination criterion not met)
  Evaluate the objective function for each individual and update the
     inertia weight
  Modify the searching points and velocities
  If solution improves, then
    Store the solution for ACO
  End if
End while
The best solutions found by PSO are passed as starting points for ACO
While (termination criterion not met)
  Generate path for each ant
  Compare response function
  If value of response function not better than earlier **then**
    Exchange with best ant's response function
    Generate path from local position to best ant
  Else
    If value of response function is better **then**
      Repeat while loop
    Else
      Wait for exchange with best ant
    End if
  End if
End while

3 Discussion

The observations that can be made regarding the performance of the PSO-DS method proposed by Victoire and Jeyakumar [87] are: the reliability of producing quality solutions, searching efficiency as iteration proceeds, accuracy of the final solution and convergence characteristics when the numbers of agents are varied. The PSO is very fast compared to other evolutionary techniques, but it does not possess the ability to improve upon the quality of the solutions as the number of generations increases. When the solution of the PSO improves in a run, the region will be fine-tuned with the DS method.

The performance of the PSO-DS method was tested with two EDP test cases and compared with the results reported in the literature for few other methods. Test result shows that the PSO-DS method is capable of handling load demand at various time intervals with no restrictions on the cost function of the units. As claimed by the authors, 77 % of the 100 trial runs produced quality solution and the convergence characteristic resembles the same for all the 100 trial runs which indicates the reliability and robustness of PSO-DS method. This hybrid method is scalable for solving the DEDP with more inequality constraints such as prohibited operating zones and spinning reserves. More accurate dispatch results can then be achieved in actual power situations.

The contribution of Coelho and Mariani [91] is the hybridization of the PSO (using Hénon map) and the Implicit Filtering (IF) direct search to solve an EDP. Chaotic PSO approach is good in solving optimization problem but often the solutions are close to but not exactly the global optimum. To get rid of this limitation, the hybrid PSO-IF approach seems to be a promising alternative.

The combination of chaotic PSO with IF is a kind of sequential hybridization directed for local search. Function of chaotic PSO is global search within a population, while function of IF is exploitation around local best solution produced by chaotic PSO in each iteration. IF explores the local search space quickly, jumps over local minima, and implicitly filters the noise.

The PSO-IF hybrid method is validated on a test bed emulating 13 thermal units. The fuel cost function considers the valve-point loading effects. The simulation result reported by the authors is comparatively better than some of the recent studies reported in the literature. The complementary search ability of chaotic PSO-IF renders its usefulness to multi-constrained optimization problems in planning and operation of power system.

In the study done by Pandian and Thanushkodi [95], a new methodology involving EP combined with EPSO has been proposed for solving non-smooth ED problem with valve-point loading. There are two parts in the proposed algorithm. The first part exploits the ability of EP in generating a near global solution. When EP meets the terminal criteria, the local search capability of EPSO is exploited to adjust the control variables so as to achieve the final optimal solution. In effect, faster convergence is obtained when EP is applied along with EPSO.

The mean cost value obtained by EP-EPSO in the ED simulation is less compared to other methods studied. Thus EP-EPSO has been established by the authors as a powerful tool for optimizing feasible solutions of the non-convex ELD problem.

The approach of PSO is similar to GA considering the fact that their search processes using probabilistic rules are based on exchange of information among the members of population. The objective of the research by Younes and Benhamida [96] is to combine PSO and GA to improve the effectiveness of the search process as a whole. The feasibility of the hybrid algorithm is tested successfully on an IEEE

25-bus system. The results show that the GA-PSO approach is quite effective in handling nonlinear ELD problems.

Long processing time and uncertainty of convergence to the global optima are the main disadvantages of GA. Again unlike GA, PSO can quickly find a good local solution but get stuck to it for rest of the iteration. The GA-PSO hybrid combination can generate a much better solution with stable convergence and appears superior over many other hybrid approaches in terms of flexibility of modeling, reliable and speedy convergence, and less processing time.

In [99], a new hybrid optimization algorithm, called FA-PSO, was presented by Soni and Pandit for solving non-convex ED problem (considering prohibited operating zones and ramp rate constraints) in power system. In FA-PSO, acceleration factors are co-evolved with the particles and the inertia weight is adjusted through fuzzy mechanism. To avoid stagnation around local optima and explore the search space effectively, this method uses a new mutation operator.

Two case studies have been employed (with systems consisting of 6 and 15 thermal units with load demand of 1263 MW and 2630 MW, respectively) to demonstrate the applicability of the proposed approach. The detailed characteristics of the units including prohibited operating zones and ramp rate limits are presented in [99] along with the convergence characteristics of the methods (FA-PSO, SOH-PSO, and PSO) tested on the systems. Test results show that FA-PSO has distinct superiority over SOH-PSO and PSO in terms of robustness, computational overhead, efficiency, and applicability to large-scale real systems.

Manteaw and Odero [100] formulated and implemented a hybridized ABC-PSO algorithm and demonstrated its successful application in the optimization of the Combined Economic and Emission Dispatch (CEED) problem. The hybrid method exploits the processing speed of PSO coupled with its convergence strength to use the results produced by ABC in yielding improved global optima.

The ABC-PSO method was tested with varying load conditions and test cases to evaluate its applicability in the CEED problem. For PSO, the maximum number of iterations and population number are considered 1000 and 15 individuals respectively, while for ABC the colony size and food number are considered 30 and 15, respectively. Though the hybrid method shows better quality solution, stable convergence characteristics and modeling flexibility with respect to other algorithms, its processing time and utility can be bettered with inclusion of mutation operators.

The hybridization of PSO with GSA reported in [103] by Dubey et al. effectively combines the exploitation ability of PSO with the exploration ability of GSA to unify their strength. The agents are initialized randomly and each agent exploring in the search space is accelerated (by means of gravity force) towards other agent having better solution (heavier mass). The agents closer to the optima proceeds slowly and assures effective exploitation. The *gbest* helps in finding the optima around a good solution. Thus the hybrid approach solved the slow speed problem of GSA algorithm on the final iterations and the problem of PSO in getting stuck at local optima.

The authors have tested the PSO-GSA hybrid algorithm in four different standard test systems, including a 6-unit system with ramp rate limits and prohibited zones, an 18-unit system with variable peak demand, a 20-unit system with transmission loss, and also a large-scale 54-unit system with valve-point loading and multiple local minima. Effect of different parameters on the performance of the algorithm was carefully studied and after considerable number of trial runs and statistical analysis the optimum parameter values selected were: $n = 100$, $a = 10$, $G_0 = 1$, $c_1 = 2.0$, $c_2 = 1.5$. Very elaborate comparative study is made with some of the recent reported methods with respect to (a) solution quality, (b) computational efficiency, and (c) robustness-test results show that PSO-GSA is superior in all aspects, i.e., lower average cost, less computational time, and greater consistency. Overall, the study is very convincing and performance-wise PSO-GSA can be rated high amongst the hybrid methodologies reviewed in this article.

In another study by Ashouri and Hosseini [102], hybrid PSO-GSA algorithm is successfully employed in ELD problem. Though the formulations are mostly the same as adopted by Dubey et al. [103], some improvisations are noticed in the expression of agent velocity-a new dynamic inertia weight is incorporated. With dynamic acceleration and weight coefficients, great exploration and exploitation happen in the first and final iterations of the algorithm respectively, resulting in better and faster solutions. A case study with 6-unit system considering transmission loss, prohibited zones, and ramp rate limits and also another study with 40-unit system with valve-point loading effect have been used to show the feasibility of the method. The parameter values used for the 6-unit case study were: $n = 30$, $a = 20$, $G_0 = 1$, $c_1 = 2.5$, $c_2 = 0.5$, and $c_3 = 0.5$. From comparison with the recent reported methods, it is observed that PSO-GSA approach ensures high quality and faster solution with stable convergence for ELD problem.

The results obtained by Amjadi and Sharifzadeh [104] prove that the proposed SA-CSA hybrid method is a more useful solution for the ED problem compared to other stochastic algorithms. Fast convergence and the ability of not being trapped in local optimums are undoubtedly the most important advantages of this new method. A comparison with methods reported in the literature speaks in favor of SA-CSA, though the result is not very encouraging in terms of fuel cost reduction of LD systems.

The authors believe that the CSA algorithm is flexible enough to hybridize with other stochastic search algorithms such as SA and PSO and each hybrid combination may be useful to solve ED problem.

Biswas et al. [105] have experimented hybridization involving DE in non-convex ELD problems taking into consideration transmission loss, ramp rate limits, and prohibited operating zones. The equality and inequality constraints are considered in the fitness function itself in the form of a penalty factor.

DE has outperformed powerful metaheuristic search algorithms like the GA and PSO but has certain limitations like stagnation before reaching global optima. The idea of computational chemotaxis of BFOA incorporated in DE greatly improvises the convergence characteristics of classical DE.

The proposed approach was tested with real data of 6, 13, 15, and 40 generator power systems. The results are comparable to those produced by other evolutionary algorithms. The solutions have good convergence characteristics. The authors infer that the DE-based hybrid algorithm is equally applicable to non-convex and non-smooth constrained ELD problems. However, it remains to be tested whether the method is applicable in practical large-sized problems with more realistic constraints.

Proposed hybrid approach by Malik et al. [107] combines GA with APO algorithm. The strength of GA is that it reaches the vicinity of global minima in relatively lesser time, but some of its weaknesses are: (1) solution is close to but not exactly the global minima hence not the optimal one (2) convergence speed slows down near the optimum point. Hybridization with APO helps to overcome these weaknesses. APO uses NSOA which is a classical approach for finding optimal solutions of power flow problems by minimizing the Lagrangian objective function.

The results of testing GA-APO hybrid approach on 3, 5, and 6-unit system show a significant reduction in generation cost with respect to GA. The cost reduces exponentially with increase in system size. Authors have demonstrated through simulation experiments that GA cost curve features peaks, dips, and flats in the wider band, whereas the GA-APO cost curve rise and fall in very narrow band and remains beneath the GA curve. Moreover, the solution time of the hybrid approach was found lesser than that of GA in all test cases.

In the DE-BBO hybrid method proposed by Bhattacharya and Chattopadhyay [112], the migration operator of BBO is combined with mutation, crossover, and selection operators of DE to maximize the good effect of all the operators—DE has good exploration ability in finding the region of global minima whereas BBO has good exploitation ability in global optimization problem. Together they enhance the convergence property to improve the quality of solution.

Proposed DE-BBO algorithm was tested with (i) 3-unit system (load 300 MW) with ramp rate limit and prohibited operating zone, (ii) 38-unit system with load 6000 MW, (iii) 40-unit system (load 10,500 MW) with valve-point loading, (iv) 10-unit system with load 2700 MW. The test results reveal that average costs obtained by DE-BBO for both convex and non-convex ELD problems are the least of all the reported methods. The computational time is also at par or better than the other methods. DE-BBO is also quite robust as it attains minimum cost 50 times out of 50 trials compared to 38 times in BBO. The tests point to the possibility that DE-BBO can be reliably tried to address optimization problems of complex power system in operation.

The HIGA algorithm proposed by Hosseini et al. [115] incorporates the concept of affinity factor and clonal selection of Immune System algorithm to improve the quality of solutions obtained by crossover and mutation operations of GA. The application of the hybrid approach is made in power generation system taking care of valve-point effects, prohibited operation zones, ramp rate constraints, and transmission losses. For two ELD test cases with different characteristics (studied by the authors), results yielded by HIGA method is better than that for the other methods in terms of cost and power loss. Based on the findings of the study, the authors conclude that the proposed HIGA algorithm can be effectively used to solve non-convex ELD problems in power systems.

Going beyond the single objective ELD problem, Koodalsamy and Simon [118] have attempted to solve multi-objective reliable emission and economic dispatch (REED) problem using the Fuzzified ABC or FABC algorithm. The ABC algorithm uses fuzzy membership approach to find the best compromise solution from the Pareto optimal set. The fuzzy membership of the reliability function is modeled while scheduling the optimum dispatch. The fuzzy membership function chosen for fuel cost and emission are same and it aids the ABC algorithm in maximizing the fitness function.

The FABC approach is tested on an IEEE 30-bus system and 3, 6, 10, 26 and 40-unit systems. Effect of different characteristics of the objective function and constraints are studied. To allow aberrations for the foraging behavior of bees, several test runs are carried out to set the optimal colony size and the limit value. From the results it is clear that the proposed hybrid method can yield a well distributed Pareto optimal set and is capable of finding a reasonably good compromise solution. The method is unique in the sense it can handle not only fuel cost aspect of ELD problem but at the same time emission and reliability factors which are equally important considering Kyoto Protocol 2008 and Energy Policy Act 2005. Moreover, the method is straightforward, easy to implement, and applicable for any large-scale power system.

FFA-ACO method proposed by Younes [119] is robust and can provide an optimal solution with fast computation time and a small number of iterations. In this hybrid method, the advantage that is exploited is that of using the metaheuristic methods which are very efficient and better than deterministic methods for the search of global solution for complex problems like ELD. The disadvantage of metaheuristic methods that is avoided is the relative long time of convergence owing to the high number of the agents and iterations. To further improve the efficiency of the hybrid method, FFA and ACO are used with as low as number of ants and fireflies as possible.

The author has studied two cases through simulation in MATLAB environment: Case 1 concerns the minimization of the cost function with constant losses, Case 2 concerns minimization of the cost function with variable loss. The results clearly show the effectiveness of performance of the FFA-ACO over other methods in terms of function cost value and convergence time.

The idea of a new hybrid evolutionary algorithm for solving ELD problem as proposed by the authors of this article is based on the well-known PSO and ACO algorithm. Though the authors have not applied the method in a simulated environment with real-system data, it seems to yield interesting results to be compared with those produced by other hybrid approaches. The PSO-ACO method is also expected to generate global optima in lesser time with less number of iterations. The basic idea behind the proposed hybrid algorithm is that the improvement of the *gbest* for each individual is according to the best path selection methodology of ACO. The intelligent decision-making structure of ACO algorithm is incorporated into the original PSO where the global best position is unique for every particle. PSO-ACO uses the random selection procedure of ACO algorithm to assign different global best positions to every distinct agent.

4 Conclusion

The ELD problem may be broadly classified as convex and nonconvex. In convex ELD input–output characteristics are assumed piecewise linear and monotonically increasing. The non-convex ED problem represents the complete, realistic problems having discontinuous and nonlinear characteristics owing to constraints (valve-point effect, transmission losses, ramp rate limits, and prohibited operating zones). The convex ED problem can be solved using mathematical programming-based optimization methods, but non-convex ED problem cannot be handled effectively by such classical approaches. Many heuristic search tools and soft computing approaches have been reported in the literature that addressed this problem with comparatively better result. But each approach has its inherent limitations-either they get stuck to local minima and fail to find the global minima, or converge very slowly to the solution or stagnate after certain iterations without improving solution. All these problems motivated the researchers to evolve hybrid approaches by combining a pair of compatible soft computing methods to minimize their individual weakness and in effect produce fast, accurate, and consistent solution.

Though in the literature quite a number of hybrid soft computing approaches are reported that address the issues of ED, this chapter reviews 14 selected hybrid approaches involving popular soft computing techniques. Out of these 14 approaches, 7 involve PSO technique. This indicates the modeling flexibility, consistency/reliability, and effectiveness of PSO as a hybrid component toward finding quality solution of ELD problem. GA features in three of the hybrid approaches whereas Fuzzy technique, ABC optimization, and DE each feature in two different hybrid approaches. It is difficult to adjudge the best or better approaches in solving ELD problem as each of them has been tested with varied characteristics of different power generation systems (simulated mostly in MATLAB) and have been compared with few (not all) other soft computing techniques or heuristic optimization tools or hybrid methods. There is no benchmark standard or common test data for comparing the simulation results obtained using different hybrid methods. Still PSO-GA hybrid combination proposed by Younes and Benhamida [96] and PSO-GSA approach by Dubey et al. and also by Ashouri and Hosseini appears superior over many other hybrid approaches in terms of lower average cost, sure and fast convergence, greater consistency, and less computational time. The PSO-ACO method proposed by the present authors is expected to overcome common limitations of many hybrid approaches involving PSO but it is yet to be tested with real data. Among methods not featuring PSO, DE-BBO method proposed by Bhattacharya and Chattopadhyay [112] and FABC algorithm proposed by Koodalsamy and Simon [118] are very promising. All the other hybrid methods discussed in this chapter are, however, unique in their approaches and indeed have some advantages over the rest.

The ELD problem for small-scale to large-scale systems has been greatly optimized by virtue of competing research in the field, evolving new optimization methods-particularly, the latest hybrid soft computing approaches. But after going through the recent and past research endeavors, the authors strongly feel that there is enough scope for research to address the challenges of future related to power system and allied aspects. Some of these challenges are: (i) Optimal Power Flow (OPF) problem, (ii) optimization of multiple objectives like Reliable Emission and Economic Dispatch (REED) problem and Combined Economic and Emission Dispatch (CEED) problem, (iii) Extended ELD problem for large number of units (40–90 units) and for more complex objective and constraint function (like exponential function and higher order polynomial).

References

1. Grigsby LL (2009) Power System Stability and Control. CRC Press, New York
2. Wood AJ, Woollenberg BF, Sheble GB (1984) Power generation operation and control. Wiley Publishers, New York
3. Yare Y, Venayagamoorthy GK, Saber AY (2009) Conference: economic dispatch of a differential evolution based generator maintenance scheduling of a power system. In: Power & Energy Society General Meeting, 2009 (PES '09), IEEE
4. Chowdhury BH, Rahman S (1990) A review of recent advances in economic dispatch. IEEE Trans Power Syst 5(4):1248–1259
5. Garg M, Kumar S (2012) Int J Electron Commun Technol IJECT 3(1)
6. Lee KY et al (1984) Fuel cost minimize at ion for both real-and reactive power dispatches. Proc Inst Elect Eng Gen Trans Distrib 131(3):85–93
7. Sailaja Kumari M, Sydulu M (2009) A fast computational genetic algorithm for economic load dispatch. Int J Recent Trends Eng 1(1)
8. Naveen Kumar KP, Parmar S, Dahiya S (2012) Optimal solution of combined economic emission load dispatch using genetic algorithm. Int J Comput Appl (0975–8887) 48(15)
9. Chopra L, Kaur R (2012) Economic load dispatch using simple and refined genetic algorithm. Int J Adv Eng Technol 5(1):584–590
10. Biswas SD, Debbarma A (2012) Optimal operation of large power system by GA method. J Emerg Trends Eng Appl Sci (JETEAS) 3(1):1–7
11. Anuj Gargeyal M, Pabba SP (2013) Economic load dispatch using genetic algorithm and pattern search methods. Int J Adv Res Electr Electron Instrum Eng 2(4) 2013
12. Arora D, Sehgal S, Kumar A, Soni A (2013) Economic load dispatch using genetic algorithm. Int J Techn Res (IJTR), 2(2)
13. Kapadia RK, Patel NK (2013) J Inform Knowl Res Electr Eng 2:219–223
14. Mansour WM, Salama MM, Abdelmaksoud SM, Henry HA (2013) Dynamic economic load dispatch of thermal power system using genetic algorithm. IRACST—Eng Sci Technol: An Int J (ESTIJ) 3(2). ISSN:2250-3498
15. Jain AK, Mandloi T Comparison of classical method and soft computing optimization algorithm applied to economic load dispatch problem. Int J Latest Trends Eng Technol (IJLTET) 3(4)
16. Aliyari H, Effatnejad R, Areyaei A (2014) Economic load dispatch with the proposed GA algorithm for large scale system. J Energy Natural Resour 3(1):1–5
17. Park J-B, Lee K-S, Shin J-R, Lee K-Y (2005) A particle swarm optimization for economic dispatch with nonsmooth cost functions. IEEE Trans Power Syst 20(1)
18. Mahadevan K, Kannan PS, Kannan S (2005) Particle swarm optimization for economic dispatch of generating units with valve-point loading. J Energy Environ 4:49–61

19. Sudhakaran M, Ajay P, Vimal Raj D, Palanivelu TG (2007) Application of particle swarm optimization for economic load dispatch problems. In: The 14th international conference on intelligent system applications to power systems, ISAP 2007, Kaohsiung, Taiwan
20. dos Santos Coelho L, Lee C-S (2008) Solving economic load dispatch problems in power systems using chaotic and Gaussian particle swarm optimization approaches. Electrical Power Energy Syst. 30:297–307
21. Baskar G, Mohan MR (2008) Security constrained economic load dispatch using improved particle swarm optimization suitable for utility system. Electric Power Energy Syst 30:609–613
22. Mahor A, Prasad V, Rangnekar S (2009) Economic dispatch using particle swarm optimization: a review. Renew Sustain Energy Rev 13:2134–2141
23. Vlachogiannis JG, Lee KW (2009) Economic load dispatch-a comparative study on heuristic optimization techniques with an improved coordinated aggregation-based PSO. IEEE Trans Power Syst 24(2)
24. Saber AY, Chakraborty S, Abdur Razzak SM, Senjyu T (2009) Optimization of economic load dispatch of higher order general cost polynomials and its sensitivity using modified particle swarm optimization. Electric Power Syst Res 79:98–106
25. Muthu Vijaya Pandian S, Thanushkodi K (2010) Solving economic load dispatch problem considering transmission losses by a Hybrid EP-EPSO algorithm for solving both smooth and non-smooth cost function. Int J Comput Electr Eng 2(3):1793–8163
26. Muthu S, Pandian V, Thanushkodi K (2011) An efficient particle swarm optimization technique to solve combined economic emission dispatch problem. Eur J Sci Res 54(2):187–192
27. Batham R, Jain K, Pandit M (2011) Improved particle swarm optimization approach for nonconvex static and dynamic economic power dispatch. Int J Eng Sci Technol 3(4):130–146
28. Sreenivasan G, Dr Saibabu CH, Dr Sivanagaraju S (2011) Solution of dynamic economic load dispatch (DELD) problem with valve point loading effects and ramp rate limits using PSO. Int J Electr Comput Eng 1(1):59–70
29. Khokhar B, Singh Parmar KP (2012) A novel weight-improved particle swarm optimization for combined economic and emission dispatch problems. Int J Eng Sci Technol (IJEST) 4 (05) 2012
30. Anurag Gupta KK, Wadhwani SK (2012) Combined economic emission dispatch problem using particle swarm optimization. Int J Comput Appl 49(6):0975–8887
31. Hardiansyah J, Yohannes MS (2012) Application of soft computing methods for economic load dispatch problems. Int J Comput Appl 58(13):0975–8887
32. Chakrabarti R, Chattopadhyay PK, Basu M, Panigrahi CK (2006) Particle swarm optimization technique for dynamic economic dispatch. IE(I) J-EL 87:48–54
33. Agrawal S, Bakshi T, Majumdar D (2012) Economic load dispatch of generating units with multiple fuel options using PSO. Int J Control Autom 5(4)
34. Linga Murthy KS, Subramanyam GVS, SriChandan K (2012) Combined economic and emission dispatch for a wind integrated system using particle swarm optimization. Int Electr Eng J (IEEJ) 3(2):769–775
35. Junaidi H, Yohannes MS (2012) Intell Syst Appl 12:12–18
36. Sharma J, Mahor A (2013) Particle swarm optimization approach for economic load dispatch: a review. Int J Eng Res Appl (IJERA) 3(1):013–022
37. Niknam T (2006) An approach based on particle swarm optimization for optimal operation of distribution network considering distributed generators. In: Proceedings of the 32nd annual conference on IEEE industrial electronics, IECON 2006, pp 633–637 1942–1948
38. Soubache ID, Ajay-D-Vimal Raj P (2013) Unified particle swarm optimization to solving economic dispatch. Int J Innov Technol Explor Eng (IJITEE) 2(4)
39. Tiwari S, Kumar A, Chaurasia GS, Sirohi GS (2013) Economic load dispatch using particle swarm optimization. Int J Appl Innov Eng Manag 2(4)

40. Singh N, Kumar Y (2013) Economic load dispatch with valve point loading effect and generator ramp rate limits constraint using MRPSO. Int J Adv Res Comput Eng Technol (IJARCET) 2(4)
41. Venkatesh B, Subbu Chithira Kala V (2013) A particle swarm optimization for multiobjective combined heat and power economic dispatch problem considering the cost, emission and losses. Int J Scient Eng Res 4(6) 2013
42. Md. Khan J, Mahala H (2013) Applications of particle swarm optimization in economic load dispatch. In: Proceedings of national conference on recent advancements in futuristic technologies (NCRAFT 13), vol 1, Issue 1, PAPER ID 0072, Oct-2013
43. Singh Maan R, Mahela OP, Gupta M (2013) Solution of economic load dispatch problems with improved computational performance using particle swarm optimization. Int J Eng Sci Invent 2(6) 01–06
44. Singh N, Kumar Y (2013) Constrained economic load dispatch using evolutionary technique. Asian J Technol Manage Res 03(02)
45. Sinha P (2013) Particle swarm optimization for solving economic load dispatch problem. Int J Commun Comput Technol 01(63)
46. Singh M, Thareja D (2013) A new approach to solve economic load dispatch using particle swarm optimization. Int J Appl Innov Eng Manage (IJAIEM) 2(11)
47. Niknam T, Amiri B, Olamaie J, Arefi A (2008) An efficient hybrid evolutionary optimization algorithm based on PSO and SA for clustering. J Zhejiang University Sci A. doi:10.1631/jzus.A0820196
48. Tikalkar A, Khare M (2014) Economic load dispatch using linearly decreasing inertia weight particle swarm optimization Int J Emerg Technol Adv Eng 4(1)
49. Abdullah MN, Bakar AHA, Rahim NA, Mokhlis H, Illias HA, Jamian JJ (2014) Modified particle swarm optimization with time varying acceleration coefficients for economic load dispatch with generator constraints. J. Electr Eng Technol 9(1):15–26
50. Sarath Babu G, Anupama S, Suresh Babu P (2014) Int J Eng Res Develop 9(11):15–23
51. Chaturvedi N et al (2014) A novel approach for economic load dispatch problem based on GA and PSO. Int J Eng Res Appl 4(3):24–31
52. Sivaraman P, Manimaran S, Parthiban K, Gunapriya D (2014) PSO approach for dynamic economic load dispatch problem. Int J Innov Res Sci Eng Technol 3(4)
53. Mistry P, Vyas S (2014) A study on: optimisation of economic load dispatch problem by PSO. Indian J Appl Res 4(6)
54. Khan S, Gupta C (2014) An optimization techniques used for economic load dispatch. Int J Adv Technol Eng Res (IJATER) 4(4)
55. Sinha N, Chakrabarti R, Chattopadhyay PK (2003) Evolutionary programming techniques for economic load dispatch. IEEE Trans Evol Comput 7(1)
56. Nomana N, Iba H (2008) Differential evolution for economic load dispatch problems. Electr Power Syst Res 78:1322–1331
57. Surekha P, Sumathi S (2012) Solving economic load dispatch problems using differential evolution with opposition based learning. WSEAS Trans Inf Sci Appl 9(1)
58. Soni SK, Bhuria V (2012) Multi-objective emission constrained economic power dispatch using differential evolution algorithm. Int J Eng Innov Technol (IJEIT) 2(1)
59. Kumar C, Alwarsamy T Solution of economic dispatch problem using differential evolution algorithm. Int J Soft Comput Eng (IJSCE) 1(6)
60. Balamurugan K, Krishnan SR (2013) Differential evolution based economic load dispatch problem. In: Proceedings of 1st national conference on advances in electrical energy applications, 3–4 Jan 2013
61. Pramod Kumar Gouda, PK, Raguraman H (2013) Economic load dispatch optimization in power system with renewable energy using differential evolution algorithm. In: Proceedings of national conference on advances in electrical energy applications, Jan 3–4 2013
62. Baijal A, Chauhan VS, Jayabarathi T (2011) IJCSI Int J Comput Sci 8(4):1
63. Tankasala GR Artificial bee colony optimisation for economic load dispatch of a modern power system. Int J Sci Eng Res 3(1)

64. Bommirani B, Thenmalar K (2013) Optimization technique for the economic dispatch in power system operation. Proceedings of national conference on advances in electrical energy applications, Jan 3–4 2013
65. Musirin I et al (2008) Ant colony optimization (ACO) technique in economic power dispatch problems. In: Proceedings of the international multi-conference of engineers and computer scientists (IMECS), vol 2I, Hong Kong 19–21 March 2008
66. Rahmat NA, Musirin I, Abidin AF (2013) Differential evolution immunized ant colony optimization (DEIANT) technique in solving weighted economic load dispatch problem. Asian Bull Eng Sci Technol (ABEST) 1(1):17–26
67. Effatnejad R, Aliyari H, Tadayyoni, Abdollahshirazi A (2013) Int J Techn Phys Probl Eng (IJTPE) 5(15):2
68. Rahmat NA, Musirin I (2013) Differential evolution immunized ant colony optimization technique in solving economic load dispatch problem. Sci Res 5(1B)
69. Vasovala PJ, Jani CY, Ghanchi VH, Bhavsar PHK (2014) Application of ant colony optimization technique in economic load dispatch problem for IEEE-14 Bus System. IJSRD —Int J Sci Res Dev 2(2)
70. Mishra NK et al (2011) Economic dispatch of electric power using clone optimization technique. Elixir Electr Eng 40:5155–5158
71. Padmini S, Sekhar Dash S, Vijayalakshmi, Chandrasekar S (2013) Comparison of simulated annealing over differential evolutionary technique for 38 unit generator for economic load dispatch problem. In: Proceedings of national conference on advances in electrical energy applications, 3–4 Jan 2013
72. Swain RK, Sahu NC, Hota PK (2012) Gravitational search algorithm for optimal economic dispatch. In: 2nd international conference on communication, computing and amp; security, vol 6, pp 411–419
73. Serapião ABS (2013) Cuckoo search for solving economic dispatch load problem. Intell Control Autom 4(4):385–390
74. Tomassini M (1999) Parallel and distributed evolutionary algorithms: a review. In: Miettinen K, M¨akel¨a M, Neittaanm¨aki P, Periaux J (eds) Evolutionary algorithms in engineering and computer science, pp 113–133. Wiley, Chichester
75. Nada MA, AL-Salami N (2009) System evolving using ant colony optimization algorithm. J Comput Sci 5(5):380–387
76. Dorigo M, Ganbardella L (1997) Ant colony system: a cooperative learning approach to the traveling salesman problem. IEEE Trans Evol Comput 1(1):53–66
77. Younes M, Hadjeri S, Zidi S, Laarioua S (2009) Economic power dispatch using an ant colony optimization method. In: The 10th international conference on sciences and techniques of automatic control & computer engineering ∼ STA, Hammamet, Tunisia, pp 20–22, Dec 2009
78. Dorigo M (1992) Optimization, learning and natural algorithms. PhD Thesis, Dipartimento di Elettronica, Politecnico di Milano, Italy
79. Sonmez Y (2011) Multi-objective environmental/ economic dispatch solution with penalty factor using Artificial Bee Colony algorithm. Sci Res Essays 6(13):2824–2831
80. Sudhakara Reddy K, Damodar Reddy M (2012) Economic load dispatch using firefly algorithm. Int J Eng Res Appl (IJERA) 2:2325–2330
81. Apostolopoulos T, Vlachos A (2011) Application of the firefly algorithm for solving the economic emissions load dispatch problem. Int J Comb 1(3):1–23
82. Wood, AJ, Wollenberg BF (1984) Example problem 4e. Power Generation, Operation and Control, pp 85–88. Wiley
83. Bhattacharya A, Chattopadhyay PK (2010) Biogeography-based optimization for different economic load dispatch problems. IEEE Trans Power Syst 25:1064–1073. http://ieexplore. ieee.org/xpl/tocpreprint.jsp
84. Rashedi E, Saryazdi S (2009) GSA: a gravitational search algorithm. Inf Sci 179:2232–2248

85. Farhat IA, Hawary E (2012) Multi-objective economic-emission optimal load dispatch using bacterial foraging algorithm. In: 25th IEEE Canadian conference on electrical and computer engineering (CCECE)

86. Passino KM (2002) Biomimicry of bacterial foraging for distributed optimization and control. IEEE Control Syst Magaz 22(3):52–67

87. Aruldoss Albert Victoire T, Ebenezer Jeyakumar A (2003) Hybrid PSO-DS for non-convex economic dispatch problems. In: Digest of the proceedings of the WSEAS conferences, Nov 2003

88. Lee SC, Kim YH (2002) An enhanced Lagrangian neural network for the ELD problems with piecewise quadratic cost functions and nonlinear constraints. Electr Power Syst Res 60:167–177

89. Won J-R, Park Y-M (2012) Economic dispatch solutions with piece-wise quadratic cost functions using improve genetic algorithm. Electr Power Energy Syst 25:355–361

90. Park JH, Yang SO, Lee HS, Park M (1996) Economic load dispatch using evolutionary algorithms. IEEE Proc 441–445

91. dos Santos Coelho L, Mariani VC (2007) Economic dispatch optimization using hybrid chaotic particle swarm optimizer. In: IEEE international conference systems, man and Cybernetics, ISIC 2007

92. Bertsekas DB (1976) On the Goldstein-Levitin-Polyak gradient projection method. IEEE Trans Autom Control 21(2):74–184

93. Hénon MA (1976) A two-dimensional mapping with a strange attractor. Commun Math Phys 50:69–77

94. Reynolds RG (1994) An introduction to cultural algorithms. In: Proceedings of the third annual conference on evolutionary programming, SanDiego, CA, USA, pp 131–139

95. Muthu Vijaya Pandian S, Thanushkodi K (2010) Int J Comput Electr Eng 2(3):1793–8163

96. Younes M, Benhamida F (2011) PRZEGLĄD ELEKTROTECHNICZNY (Electrical Review), ISSN:0033-2097, R. 87 NR 10/2011

97. Younes M, Hadjeri S, Zidi S, Houari S, Laarioua M (2009) Economic power dispatch using an ant colony optimization method. In: 10th International conference on sciences and techniques of automatic control & computer engineering, Hammamet, Tunisia, pp 785–794 Dec 20–22

98. Bouzeboudja H, Chaker A, Allali A, Naama B (2005) Economic dispatch solution using a real-coded genetic algorithm. Acta Electrotechnica et Informatica 5(4)

99. Soni N, Dr Pandit M (2012) A fuzzy adaptive hybrid particle swarm optimization algorithmt to solve non-convex economic dispatch problem. Int J Engineering Innov Technol (IJEIT) 1 (4)

100. Manteaw ED, Abungu Odero N (2012) Multi-objective environmental/economic dispatch solution using ABC_PSO hybrid algorithm. Int J Sci Res Publ 2(12)

101. Bharathkumar S, Arul Vineeth AD, Ashokkumar K, Vijay Anand K (2013) Multi objective economic load dispatch using hybrid fuzzy, bacterial Foraging-Nelder–Mead algorithm. Int J Electr EngTechnol 4(3):43–52

102. Ashouri M, Hosseini SM (2013) Application of new hybrid particle swarm optimization and gravitational search algorithm for non convex economic load dispatch problem. J Adv Comput Res Quart 4(2):41–51

103. Dubey HM, Pandit M, Panigrahi BK, Udgir M (2013) Economic load dispatch by hybrid swarm intelligence based gravitational search algorithm. Int J Intell Syst Appl 5(08):21–32

104. Amjadi N, Sharifzadeh H (2010) Solution of nonconvex economic dispatch problem considering valve loading effect by a new modified differential evolution algorithm. Electr Power Energy Syst

105. Biswas A (2009) Hybrid artificial intelligence systems lecture notes in computer science 5572:252–260

106. Lampinen J, Zelinka I (2000) On stagnation of the differential evolution algorithm. In: Proceedings of MENDEL, 6th International Mendel Conference on Soft Computing, Brno, Czech Republic

107. Nadeem Malik T, ul Asar A, Wyne MF, Akhtar S (2010) A new hybrid approach for the solution of nonconvex economic dispatch problem with valve-point effects. Electr Power Syst Res 80(9):1128–1136

108. Malik TN, Abbasi AQ, Ahmad A (2006) Computational framework for power economic dispatch using genetic algorithm. In: Proceeding of the third international conference on informatics in control, automation and robotics (ICINCO), pp 191–194. Stubal, Portugal, Aug 1–5

109. Wood AJ, Wollenberg BF (1996) Power Generation Operation and Control. John Wiley, New York

110. Pai MA (2006) Computer techniques in power system analysis. Tata McGraw-Hill, New Delhi

111. Narayana PP, Latha K (2004) Evolutionary programming based economic power dispatch solutions with independent power producers. In: IEEE international conference on electric utility deregulation, restructuring and power technologies (DRPT), pp 172–177

112. Bhattacharya A, Chattopadhyay PK (2010) Hybrid differential evolution with biogeography based optimization for solution of economic load dispatch. IEEE Trans Power Syst 25(4)

113. Selvakumar I, Thanushkodi K (2007) A new particle swarm optimization solution to nonconvex economic dispatch problems. IEEE Trans Power Syst 22(1):42–51

114. Chiang CL (2005) Improved genetic algorithm for power economic dispatch of units with valve-point effects and multiple fuels. IEEE Trans Power Syst 20(4):1690–1699

115. Hosseini MM, Ghorbani H, Rabii A, Anvari Sh (2012) A novel heuristic algorithm for solving non-convex economic load dispatch problem with non-smooth cost function. J Basic Appl Sci Res 2(2):1130–1135

116. Soroudi A, Ehsan M, Zareipour H (2011) A practical eco-environmental distribution network planning model including fuel cells and non-renewable distributed energy resources. Renew Energy 36:179–188

117. Jain LC, Palade V, Srinivasan D (2007) Advances in evolutionary computing for system design. In: Studies in computational intelligence, vol 66. Springer

118. Koodalsamy C, Simon SP (2013) Fuzzied artificial bee colony algorithm for nonsmooth and nonconvex multiobjective economic dispatch problem. Turkish J Electr Eng Comput Sci 21:1995–2014

119. Younes M (2013) A novel hybrid FFA-ACO algorithm for economic power dispatch. Control Eng Appl Inf 15(2):67–77

120. Sayah S, Zehar K (2008) Modified differential evolution algorithm for optimal power flow with non-smooth cost functions. Energy Convers Manag 49:3036–3042

Unsolved Problems of Ambient Computationally Intelligent TBM Algorithms

Ashish Runthala and Shibasish Chowdhury

Abstract Structural and functional characterization of protein sequences is one of the important areas of biological research. Currently, a small number of experimentally solved protein structures exist in Protein Data Bank (PDB) in comparison to their considerably higher count of sequence available in UniProtKB/Swiss-Prot. Ambient template-based modelling (TBM) algorithms computationally predict conformational details of a protein sequence on the basis of its evolutionary related similarity with other experimentally solved protein structures. Despite several improvements, shortcomings still obstruct the accuracy of every single step of a TBM algorithm. In this study, we discuss the shortcomings as well as probable corrective measures of major TBM algorithm steps like search and selection of the reliable templates, construction of an accurate target–template alignment, model building, and sampling, and model assessment for selecting the best conformation.

Keywords CASP · TBM · Domain · HMM · MODELLER · TM_Score · GDT

1 Introduction

Proteins, the workforce of a cell structure, are encoded by a set of amino acids. The amino acids encoded in a primary protein sequence interact among themselves and the surrounding environment to fold into a functionally active three-dimensional conformation. Structural detail of a protein is very important to understand the complete biological information of a cell.

Currently (as of July 18th, 2015), 110,471 experimentally solved protein structures have been released by the Protein Data Bank (PDB) [5] which is much smaller than 92,000,000 sequences existing in UniprotKB (UNIversal PROTein

A. Runthala · S. Chowdhury (✉)
Department of Biological Sciences, Birla Institute of Technology
and Science, Pilani, Rajasthan 333031, India
e-mail: shiba@pilani.bits-pilani.ac.in

© Springer India 2016
S. Bhattacharyya et al. (eds.), *Hybrid Soft Computing Approaches*,
Studies in Computational Intelligence 611,
DOI 10.1007/978-81-322-2544-7_3

resource KnowledgeBase)/TrEMBL (Translated European Molecular Biology Laboratory) (http://www.uniprot.org/). Even after excluding homologous sequences from this set, the residual 46,950,870 sequences also surpass the count of experimentally solved structures. For a large number of protein sequences, experimentally solved structures are still not solved.

Several experimental methodologies like X-ray, Nuclear Magnetic Resonance (NMR), and Cryo-Electron Microscopy are currently available to determine protein structure. However, for technical complexity and limitations, the gap between the number of sequenced proteins and experimentally solved protein structures is constantly increasing. This sequence–structure gap needs to be bridged to better understand functional details of protein sequences. Computational structure prediction seems to be a realistic method to quickly understand protein structure without any complex experimental procedure. All these computational algorithms predicting a protein structure from its sequence are termed as modeling tools [27].

2 Protein Structure Prediction Algorithms

In the past few decades, several algorithms have been developed to predict protein structures from their primary sequences. These algorithms are evaluated every 2 years for prediction accuracy by a team of researchers in a global blind test termed as Critical Assessment of Structure Prediction (CASP). Started in 1994, CASP efforts have significantly improved the accuracy of modeling algorithms [5, 27, 92]. The protein structure prediction methodologies are categorized into three groups, viz. ab initio or de novo, comparative or homology modeling, and threading.

The ab inito algorithm, being the most challenging problem, attempts to construct a protein model from its sequence details. These algorithms do not employ any structural information available in the form of solved protein structures. It is assumed that the native structure of a protein sequence exists at the lowest energy conformation (*global minima of the energetic landscape*). To search this native conformation, the structure prediction algorithms span the energetic landscape of a protein sequence (target) for searching the most stable conformation with the lowest free energy. These methodologies employ an energy function to define the free energy of a protein structure and also use a well-defined step parameter for quickly spanning the conformational landscape of a target sequence [7]. This methodology is normally employed in several model predictions and refinement algorithms like minimum perturbation, Monte Carlo (MC), Simulated Annealing (SA), Molecular Dynamics (MD) simulations, Genetic Algorithm, and graph theory-based algorithms [7, 32, 59, 88].

Despite several methodological improvements, ab intio algorithms are still not capable of accurately predicting the protein structures. A protein sequence normally has an extremely large and rugged conformational space, wherein the search for the most stable structure becomes a challenging task. On the other hand, as this methodology predicts the protein structure solely from the sequence details, it can predict a novel fold and is normally employed to predict structure for a target sequence that does not share significant sequence similarity with the experimentally solved protein structures.

Homology or comparative modeling on the other hand employs the already solved structures (*templates*) to predict the conformation of a target sequence [59]. It assumes that similar protein sequences encode similar structures and is mostly helpful when the target shares a significant sequence homology with the templates. Threading algorithms are in between these two broad categories and employ both ab initio and comparative algorithms for predicting the best possible optimal structure of a target [59]. Threading predicts reasonably correct structure for a target that shares sequence similarity only for some local segments scattered across its length. Altogether, these three modeling methodologies are grouped under template-based modeling (TBM) and free modeling (FM) [32]. TBM includes the comparative modeling and threading and exploits the evolutionary relationship

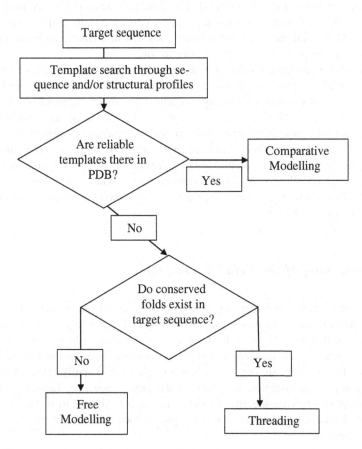

Fig. 1 Various groups of protein structure prediction algorithms

between target and template sequences, whereas the ab initio algorithms are grouped under FM category [88]. All these three modeling methodologies predict models with varying accuracy and modeling reliability [59]. These methodologies are diagrammatically represented below in Fig. 1.

3 TBM Algorithms

The protein structure can be easily predicted for a sequence that shares a sequence or structural similarity with the templates in PDB. Although always correct, a sequence similarity normally implies an evolutionary relationship between the target and template sequences. Hence, a target sequence can be easily modeled if its functionally similar homologous or orthologous structures are available in the PDB and can be accurately aligned with the target.

Ideally, the proteins are often encoded by a set of structurally conserved folds (*sequence segments of a few residues*). The threading methodology computes the best set of structural segments, something like 3D jigsaw puzzle, to model the target sequence. Hence, all the structural domains encoded in a target can be accurately modeled in an overall correct target conformation.

Although comparative modeling and threading algorithms are grouped within the broader TBM category, clearly demarcating them by a reliable boundary is difficult, especially while employing evolutionarily distant templates [59, 89]. Hence, the current TBM algorithms employ both threading and comparative modeling for predicting the protein conformation. A typical TBM algorithm usually consists of steps like template search, template selection, aligning target and template sequences, model construction, model assessment, and model refinement. All these steps are equally responsible for constructing a reliable model for the target sequence. However, the first four steps are the most important steps for predicting an accurate protein model.

3.1 Screening of the Reliable Templates

For a target sequence, the first step is to search the correct template(s) that maximally spans the target sequence [60]. It is observed that a target sharing a high sequence identity (~ 40 %) to the template can be accurately modeled with an accuracy level of a medium-resolution NMR structure or a low-resolution X-ray structure. This is normally the case when a single evolutionarily related template provides an almost complete coverage of the target sequence. However, as the target–template sequence identity decreases, the models diverge from the templates and a set of related templates is required to predict the accurate conformation of the target sequences.

Templates are normally searched through several programs. These programs are categorized into two groups as pairwise comparison and profile search. While the former category includes Basic Local Alignment Search Tool (BLAST) and FASTA [52] that compute a pairwise alignment of the target sequence against the PDB, the latter algorithms like Position Specific Iterative Basic Local Alignment Search Tool (PSI-BLAST) [1], HMMER [20], Intermediate Sequence Search (ISS) [72], Comparison Of Multiple Alignments (COMA) [46], Domain Enhanced Lookup Time Accelerated BLAST (DELTA-BLAST) [8], and HHPred [67] compare the target sequence profile against the template sequence profile. Pairwise sequence comparison methods heuristically search the target and template sequences for the high-scoring segment pairs to compute their statistically significant alignment. Pairwise- and profile-based template search tools utilize PDB along with databases like Structural Classification of Proteins (SCOP) and Protein FAMily (PFAM) [68] and are able to find conserved structural segments from several templates to maximally cover the target sequence. The profile-based methods construct and compare the target and template profiles to efficiently exclude the distant templates sharing an insignificant sequence similarity with target. It sometimes results in the low-scoring and unreliable hits that do not share a significant sequence similarity with the target [19]. It normally occurs due to the consideration of a low E-value threshold for constructing sequence profile. The profile-based methods also compare the PSI-blast-based secondary structure PREDiction (PSIPRED) [10] of the target against the templates to probabilistically select the correct ones [33]. Recently, it was realized that this algorithm considers the unreliable target sequence segments on either side of the high-scoring target chunk that is well aligned with the template(s), and it has therefore encountered a greedy-natured template consideration error in PSI-BLAST [26].

The template search algorithms employ several scoring measures like residue substitution matrices, sequence identity, gap penalties, sequence similarity, and predicted structural similarity to rank a correct template against the target sequence [59]. Profile-based methods extract features like structural context, topological orientation, and solvent accessibility for every amino acid position of the target sequence. Structural context and topological orientation are especially very crucial to construct biologically correct profiles of transmembrane protein sequences.

It is also observed that some templates give false positive results against several targets due to the availability of short conserved segments [40]. The low-scoring templates, searched for a target sequence, are scored through other measures to affirm their modeling credibility [19, 76]. In order to avoid the false positive template matches, the template dataset is generally culled at a specific sequence identity threshold so that no two hits share a considerable structural similarity [60]. It provides the conformational diversity of templates for reliably modeling of the target sequence.

All searched templates are further scored through several other measures like substitution matrix BLOcks SUbstitution Matrix (BLOSUM) [29] and E-value to measure their statistical reliability. The BLOSUM62 matrix, constructed through the sequences not sharing more than 62 % sequence similarity, is normally the default scoring scheme employed by the template search algorithms. It is highly competent to detect even the distant sequence similarity relationship. The additional scoring measures traditionally employ the same residue substitution scores at different locations. This type of traditional template scoring, employed by PSI-BLAST, has been extensively improved and implemented by methods like Context-Specific BLAST (CS-BLAST) [2, 6] and HHPred [67]. These methods consider the target–template residue substitution score, as per the constructed alignment, along with its implied structural context. These algorithms probabilistically score the target residue INsertion or DELetion (INDEL) against the templates to find the reliable hits.

3.2 Template Selection

The objective of the protein modeling is to predict an accurate protein structure and so the best set of templates must be employed. Even when the correct templates are available for a target sequence, the template search algorithms fail to screen them for the target sequences. Still, if the correct templates are properly screened, consistently selecting their best set through consideration of functional linkage, the evolutionary and phylogenic relationship of the well-annotated functional motifs is yet another problem. However, if several continuous or discontinuous structural domains are available in several top-ranked templates, the selection of their best becomes a complicated problem to accurately predict the complete target conformation.

The phylogenic relationship of the target and template sequences; structural accuracy of templates in terms of B-factor, R-factor, R_{free}, and resolution; sequence identity; statistically high-alignment score; and E-value between target and template sequences are normally considered to select the reliable templates from the searched hits [63]. In CASP10, template ranking was employed by several TBM algorithms like RAPTOR-X, HHPred, and LEE to select the top-ranked hits for a target sequence [35]. CpHmodels algorithm ranks templates on the basis of length and substitution matrix scores. While the target–template residue composition difference [61] is considered by COMA [46], DCLAB [44], and ZHOU-SPINE-D [85], the sequence identity, similarity score, and coverage span are considered by Jones UCL and FAMSD algorithms [11]. In FOLDPRO algorithm, alignment score and E-value are considered to be rank templates [90]. PSIPRED [33] predicted that structural similarity compared against the templates is also employed by Zhang [38], Chuo_fams, and CIRCLE algorithms [90]. In these global structure prediction CASP tests, the reliable templates were also selected through the HMM sequence profile employed along with a conservative E-value threshold of 10^{-20} and target sequence coverage of 75 % by the MULTIple-level COMbination (MULTICOM) CLUSTER ([90], www.predictioncenter.org). BIOICM considered a significant E-value threshold of 0.005 to select the templates or homologues closest to the target sequence (www.predictioncenter.org). FALCON TOPO considered templates that were encoding functionally similar annotated structural domains and were structurally similar to the top-ranked selected templates within a specific Root-Mean-Square Deviation (RMSD) threshold (www.predictioncenter.org). Some algorithms like LEE, PLATO, Chou-fams, ZHOU-SPINE-DM, and MeilerLab have even used the PDB culling at specified sequence identity thresholds (www.predictioncenter.org). These methodologies generally employed PDB95 culling to select templates and here only a representative hit was considered from the set of templates sharing more than 95 % sequence identity (www. predictioncenter.org). Sadowsky on the other hand excluded the unreliable hits that did not span at least 75 % target sequence for his PROTAGORAS algorithm (www.predictioncenter.org). Selection of the best set of templates for an accurate model prediction then requires consideration of all the available template scoring measures. The modeling algorithm LEE considers the top 20 templates or alignments for predicting an accurate near-native model of the considered target sequence. Quite similar to this LEE algorithm, MULTICOM employs only the top 10 scoring hits (www.predictioncenter.org, [90]). These algorithms, respectively, select the top 20 and top 10 templates to employ the minimal possible number of high-scoring hits for maximally covering the target sequence. These methodologies did not consider the statistically insignificant hits for spanning the target as it may decrease the modeling accuracy by diluting the structural topology extracted from the selected high-scoring templates, reliably spanning the target sequence.

3.3   Target–Template Alignment

An accurate alignment is solely responsible for predicting the accurate target model through the correctly selected template(s). The selected templates are aligned with the target sequence. The target–template alignment employed by the template search tool can also be normally employed to model the target sequence. However, the initial alignment accuracy can be improved by employing an additional alignment construction step. The additionally employed algorithms are specifically trained to detect distant evolutionary relationship that is normally observed only for short protein sequences and may therefore be useful to improve modeling accuracy especially for bigger or multi-domain targets or when sequence identity between the target–template sequences falls below 30 %.

Several programs computing a domain-specific local alignment (*based on Smith–Waterman algorithm*) and global alignment (*based on Needleman–Wunsch algorithm*) are employed [21, 37, 50, 53, 54, 59, 73, 91]. For improving the accuracy of an alignment, these algorithms have been constantly developed and defined as TCOFFEE [50], Sequence and secondary structure Profiles Enhanced Multiple alignment (SPEM) [91], PRALINE (Profile ALIgNmEnt) [54], Multiple Alignment using Fast Fourier Transform (MAFFT) [37], MUltiple Sequence Comparison by Log- Expectation (MUSCLE) [21], PROfile Multiple Alignment with predicted Local Structure (PROMALS) [53], CLUSTALW [73], HMMER [20], STructural

Alignment of Multiple Proteins (STAMP), and Combinatorial Extension (CE) [65] to improve the accuracy of Multiple Sequence Alignment (MSA) [30].

Alignment algorithms, normally scored through affine gap penalties and residue substitution scores, are employed for constructing a pairwise target–template alignment or a MSA of several templates with the target sequence. MSA can be constructed through a progressive or a structural alignment of templates. The progressive alignment methodology successively aligns the templates before aligning it further with the target sequence. It computes all combination of pairwise alignments of the selected templates and then progressively adds the next most similar template starting with the best pair. However, if the initial template employed to seed this alignment or the progressive order is incorrect, it produces an inaccurate alignment which may be difficult to accurate manually. Other MSA construction strategy employs the structural information of templates to construct the best possible alignment. It evaluates the structural similarity of templates in terms of RMSD by structural superimposition, and it maps all the conformational motifs within a pre-defined RMSD cutoff of the templates into an alignment. While aligning distantly related template sequences with similar structures, this alignment results in a more accurate alignment as it reasonably keeps the conserved structural folds of templates at the same loci in the MSA [31]. Although these algorithms have been extensively developed through computationally complex measures, the constructed pairwise and multiple sequence alignments are not consistently correct. Therefore, the biological credibility of the aligned structural folds is sometime screened through the Dictionary of Protein Secondary Structure (DSSP) database [36].

The redundant or inefficient templates are also routinely removed to improve the accuracy of the MSA constructed through the left over representative templates. LEE algorithm employs this methodology to select the representative templates for constructing a reliable MSA by removing all the redundant and insignificant hits with TM score falling outside 0.6 and 0.975 [35]. Other than these alignment measures, the sequence alignment of templates or sub-structural segments is also considered to map the target sequence through PFAM. As structural alignment of templates may enforce the alignment construction as per the employed structural details of templates, it may consequently dilute and inaccurately localize the essential structural segment of some templates.

Recently, molding approach was also employed to predict the target models through several alternative target–template alignments for selecting the best predicted conformation [16]. Templates have been even aligned globally to maximally span the target sequence. Several algorithms have even employed the complete target alignment of the top-ranked template for spanning the additional target segments through the other high-scoring hits (www.predictioncenter.org). The minimal number of top-ranked templates has also been employed to best span the target sequence for constructing a reliable model in Protein Homology/analogY Recognition Engine (PHYRE2) and HHPred algorithms [90].

Zhang server employs globally aligned templates for accurate TBM prediction and local structural segments for FM [87]. FAIS-Server has even employed all templates and alignments to extract the best available information for the entire target sequence. For constructing an improved alignment, several algorithms including MARINERII, PROTAGORAS, and CaspIta have employed pairwise structural alignment and secondary structure information (www.predictioncenter. org). To construct an accurate MSA, the top-ranked template has been employed by MULTICOM-CLUSTER as a seed to construct a reliable MSA and ultimately build a correct protein model topology [90]. Among these alignment methodologies, JIANG_THREADER employed the MODELLER SALIGN algorithm to construct MSA of the considered templates on the basis of their structural topology (www. predictioncenter.org). Some prediction methodologies have even considered a cutoff on the maximal allowed overlap of the templates, like the DISTILL algorithm has employed a maximal overlap cutoff of nine residues among all of its selected top-ranked hits for accurately spanning the target sequence [48, 90].

3.4 Model Construction

Once the reliable alignment is constructed for the target sequence, it can be employed to model the target sequence. The structural information extracted from the template residues is applied to the aligned target residues. The template scaffolds covering the target sequence are then employed to construct the complete all-atom target structure by modeling the unaligned target segments through the ab initio algorithm. The model construction algorithms are generally grouped in the following six categories.

3.4.1 Rigid-Body Assembly

This algorithm considers the target–template aligned structural segments as rigid bodies and integrates them by considering the unaligned target segments as the loop regions or linker segments. It models the loop segments either by the ab initio algorithm or by employing the PDB database segments that best fit the structural topology of the target segments harnessed from the templates. The algorithms COMPOSER and Swissmodel employ this algorithm in a semiautomatic and automatic manner, respectively [39, 64].

3.4.2 Hexapeptide Assembly

This algorithm considers the target sequence as a set of hexapeptides and screens PDB for the best set of structurally fitting and matching Cα segments. This segment matching or coordinate reconstruction algorithm models the unaligned target segments either by the structures available within PDB database or through ab initio modeling. SEGMOD [42] and 3D-JIGSAW [4] algorithms employ this methodology. Even the loop constructing algorithms are considered to be member of this group as loops also fit the available structural segments of templates, which is the best possible way for mapping the target sequence [34].

3.4.3 Structural Restraint Satisfaction

This modeling methodology optimally satisfies the structural restraints of a target sequence through the considered templates. In this category, MODELLER has been extensively employed to construct a protein model [62]. It attempts to best fit the structural restraints harnessed through the templates for the mapped equivalent residues for a target sequence. As per the alignment, the distance and angular restraints taken from template residues are applied onto the equivalent target residues. The geometric restraints are then employed in an energy function through force field equations, normally expressed as a Probability Density Function (PDF), to construct a target model with a proper topology [9, 62]. This initial model normally encodes several non-physical local atomic clashes and that needs further structural refinement [60]. Hence, the models are energetically relaxed through MD and SA [15] to generate a set of alternative models. The best near-native conformation can then be selected among these models through several assessment measures, as explained later in the text. While these algorithms predict the complete all-atom target model, several others specifically predict Cα backbone, side chains, and loop conformation. Accurate modeling of the loop segments or INDELS of a target–template alignment is very important to construct an accurate protein model.

3.4.4 Loop Modeling

Loops are normally predicted through ab initio and knowledge-based methods [19, 24]. Knowledge-based methods extract the best fitting structural segments of loops from the PDB database to structurally fit the topology of the proximal target residues (*Loop stems*). This method predicts accurate model only if the best fitting and correct loop segment is available in the PDB. This algorithm has been implemented in several programs like MODLOOP [25] and Superlooper (www.predictioncenter. org). However, it can also give wrong results if loop incorrectly localizes the stem segment and disturbs the complete model. The ab initio modeling of the loop segments is worked out normally to construct the complete structure of the considered target sequence. If this loop structure is incorrectly predicted, its topology can disturb the target model with its atomic interactions or clashes.

3.4.5 Side-Chain Modeling

Side-chain orientations are also separately modeled in a predicted protein conformation to improve their structural localization without destabilizing the complete target model. Side-chain orientations are normally fixed through two methodologies. The first category employs different conformations of side chains (*rotamers*) from high-resolution experimental structures to model the side chains [79]. The second methodology extracts the template residue topology, as per the considered alignment, for the corresponding target residues without altering the topological orientation of backbone residues.

The first method is easier to apply on the protein models as it superimposes the side chains on the predicted backbone topology of the target model. So it optimally fixes the side-chain rotamers of the predicted models without altering the conformation of the backbone residues. It can reliably predict the side chains even for an incorrect structural topology of the model. This rotamer-based algorithm predicts the side-chain conformation along with an energetic assessment to predict an accurate model. While adding the side-chain rotamers, the van der Waals (VDW) surface is normally employed as the dielectric boundary as it provides a simple way to approximately consider the plausible effects of structural fluctuations for the static structure of the predicted model [69]. This rotamer-based side-chain modeling methodology is employed in the Side Chains with a Rotamer Library (SCWRL) program that has a repository of the side-chain rotamers normally available in high-resolution experimentally solved PDB structures. SCWRL incorporates the energetically favored side-chain rotamers to the modeled protein backbone of the considered model for constructing the complete model [77]. Several studies have proved that correctly localized side chains are very important for a biologically accurate model and are often involved in studying the protein–protein interactions [57].

3.4.6 Modeling Through the Cα Backbone

Modeling the target structure through Cα backbone harnesses the backbone topology of templates and employs it as a chassis to construct the complete model. This methodology is employed by several programs like PowerfUL CHain Restoration Algorithm (PULCHRA) [22], Maxsprout [74], and REconstruct atomic MOdel (REMO) [43]. In these methods, Cα backbone of the target sequence is constructed using reliable structural segments of the top-ranked templates [70]. The constructed protein backbone models are devoid of side chains which are then additionally incorporated through the side-chain modeling algorithms.

3.5 Model Assessment

The predicted set of models for a target sequence is assessed through several measures to select the best possible conformation. This model assessment step evaluates all the predicted models for the errors that are normally incorporated during the modeling process. All the predicted models are either individually assessed or mutually compared (*termed as clustering*) [90] to select the best predicted conformation. The clustering algorithms assess the stereo-chemical correctness of the protein models through energetics and statistical potentials [41].

Several physico-chemical energetic and knowledge-based assessment measures are used to select the best predicted model for a target sequence. For example, AL0 assesses the accuracy of the employed target–template alignment. It normally estimates the alignment errors in the constructed model. The alignment shift error is estimated in the constructed models through the alignment and template information. AL0 estimates the alignment accuracy with no residue shifts and it reflects the correct alignment. Similarly AL1, AL2, AL3, and AL4 estimate the alignment accuracy with an allowed residue shift threshold of 1, 2, 3, and 4, respectively. The sequence-guided structural assessment also employs Levitt Gerstein Algorithm (LGA)-guided structural superimposition within a 5 Å deviation against the selected template [84].

Predicted residue topology is considered to be correctly modeled if it falls within 3.8 Å distance deviation against the equivalent residue in the considered template. The second set of measures evaluating the model accuracy assesses correct topology of secondary structural elements, solvent exposed surface area, residue interactions, hydrogen-bonding network and molecular packing, molecular environment, and charge stability [41, 59]. All the predicted sets of model decoys are ranked through these scores to select the best conformation. Some of the other assessment measures are GA341, Molecular Probability Density Function (MOLPDF), Discrete Optimized Potential Energy (DOPE), and Normalized_ DOPE_Score (Z_Score), and are normally employed by MODELLER.

Additional algorithms like Atomic NOn-Local Environment Assessment (ANOLEA), PROtein structure CHECK (PROCHECK), Alignment-based Quality Assessment (AQUA), PROtein Structure Analysis (PROSA), BIOTECH, VERIFY3D, ERRAT, and EVAluation of protein structure prediction servers (EVA) are also employed to assess the accuracy of predicted models. All these measures are grouped into a single category as Model Quality Assessment (MQA) by CASP [41, 80, 84]. Distance-scaled, Finite Ideal-gas REference (DFIRE), and DISOrder PREDiction (DISOPRED) algorithms have been additionally employed under this category in CASP9 and CASP10 to select the best predicted conformation [82, 90]. The CASP also employed several different algorithms like LEE, ZICO, DOMFOLD, RAPTOR, FIEG, MUFOLD, PRECORS, FAIS@HGC, PLATO, GS-METAMQAP, DISOCLUST, ZHOU-SPARX, Threading ASSEmbly Refinement (TASSER), SphereGrinder, and 3D-JIGSAW ([90], www.predictioncenter.org) to reliably select the best target models.

Several other measures like Longest Continuous Segment (LCS) [84], TM_Score, Global Displacement Test Total Score (GDT-TS), Global Displacement Test High Accuracy (GDT-HA), and RMSD are also normally employed to structurally screen the models and select the best conformation [59]. LCS measure employs LGA method and calculates the longest segment of model conformation that falls within a specific structural deviation of Cα RMSD cutoff against the selected templates [84]. TM_Score computes the structural similarity of the two models as per their sequence alignment-guided optimal structural superimposition. Through alignment information, TM_Score maps the equivalent residues from target and template residues and optimally superimposes these structures through the features extracted from Voronoi tessellation:

$$
\text{TM\_Score} = \frac{1}{\text{Target\_length}} \sum_{i=1}^{\text{Alignment\_Length}} \left(\frac{1}{1 + \left(\frac{\text{Distance}^i}{\text{Distance}^o} \right)^2} \right). \tag{1}
$$

The TM_Score, represented as Eq. 1, varies from 0 to 1 with 0.5 being the lower cutoff affirming the structural similarity of the two structures [80].

TM_Score is computed on the basis of the target sequence length and it calculates the proportion of aligned residues that are within a minimal distance deviation. It is normalized by a Distance^0 parameter that is computed as $1.24\sqrt[3]{\text{LM} - 15} - 1.8$, and it makes this score independent of the sequence lengths. Due to this length normalization, two protein models of a considered target sequence, predicted through different templates or alignments, can be accurately compared.

GDT evaluates the average proportion of Cα residues of a predicted model that are within a predefined distance deviation from the equivalent residues of the considered template in a sequence-independent optimal structural superimposition [32]. GDT-TS score computes the average fraction of Cα residues of model structure that are present within a predefined distance threshold of 1, 2, 4, and 8 Å. On the other hand, GDT-HA measures average fraction of Cα residues that are present within a predefined distance threshold of 0.5, 1, 2, and 4 Å. As GDT-TS employs a higher distance deviation to calculate an average similarity score, it correctly distinguishes two different models that are structurally deviant. However, as the GDT-HA measure evaluates the same average score at lower distance deviations, it efficiently distinguishes even pretty close models.

Another assessment measure RMSD computes the squared average distance deviation between the equivalent residues of the model and the selected template. Its optimal value is normally calculated through the optimal superimposition of the model and template structures by employing the rotational and translational shifts of one structure. It can be employed for estimating the accuracy of Cα backbone atoms or the complete model. Although it numerically makes some sense, it is not a biologically reliable measure. It is possible that a local structural segment is deviant in the structures which may still share a similar overall conformation [80]. Likewise, if an incorrect template is selected to model a target sequence, the constructed model will not be an accurate conformation even if it shows a lower RMSD score against the selected template.

3.6 Model Refinement

The model predicted through the reliable and top-ranked template(s) frequently suffers from several atomic clashes which disturb topology and hydrogen-bonding network of the overall structure. To further improve these algorithms, all the energetically or biochemically disallowed contacts can be represented as a distance matrix calculated among all the atoms of the model [28]. A well-predicted model is therefore expected to be free of all the atomic clashes. Energetic refinement methods are therefore employed for energy minimization of the predicted conformation to improve its structural topology [86]. The algorithms employed by these refinement methodologies are similar to those employed for the ab initio modeling of the protein sequences. If the target–template alignment is correctly constructed, it can be employed to structurally fix the well-aligned as well as structurally conserved segments of the target sequence. The unaligned area of the target sequence or the target residues which are not aligned against the considered templates can then be accurately modeled by screening its conformational space. As this strategy samples a small number of the unaligned target residues in the conformational space, it decreases the sampling complexity to a great extent.

In the energetic refinement algorithms, two parameters, a correct energy equation representing the native energy of a protein model and an efficient step parameter, employed to screen the conformational space, and specifically play an important role to predict the accurate model structure. MC and MD samplings have been employed here for energetically refining the protein models and have correctly predicted the backbone topology of the models with accurately localized side chains [13]. These algorithms are implemented in force field equations, and PDF functions are employed in many tools including Assisted Model Building with Energy Refinement (AMBER) [51], GROningen MAchine for Chemical Simulations (GROMACS) [23], and Chemistry at HARvard Molecular Mechanics (CHARMM) [78].

Structurally incorrect segments of the protein models have been explicitly sampled to further improve the sampling accuracy [55]. Predicted target models have also been structurally refined through the model clustering methodology by the 3D-JIGSAW algorithm. Through clustering, this algorithm employs structurally variant segments of the predicted models as hotspots and predicts the consensus model with a possibly improved topology for these hotspot segments.

Although there has been extensive work to improve the sampling accuracy, several shortcomings still obstruct the accuracy of these energetic refinement algorithms [86]. The current energetic refinement methodologies normally fail to drive the predicted models away from the topology of the considered templates toward the actual native conformation of the target sequence and do not consistently improve the accuracy of the predicted conformations [48].

4 Unsolved Problems of a TBM Algorithm

As TBM algorithms can pave way to develop efficient computational algorithms that can certainly bridge ever-increasing gap between the count of protein sequences and experimentally solved structures, major unsolved problems of the TBM algorithm are individually explained below for each of the protein modeling steps and few algorithm is discussed which can further develop new robust efficient methodologies.

4.1 Screening and Selecting Reliable Templates

While searching reliable templates for a target sequence, several problems are occasionally encountered. Secondary structure prediction tools are normally employed to accurately align the target against the templates for selecting the reliable structures. When target sequence is assigned an incorrect secondary structure, template search algorithms also screen unreliable templates for the considered target sequence. These errors normally result in an incorrect placement of gaps in the target–template alignment and it constructs an inaccurate and biologically futile alignment. Wrongly assigned gap can dissect the secondary structure segments of the target. Moreover, if the insertion or deletion is not a loop conformation, which is usually and wrongly considered by researchers, it may lead to fatal topological errors in the predicted model. It is because INDEL segments may actually encode structurally important secondary structure information.

Although several algorithms attempt to construct a correct target–template alignment by comparing their sequence profiles, reliable alignments are not consistently constructed. Hence, there is still a requirement of a robust algorithm to harness reliable structural topology of INDELS by database, predictive or conformational search methodologies for correctly aligning the target–template sequences.

PDB culling is normally employed by modeling algorithms. It screens the structurally most similar hit among the related templates to select the most related representative structure. This culling step does not evaluate the similarity of templates against target and it may choose a distant hit as a representative structure. It is because this selected representative hit may be structurally similar to all of its other considered hits, although it could be an evolutionarily distant and unreliable template for target. Algorithms still need to be computationally more intelligent and quick to perform sequence similarity check of clustered hits among themselves and also against target.

Target segments unaligned with the top-ranked template(s) are sometime pruned to construct satisfactory structure of the well-aligned target residues. Complete target model is sometimes constructed by threading the unaligned target sequence through additional reliable templates, although it might not predict accurate models consistently. To construct an accurate target model, target sequence should be maximally covered through reliable templates. Template scoring system should consider the target to accurately evaluate template ranking score. Templates that additionally span the target over the considered top-ranked template(s) and construct a correct target model topology, as estimated by its structural assessment against the selected template, are more reliable. Currently, the templates searched through different searched algorithms are considered as different entities and their close relationship in terms of structural similarity is not at all considered. This mutual structural similarity can reliably select additional templates over the selected ones to reliably construct an improved model.

Different template search algorithms are routinely employed to select reliable templates for a target. However, it is still possible that correct templates are not properly screened. For a target sequence, it is expected from a template search algorithm that it will screen all the reliable hits. Therefore, to answer this normally posed query, we screened efficiency of the usually employed template search measure HHPred to select the best hits for a CASP10 target sequence T0752 with 156 amino acid residues. It is a TBM-HA target encoding a single structural domain, as assessed by CASP. To save computational time, we considered MODELLER PDB95 dataset (*PDB culled at* 95 % *sequence identity*) for selecting a reduced PDB database of 41,967 structures as representative candidates for a total of 96,920 available structures, as on January 16, 2014. These 41,967 structures were then individually employed as templates to model the T0752 target for assessing their modeling accuracy in terms of TM_Score against the actual native structure.

As employed by CASP10 assessors, we also assessed these T0752 models for the structurally conserved 148 residue segments covering 2–149 amino acids. This modeling step yielded structures with TM_Score accuracy ranging from 0 to 1, as shown in Table 1. All the 41,967 models, constructed through the selected representative templates, are grouped at every successive 0.1 TM_Score intervals, as enlisted in Table 1. Most of these models show an insignificant modeling accuracy with TM_Score lesser than 0.3. As expected, only 42 templates showed the

Table 1 TM_Score range of the culled template set models

TM_Score range	Number of models
0–0.1	692
0.1–0.2	39,776
0.2–0.3	1430
0.3–0.4	27
0.4–0.5	28
0.5–0.6	8
0.6–0.7	3
0.7–0.8	2
0.8–0.9	0
0.9–1	1
Total	41,967

TM_Score accuracy higher than 0.4, and therefore all of their structurally similar hits, rejected earlier, were now considered. The single-template models constructed through these templates yielded only 24 models, with TM_Score higher than 0.7, as enlisted in Table 2. Table 2 represents all these accurate templates as Y (Yes) or N (No), respectively, for their presence or absence in the PSI-BLAST or HHPred results. The sequence identity and BLOSUM62 scores, computed through the employed pairwise alignments of these templates against the target sequence, are also listed here. It attempts to show that the correct templates are not efficiently screened by both the current best PSI-BLAST and HHPred template search and selection tools. Among these 42 templates and 141 structurally similar hits and as highlighted in Table 2, only two templates 4GB5_A (*Actual experimental structure of target sequence*) and 3B8L_A (*Top-ranked HHPred template*) were correctly screened by both MODELLER's PDB95 and HHPred's PDB70 culled datasets. The hits screened by PSI-BLAST and HHPred algorithms are enlisted as Y and N in Table 2 to show their presence or absence in the search results. Moreover, quite intriguingly, 22 templates other than the two HHPred resultant hits also share a

Table 2 Modeling accuracy of 24 hits with TM_Score more than 0.7

Sl. no	Hit	TM_Score ≥ 0.7	GDT-TS	Hits resulted in		Sequence identity	BLOSUM62 Score
				PSI-BLAST	HHPred		
1	4GB5_A	0.999	10000	Y	Y	97.37	5.25
2	4STD_C	0.836	72.774	N	N	18.12	0.34
3	4STD_B	0.786	67.295	N	N	19.38	3.12
4	4STD_A	0.782	65.753	N	N	19.38	3.12
5	7STD_B	0.775	64.897	N	N	19.38	3.12
6	3STD_A	0.774	63.699	N	N	19.5	3.14
7	3STD_B	0.774	64.555	N	N	19.5	3.14
8	6STD_A	0.774	64.384	N	N	19.38	3.12
9	5STD_A	0.772	63.185	N	N	19.38	3.12
10	1STD_A	0.771	65.068	N	N	19.5	3.14
11	5STD_B	0.77	64.726	N	N	19.38	3.12
12	6STD_B	0.77	64.726	N	N	19.38	3.12
13	7STD_A	0.77	62.842	N	N	19.38	3.12
14	3STD_C	0.768	62.842	N	N	19.5	3.14
15	5STD_C	0.767	63.699	N	N	19.38	3.12
16	6STD_C	0.767	63.014	N	N	19.38	3.12
17	7STD_C	0.766	63.699	N	N	19.38	3.12
18	2STD_A	0.763	63.356	N	N	20.75	3.77
19	3B8L_D	0.735	64.384	N	N	25.33	0.83
20	3B8L_A	0.733	63.87	N	Y	25.08	0.83
21	3B8L_C	0.733	63.87	N	N	25.33	0.8
22	3B8L_B	0.73	62.5	N	N	25.33	0.83
23	3B8L_E	0.73	63.87	N	N	25.33	0.8
24	3B8L_F	0.73	64.384	N	N	24.75	0.83

significant similarity with the target sequence. Despite this high similarity, as compared to 3B8LA, all the correct templates were not properly ranked and selected by HHPred. This experiment quite clearly shows that the accurate templates much better than 3B8L_A were also available during CASP10. However, these templates were not screened by different algorithms including HHPred.

4.2 Target–Template Alignment

A target–template alignment is expected to slither and map the correlating residues that perform or are expected to execute the same functional role in the target as well as the template structures. This target–template sequence or structural similarity is most plausibly due to their evolutionarily related relationship. Several algorithms are normally employed to construct the best possible target–template alignment. Target–template sequence alignment needs to be inspected for the biologically significant structural similarity of all aligned residues. Alignment shift or incorrect residue assignment error leads to an incorrect target model topology that cannot be properly rectified in model building, sampling, or refinement stages. For a difficult TBM target, several local target sequence areas show a considerable structural divergence against several templates. Hence, for constructing a correct target–template alignment, target and template profiles are, respectively, constructed through hhblits [56] and HHSearch [67].

While hhblits employs Uniprot20 to construct sequence profiles, HHSearch further employs hhfliter to construct profile alignments. Several alternative target–template alignments are normally constructed through varying scoring parameters like residue depth, secondary structure, biologically incompatible alignment of charged residues, solvent accessibility, and unfavorable burial of charged residues to construct an accurate model [3]. Despite employing these rigorous measures, employed target–template profiles routinely consider even evolutionarily distant or unreliable hits. Target sequence profile spanned with the template profile alignment can pose a major computational challenge to select top-ranked reliable templates in a biologically significant alignment for constructing an inaccurate target model.

One of the unsolved alignment problems is the prediction of discontinuous domains (*discontinuous target*) encoding the structural segments. As domains are structural subunits that can fold and evolve independently, correct identification of domain boundaries is important to precisely understand evolutionarily related functional mechanism of a protein. It is also important to correctly model target sequence through threading with accurate template segments. Several methods harnessing domain boundary information of target through an alignment with homologous templates and families have been developed. However, when close templates do not maximally cover target sequence, or when the target–template sequence identity falls below 30 %, domain boundary prediction becomes very difficult.

As protein structures are more conserved than sequences, several algorithms even compare predicted conformation of target sequence with templates to accurately parse its domain boundaries. A target sequence normally encodes multiple structural domains, and hence if its domain boundaries can be correctly parsed, its conformation can be accurately predicted. The currently employed automated algorithms do not accurately parse target sequence against templates and so an algorithm that involuntarily parses target domains accurately is mandatorily required. It would be further expected to splendidly work for both small single-domain as well as large multi-domain targets. Although there are many algorithms like LOcal MEta Threading Server (LOMETS), threading target sequences through several other complementary approaches like MUlti-Sources ThreadER (MUSTER), Profile Profile Alignment (PPA-I), Sequence Alignment and Modeling system (SAM-T02), PROtein Structure Prediction and Evaluation Computer Toolkit (PROSPECT2), HHPred, Sequence, secondary structure Profiles And Residue-level Knowledge-based energy Score (SPARKS2), and SP3 have been employed for this purpose (www.predictioncenter.org). While these MUSTER, PPA-I, SAM-T02, HHPred, and PROSPECT2 algorithms employ an internal ITASSER PDB70 library, the SPARKS and the SP3 algorithms employ PDB40 library [81]. These libraries are normally utilized along with class, architecture, topology and homology (CATH) to accurately predict domain boundaries for a target sequence. Reliable parsing of a domain employs the analysis of linker segments connecting evolutionarily conserved template domains. Despite considering all the complex sequences or structural databases through computationally heavy algorithms, accurate localization of domain boundaries of all structurally conserved elements encoded in a target sequence still remains a major challenge.

Accepting all these challenges for constructing a reliable target–template alignment and considering the fact that multiple templates are mostly employed to model a target sequence, construction of a reliable MSA becomes a mathematical nondeterministic polynomial time complexity problem. Correct target–template alignment is normally computed through dynamic programming measure which usually employs Smith–Waterman and Needleman–Wunsch alignment algorithms for protein and DNA sequences, respectively. Dynamic programming-based MSA construction is ideally a complex problem. The computational complexity of this problem is based on the total number of considered sequences and the average length of each of the considered sequences [49]. However, constructing MSAs through dynamic programming becomes practically impractical as computational complexity of dynamic programming increases exponentially with the increase of sequence number and sequence length.

Alternately, progressive alignment algorithms like CLUSTALW have been developed for constructing MSAs through similarity-based guided tree [73]. However, it fails when errors occurring in early alignment stages cannot be fixed later and, as an attempt to solve them, several other algorithms like TCOFFEE [50], MUSCLE [21], and SPEM [91] have been developed. Most of these algorithms attempt a rigorous optimization of the consistency-based scoring function instead of employing a heuristic progressive alignment approach. A correct relationship between optimal scoring functions and alignment accuracy needs to be properly checked for constructing a biologically significant alignment.

Optimization algorithms attempt to construct the best alignment through global optimization, as employed by Simulated Annealing (SA) and Simulated Annealing Genetic Algorithm (SAGA) scoring functions [71]. Although these SA-based methods are quite versatile, their computational efficiency is normally lower than the problem-specific optimizations. Conformational Space Annealing (CSA) algorithm is developed to further solve this problem and it combines SA, GA, and MC minimization of the alignment score function. It employs distance measures between two templates and controls the conformational diversity of protein structures for constructing their optimal alignment. It can also construct alternative alignments that might correspond to biologically significant MSAs. This algorithm is termed as MSACSA methodology and contrary to other progressive methods it is expected to be more accurate. Despite all these computational challenges, an optimal alignment might be mathematically high scoring, although it may not be biologically correct. It is because, unlike the mathematically top-scored optimal alignment, a biologically correct alignment has accurately aligned residues that are expected to share a structural and functional similarity. Structural alignment of proteins is termed as "Gold Standard" of the alignment accuracy. It is because protein structures are much more evolutionarily conserved than the sequences. Alignment scoring functions are normally tuned to optimize the gap penalties and amino acid substitution scores.

The optimal alignment of templates can be substantially different from their suboptimal structural alignments. Due to marginal difference in the employed gap penalty or substitution matrix scoring measures, a suboptimal alignment can have a low score than the optimal alignment. The scoring changes often generate a set of suboptimal alignments depending on the alignment paths selected during the alignment construction stage. The number of suboptimal alignments increases when the distantly related proteins are considered for constructing an optimal alignment. The number of suboptimal or alternative alignments grows quickly when the alignment construction steps move away from the top-scoring optimal path. These suboptimal alignments can efficiently sample the degree of freedom in an alignment space to make it computationally tractable. Suboptimal alignments are properly represented by Vincenzo Cutello's group [17]. As the loops are also fixed as per employed scoring constraints in an optimal alignment, a suboptimal alignment can plausibly consider geometric distance between considered loops in a structural alignment. Hence, suboptimal or near-optimal alignments may be biologically more accurate. These near-optimal alignments can then be scored through a suboptimal alignment diversity score [12] and the top-scoring alignment can then be selected for modeling the target sequence. These suboptimal alignments can increase modeling accuracy and can sometimes encode a target model with a significantly higher number of topologically correct residues [66]. This structurally corrected target model has biologically acceptable distance deviations among all of its residues, as normally evaluated through a contact map matrix for a protein model. A predicted model with the same contact map, as compared to the respective map of considered template, is therefore considered to be more accurate. The contact map similarity or contact map overlap has been evaluated in a stochastic algorithm PROPOSAL [47]. This recently developed algorithm probabilistically scores the local alignment on the basis of structural similarity of templates and constructs the optimally scored MSA of considered protein structures. So alignment construction tool should construct an array of suboptimal alignments by employing the target and template information and then select the best possible alignment through efficient ranking to construct the accurate target model.

4.3 Model Construction

The protein models are normally constructed through several algorithms. However, some problems still obstruct their path to predict accurate conformations. For selecting the best set of representative templates, among all the structurally similar template sets, a reliable MSA becomes a cumbersome exercise [59]. Several algorithms including MULTICOM have been developed. These algorithms screen an initial set of considered templates for selecting the top-ranked hits to maximally cover the target sequence and construct an accurate model of target sequence [90]. These algorithms employ iterative and computationally complex steps. However, as an attempt to reliably solve a few modeling problems, the protein modeling server or tool should attempt to consider the following constraints for developing an improved and reliable modeling algorithm.

If the selected templates are properly scrutinized, the modeling errors normally caused due to selection of unreliable or distantly related templates can be efficiently solved. The top-ranked templates will certainly construct an accurate MSA. Furthermore, consideration of optimal and suboptimal MSAs will significantly decrease computational degree of freedom to select the best templates through an accurate guide tree (*Phylogenic tree of the selected templates for constructing an accurate structural topology-based sequence alignment*). Sometimes, a biologically incorrect suboptimal alignment can be scored quite significantly to result in an erroneous target model and so HMM-based alignment scoring can help us tackle this problem to some extent.

4.4 Model Assessment

Currently employed measures, viz. MOLPDF DOPE score, GA341, Normalized DOPE score (*Z_Score*), TM_Score, GDT-TS, GDT-HA, RMSD, and structurally correct residues falling within a distance deviation, do not rank the best single- or multi-template model consistently [58].

To select the reliable assessment measure(s), we have compared protein models through MaxCluster to evaluate structural similarity scores [66]. MaxCluster tool considers the target model residues as topologically correct only if their distance deviation against the structurally equivalent residues of the considered template falls within 3.5 Å. Although these assessment measures are mutually interrelated, a single model is not unanimously selected as the most accurate conformation. The considered scoring measures show some sort of non-linearity among their scores. These measures mostly do not reliably select a single-target model as the correct structure for any of the individual domains or an overall accurate topology. These measures are designed to evaluate different structural features of the generated set of protein models for selecting the best conformation.

The referred assessment measures can easily distinguish between a bad model and a correct conformation, and often fail to select the best structure among the close near-native models. Therefore, an increased model sampling becomes yet another confusing task sometimes, especially when the target sequence encodes multiple structural domains. To assess scoring reliability of these measures, we employed a well-studied 182 residue CASP8 target T0390, which is an Ephrin type-A receptor 2 from *Homo sapiens* (*PDB ID: 3CZU*). This single-domain target is modeled using EphB2/EphrinA5 complex structure (*PDB ID: 1SHW*) as a template through MODELLER9.9 for a short 100 model sampling run. It yielded diverse models with varying accuracy for all the considered scores, as graphically represented in Table 3. Table 3 shows the individual undulations of considered assessment measures for the short 100 model sampling run.

As per our results, graphically represented in Table 3, GA341 and MOLPDF appear to be completely unreliable measures. It is because MOLPDF shows significant crinkling deviations even for the close models across the sampled set of 100 predictions. Contrarily, GA341 equally scores each of these models. Although GA341 can distinguish between a good and a bad model, it fails to correctly differentiate between the correct near-native models. Therefore, the MOLPDF and GA341 scores do not reliably correlate with the TM_Score measure. Quite intriguingly, Z_Score measure shows a consistent undulation curl, with negligible deviations to the TM_Score scoring estimations. The energetic assessment or DOPE on the other hand is contrarily untrustworthy. It is because the models with the best DOPE scores are not found to be the best conformations as per the TM_Score measure. DOPE parameter becomes futile to select the accurate model. Contrarily, the normalized energetic assessment fairly distinguishes two models with same TM_Score. Hence, to make the model assessment measure highly effective and robust for selecting the best predicted model, the TM_Score and Z_Score are used together. Hence, for each sampled set of predicted models for a target sequence, the top five models with highest TM_Score are screened to finally select the conformation with the lowest Z_Score and we have found this assessment set very effective. As per Table 3, any assessment measure that preferentially selects the models with a more accurate overall topology, or more residues falling within a minimal distance deviation against considered template, is correct. Here, accurate models should therefore have 121 residues within the considered 3.5 Å distance deviation against the corresponding residues of the actual native structure of T0390 (*Shown as SCR in* Table 3).

Table 3 Different assessment measures of the 100 models sampled for the target T0390

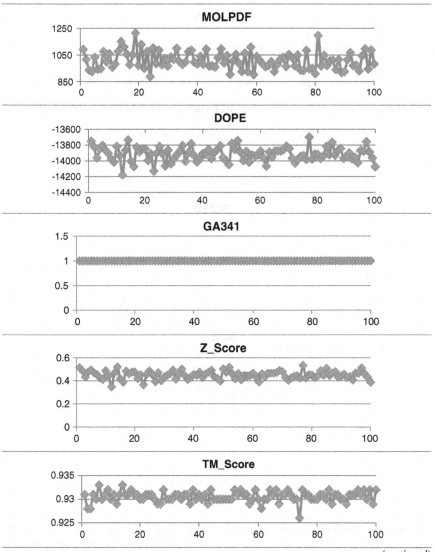

(continued)

Table 3 (continued)

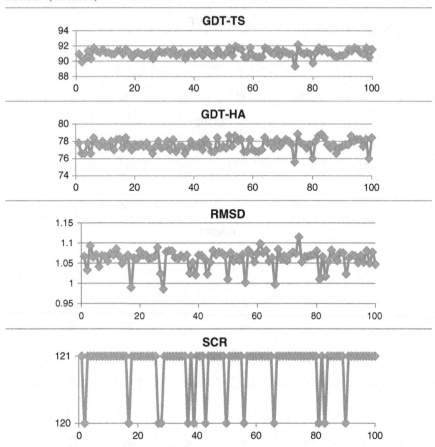

Currently, TM_Score is considered as a highly credible assessment measure. It considers minimal distance deviation between the equivalent residues of two considered structures and emphasizes the count of topologically correct residues. Hence, selecting it as primary assessment measure, the top ten models among the sampled decoys are selected and employed to evaluate mutual non-linearity of other considered assessment measures, as enlisted in Table 4. Here, all the selected models are structurally assessed as per the CASP employed discontinuous domain information spanning the 126 residue segments from 29–123 to 130–160 amino acids [58, 66].

The protein model assessment has become one of the major modeling issues. Although this step has been significantly improved compared to other steps, some problems are still there. It is because the best available and widely used TM_Score

Table 4 Assessment scores of the top ten models for the short 100 model sampling run of T0390

Model	MOL PDF	DOPE score	GA 341	Z_Score	TM_Score	GDT-TS	GDT-HA	RMSD	SCR
6	889.00	−14187.18	1	0.3441	0.933	91.734	78.427	1.041	121
14	1171.95	−13712.27	1	0.5289	0.933	91.331	78.226	1.049	121
9	918.82	−14143.62	1	0.3610	0.932	91.532	78.024	1.055	121
17	1014.07	−13953.83	1	0.4349	0.932	91.532	78.427	0.99	120
28	942.70	−14080.99	1	0.3854	0.932	91.331	78.024	0.986	120
38	957.79	−14170.31	1	0.3507	0.932	90.927	77.218	1.052	121
44	1012.26	−14171.64	1	0.3501	0.932	91.532	78.226	1.052	121
52	965.55	−14068.48	1	0.3903	0.932	91.734	78.629	1.055	121
54	992.71	−14204.79	1	0.3372	0.932	91.935	78.629	1.057	121
64	888.58	−14138.81	1	0.3629	0.932	91.734	78.427	1.055	121

measure also suffers from several logical problems. It does not select the best model topology from the structurally similar near-native decoys. It does not properly assess structural orientation of the individual residues of a protein model. Despite this, the MQA measures are tested and benchmarked in correlation with TM_Score and this further complicates the model assessment step [45].

Recently, Random Forest-Based Protein Model Quality Assessment (RFMQA) algorithm is developed to assess quality of the model structure. To select the best conformation, it statistically compares the structural features of a model in comparison to those predicted for the considered target sequence. Here, for a target sequence, structural features of a predicted model, viz. secondary structure (*predicted through PSIPRED*), solvent accessibility (*predicted through DSSP*), and the energetic potential (*estimated by DFIRE,* Generalized Orientation-dependent All-atom Potential (GOAP) *and RWplus*), are considered [45]. Although it attempted to assess the predicted models very well, it failed to meet expectations due to some logistic reasons. This approach does not measure the depth or distance of model residues from the superficial surface or centroid. So a tightly folded protein model may not have an additional edge over the other predicted decoys. Hence, even during measuring solvent accessibility topology of buried atoms or atoms excluded in computing the solvent accessibility is not considered by RFMQA. This solvent accessibility can also be incorrectly correlated, if the PSIPRED secondary structure assignment fails, even when it correctly played its part and accurately predicted its solvent accessibility score. Besides this, if a predicted model structurally encodes a local fold whose mirror image copies are predicted in several other incorrect models, this algorithm probably fails to select the correct conformation. It is because the fold topology can be quite differently located on the surface or inside the protein core. It may result in varying scores and makes the model selection step very confusing.

Moreover, all the three considered factors employed by RFMQA measure can show mutually nonlinear scorings for a single model, although an actually correct conformation. So the employed random forest regression may wrongly predict its average value during optimization and it may imperfectly over-score some other comparatively incorrect models showing a high information gain (*Gini impurity*) for the randomly selected features. Moreover, energetic assessment of predicted models, as per the DFIRE potential, also sometimes wrongly ranks the correct models. It is because if the correct near-native model is actually predicted at a relatively higher energy level, it is considered as a low-accuracy prediction by DIFRE. Additionally, if incorrect models majorly constitute the considered set of models, the employed random forest algorithm fails altogether as it might not succeed to reliably select the correct models through the evaluation of two randomly selected features. Although these three considered factors are mathematically weighed by RFMQA, it could have failed to reliably select the best predicted model. Therefore, a correct assessment measure to reliably select the best predicted conformations consistently is still required.

4.5 Model Refinement

Although model refinement step has been improved to a great extent, it is still obstructed by some logical problems. It is sometimes observed that structurally refined models do not show an improved model topology, probably being closer to the actual native conformation of target sequence. However, it is also possible that the best refined or probably improved model is not carefully selected from the sampled set of conformational decoys [59]. Recently, side chains were additionally sampled for the considered protein model to further construct an improved model through ProQM assessment. Here, side-chain rotamers were sampled by fixing the backbone topology of the considered protein model. While it sometimes incorrectly packs side chains in the selected model, it retains the correct backbone topology and this reconstruction of side chains has shown an improved accuracy of the predicted models [75]. Although several refinement methodologies have been developed, the correct natural in vivo protein folding process is still not correctly implemented in our protein folding and energetic refinement algorithms. Quite intriguingly, nature does the same in a fraction of second and that pinnacle of accuracy is still not achieved by the current structure prediction algorithms.

5 Conclusion

Different steps of protein structure prediction algorithms fall broadly under soft computing [83] field as these algorithms are tolerable to uncertainty and imprecise answers. These algorithms are quite robust as these algorithms heuristically work out to obtain an optimal solution in a short computational time. To obtain reliable templates and to build model structure, several algorithms are stitched together. Even energetic refinement methodologies involve several approximations to search quickly the conformational landscape of a target sequence. These refinement methodologies are programmed to optimize the scoring functions to compute an optimal solution [14, 18]. Although the soft computing algorithms are highly robust, obtaining an accurate solution is still a major challenge.

Templates are an exquisite source of structural information and are chassis of TBM algorithms which help us predict accurate protein models. Although the homology modeling or TBM algorithms have been improved, the available structural information of templates is still not currently employed correctly. Due to a high degree of freedom to choose templates for employing them in the convincingly high-scoring alignment, their best set cannot be consistent as well as reliably selected for modeling a target. Problems still exist at every single modeling step and all the steps need to be improved altogether to predict correct models not only for the single-domain but also for the multi-domain targets including the transmembrane proteins as well. Some of our corrective measures for the template selection, model sampling, and assessment steps showed a significantly higher modeling accuracy compared to the other available methodologies.

Threading algorithms currently hold the key to successfully predict near-native protein models. Currently, TBM algorithms are highly intermixed with threading methodologies and have become successful for two reasons. Firstly, these algorithms employ a large set of threading templates or structural segments for constructing a top-scoring protein model. Moreover, the structural segments of different templates are considered to model the target without being obstructed with the extremely large conformational space of a target. Secondly, as the best sets of mutually complementary and most favorable structural segments are employed, these algorithms can efficiently predict correct topology for the individual domains as well as the complete target model. However, a highly accurate protein model can only be predicted if the target sequence segments linking the conserved structural elements show a negligible structural diversity and are easily available in PDB.

Although TBM algorithms are popular and predict protein structure quite accurately, we are still far away from the experimental accuracy. These algorithms need to be further studied to predict accurate conformations closer to the native protein structures.

References

1. Altschul SF, Madden TL, Schaffer AA, Zhang J, Zhang Z, Miller W, Lipman DJ (1997) Gapped BLAST and PSI-BLAST: a new generation of protein database search programs. Nucleic Acids Res 25(17):3389–3402
2. Angermüller C, Biegert A, Söding J (2012) Discriminative modeling of context-specific amino acid substitution probabilities. Bioinformatics 28(24):3240–3247
3. Barbato A, Benkert P, Schwede T, Tramontano A, Kosinski A (2012) Improving your target-template alignment with MODalign. Bioinformatics 28(7):1038–1039
4. Bates PA, Kelley LA, MacCallum RM, Sternberg MJE (2001) Enhancement of protein modeling by human intervention in applying the automatic programs 3D-JIGSAW and 3D-PSSM. Proteins Struct Funct Genet 5(5):39–46

5. Berman H, Bourne P, Gilliland G, Westbrook J, Arzberger P, Bhat T (2000) Protein Data Bank. http://www.rcsb.org/pdb/home/home.do, 08 Sept 2014
6. Biegert A, Söding J (2009) Sequence context-specific profiles for homology searching. Proc Nat Acad Sci USA 106(10):3770–3775
7. Bonneau R, Baker D (2001) Ab-initio protein structure prediction: progress and prospects. Annu Rev Biophys Biomol Struct 30:173–189
8. Boratyn GM, Schäffer AA, Agarwala R, Altschul SF, Lipman DJ, Madden TL (2012) Domain enhanced lookup time accelerated BLAST. Biolo Direct 7, 12
9. Brooks BR, Bruccoleri RE, Olafson BD, States DJ, Swaminathan S, Karplus M (1983) CHARMM—a program for macromolecular energy, minimization, and dynamics calculations. J Comput Chem 4(2):187–217
10. Buchan DW, Minneci F, Nugent TC, Bryson K, Jones DT (2014) Scalable web services for the PSIPRED protein analysis workbench. Nucleic Acids Res 41:W349–W357
11. CASP Home Page. www.predictioncenter.org
12. Chen H, Kihara D (2011) Effect of using suboptimal alignments in template-based protein structure prediction. Proteins: Struct, Funct, Bioinf 79(1):315–334
13. Chen J, Charles L, Brooks CL III (2007) Can molecular dynamics simulations provide high-resolution refinement of protein structure?. Proteins: Struct, Funct, Bioinf 67(4):922–930
14. Clerc M, Kennedy J (2002) The particle swarm: explosion stability and convergence in a multi-dimensional complex space. IEEE Trans Evol Comput 6(1):58–73
15. Clore GM, Brunger AT, Karplus M, Gronenborn AM (1986) Application of molecular dynamics with interproton distance restraints to three-dimensional protein structure determination, A model study of crambin. J Mol Biol 191(3):523–551
16. Cozzetto D, Giorgetti A, Raimondo D, Tramontano A (2008) The evaluation of protein structure prediction results. Mol Biotechnol 39(1):1–8
17. Cutello V, Nicosia G, Pavone M, Prizzi I (2011) Protein multiple sequence alignment by hybrid bio-inspired algorithms. Nucleic Acids Res 39(6):1980–1992
18. Dozier G, Bowen J, Homaifar A (1998) Solving constraint satisfaction problems using hybrid evolutionary search. IEEE Trans Evol Comput 2(1):23–33
19. Dunbrack RL Jr (2006) Sequence comparison and protein structure prediction. Curr Opin Struct Biol 16(3):374–384
20. Eddy SR (1998) Profile hidden Markov models. Bioinformatics 14(9):755–763
21. Edgar RC (2004) Muscle: multiple sequence alignment with high accuracy and high through-put. Nucleic Acids Res 32(5):1792–1797
22. Feig M, Rotkiewicz P, Kolinski A, Skolnick J, Brooks CL 3rd (2000) Accurate reconstruction of all-atom protein representations from side-chain-based low-resolution models. Proteins: Struct, Funct, Bioinf 41(1):86–97
23. Fernández-Pendás M, Escribano B, Radivojević T, Akhmatskaya E (2014) Constant pressure hybrid Monte Carlo simulations in GROMACS. J Mol Model 20:2487
24. Fiser A, Fieg M, Brooks CL 3rd, Sali A (2002) Evolution and physics in comparative protein structure modeling. Acc Chem Res 35(6):413–421
25. Fiser A, Sali A (2003) ModLoop: automated modeling of loops in protein structures. Bioinformatics 19(18):2500–2501
26. Gonzalez MW, Pearson WR (2010) Homologous over-extension: a challenge for iterative similarity searches. Nucleic Acids Res 38(7):2177–2189
27. Guo JT, Ellrott K, Xu Y (2008) A historical perspective of template-based protein structure prediction. Methods Mol Biol 413:3–42
28. Hao F, Xavier P, Alan EM (2012) Mimicking the action of folding chaperones by Hamiltonian replica-exchange molecular dynamics simulations: application in the refinement of de-novo models. Proteins: Struct, Funct, Bioinf 80(7):1744–1754
29. Henikoff S, Henikoff JG (1992) Amino acid substitution matrices from protein blocks. Proc Nat Acad Sci USA 89(22):10915–10919
30. Huang IK, Pei J, Grishin NV (2013) Defining and predicting structurally conserved regions in protein superfamilies. Bioinformatics 29(2):175–181

31. Jaroszewski L, Rychlewski L, Godzik A (2000) Improving the quality of twilight-zone alignments. Protein Sci 9(8):1487–1496
32. Jauch R, Yeo HC, Kolatkar PR, Neil DC (2007) Assessment of CASP7 structure predictions for template free targets. Proteins: Struct, Funct, Bioinf 69(8):57–67
33. Jones DT (1999) Protein secondary structure prediction based on position-specific scoring matrices. J Mol Biol 292(2):195–202
34. Jones TA, Thirup S (1986) Using known substructures in protein model building and crystallography. EMBO J 5(4):819–822
35. Joo K, Lee J, Sim S, SY Lee, Lee K, Heo S, Lee I, Lee SJ, Lee J (2014) Protein structure modeling for CASP10 by multiple layers of global optimization. Proteins: Struct, Funct, Bioinf 82(2):188–195
36. Karchin R, Cline M, Mandel-Gutfreund Y, Karplus K (2003) Hidden Markov models that use predicted local structure for fold recognition: alphabets of backbone geometry. Proteins: Struct, Funct, Bioinf 51(4):504–514
37. Katoh K, Kuma K, Toh H, Miyata T (2005) MAFFT version 5: improvement in accuracy of multiple sequence alignment. Nucleic Acids Res 33(2):511–518
38. Kedarisetti BKD, Mizianty MJ, Dick S, Kurgan L (2011) Improved sequence-based prediction of strand residues. J Bioinf Comput Biol 9(1):67–89
39. Kopp J, Schwede T (2004) The SWISS-MODEL Repository of annotated three-dimensional protein structure homology models. Nucleic Acids Res 32(1):D230–D234
40. Kristensen DM, Chen BY, Fofanov VY, Ward RM, Lisewski AM, Kimmel M, Kavraki LE, Lichtarge O (2006) Recurrent use of evolutionary importance for functional annotation of proteins based on local structural similarity. Protein Sci 15(6):1530–1536
41. Kryshtafovych A, Fidelis K (2008) Protein structure prediction and model quality assessment. Drug Discov Today 14(7–8):386–393
42. Levitt M (1992) Accurate modeling of protein conformation by automatic segment matching. J Mol Biol 226(2):507–533
43. Li Y, Zhang Y (2009) REMO: a new protocol to refine full atomic protein models from C-alpha traces by optimizing hydrogen-bonding networks. Proteins: Struct, Funct, Bioinf 76 (3):665–676
44. MacCallum JL, Hua L, Schnieders MJ, Pande VS, Jacobson MP, Dill KA (2009) Assessment of the protein-structure refinement category in CASP8. Proteins: Struct, Funct, Bioinf 77 (9):66–80
45. Manavalan B, Lee J, Lee J (2014) Random forest-based protein model quality assessment (RFMQA) using structural features and potential energy terms. PLoS ONE 9(9):e106542
46. Margelevicius M, Venclovas C (2010) Detection of distant evolutionary relationships between protein families using theory of sequence profile-profile comparisons. BMC Bioinf 11:89
47. Micale G, Pulvirenti A, Giugno R, Ferro A (2014) Proteins comparison through probabilistic op-timal structure local alignment. Frontiers Genet 5:302
48. Moult J, Fidelis K, Kryshtafovych A, Rost B, Hubbard T, Tramontano A (2007) Critical assessment of methods of protein structure prediction—Round VII. Proteins: Struct, Funct, Bioinf 69(8):3–9
49. Nguyen KD, Pan Y, Nong G (2011) Parallel progressive multiple sequence alignment on reconfigurable meshes. BMC Genom 12(5):S4
50. Notredame C, Higgins DG, Heringa J (2000) T-COFFEE: a novel method for fast and accurate multiple sequence alignment. J Mol Biol 302(1):205–217
51. Pany Y (2014) Low-mass molecular dynamics simulation: a simple and generic technique to enhance configurational sampling. Biochem Biophys Res Commun 452:588–592
52. Pearson WR (2014) BLAST and FASTA similarity searching for multiple sequence alignment. Methods Mol Biol 1079:75–101
53. Pei J, Kim BH, Tang M, Grishin NV (2007) PROMALS web server for accurate multiple protein sequence alignments. Nucleic Acids Res 35:W649–W652
54. Pirovano W, Feenstra KA, Heringa J (2007) PRALINE™: a strategy for improved multiple alignment of transmembrane proteins. Bioinformatics 24(4):492–497

55. Qian B, Raman S, Das R (2007) High-resolution structure prediction and the crystallographic phase problem. Nature 450(7167):259–264
56. Remmert M, Biegert A, Hauser A, Söding J (2012) HHblits: lightning-fast iterative protein sequence searching by HMM-HMM alignment. Nat Methods 9:173–175
57. Repiso A, Oliva B, Vives Corrons JL, Carreras J, Climent F (2005) Glucose phosphate isomerase deficiency: enzymatic and familial characterization of Arg346His mutation. Biochimica et Biophysica Acta (BBA)—Molecular Basis of Disease 1740(3):467–4471
58. Runthala A, Chowdhury S (2014) Iterative optimal TM_Score and Z_Score guided sampling significantly improves model topology. In: Proceedings of the International MultiConference of Engineers and Computer Scientists (Lecture Notes in Engineering and Computer Science), March 12–14 Hong Kong, pp 123–128
59. Runthala A, Chowdhury S (2013) Protein structure prediction: are we there yet?, SCI 450. In: Pham TD, Jain LC (eds) Innovations in Knowledge-based Systems in Biomedicine and Computational Life Science, Springer-Verlag Monograph Volume, pp 79–115
60. Runthala A (2012) Protein structure prediction: challenging targets for CASP10. J Biomol Struct Dyn 30(5):607–615
61. Rykunov D, Fiser A (2007) Effects of amino acid composition, finite size of proteins, and sparse statistics on distance-dependent statistical pair potentials. Proteins: Struct, Funct, Bioinf 67(3):559–568
62. Sali A, Blundell TL (1993) Comparative protein modelling by satisfaction of spatial restraints. J Mol Biol 234(3):779–815
63. Sanchez R, Sali A (1997) Evaluation of comparative protein structure modelling by MODELLER-3. Proteins: Struct, Funct, Bioinf 1:50–58
64. Schwede T, Kopp J, Guex N, Peitsch MC (2003) SWISS-MODEL: an automated protein homology-modelling server. Nucleic Acids Res 31(13):3381–3385
65. Shindyalov IN, Bourne PE (1998) Protein structure alignment by incremental combinatorial extension of the optimum path. Protein Eng 11(9):739–747
66. Siew N, Elofsson A, Rychlewski L, Fischer D (2000) MaxSub: an automated measure for the assessment of protein structure prediction quality. Bioinformatics 16(9):776–785
67. Söding J (2005) Protein homology detection by HMM-HMM comparison. Bioinformatics 21:951–960
68. Söding J, Biegert A, Lupas AN (2005) The HHpred interactive server for protein homology detection and structure prediction. Nucleic Acids Res 33:W244–W248
69. Song Y, Mao J, Gunner MB (2009) MCCE2: Improving protein pKa calculations with extensive side chain rotamer sampling. J Comput Chem 30(14):2231–2247
70. Subramaniam S, Senes S (2014) Backbone dependency further improves side chain prediction efficiency in the Energy-Based Conformer Library (bEBL). Proteins: Struct, Funct, Bioinf
71. Takaya D, Takeda-Shitaka M, Terashi G, Kanou K, Iwadate M, Umeyama H (2008) Bioinformatics based Ligand-Docking and in-silico screening. Chem Pharm Bull 56 (5):742–744
72. Teichmann SA, Chothia C, Church GM, Park J (2000) Fast assignment of protein structures to sequences using the intermediate sequence library PDB-ISL. Bioinformatics 16(2):117–124
73. Thompson JD, Higgins DG, Gibson TJ (1994) CLUSTALW: improving the sensitivity of progressive multiple sequence alignment through sequence weighting, position-specific gap penalties and weight matrix choice. Nucleic Acids Res 22(22):4673–4680
74. Tosatto S (2006) Spritz: a server for the prediction of intrinsically disordered regions in protein sequences using kernel machines. Nucleic Acids Res 34:W164–W168
75. Wallner B (2014) ProQM-resample: improved model quality assessment for membrane proteins by limited conformational sampling. Bioinformatics 30(15):2221–2223
76. Wang G, Dunbrack RL Jr (2003) PISCES: a protein sequence culling server. Bioinformatics 19(12):1589–1591
77. Wang Q, Canutescu AA, Dunbrack RL Jr (2008) SCWRL and MolIDE: computer programs for side-chain conformation prediction and homology modeling. Nat Protoc 3(12):1832–1847

78. Wu EL, Cheng X, Jo S, Rui H, Song KC, Dávila-Contreras EM, Qi Y, Lee J, Monje-Galvan V, Venable RM, Klauda JB, Im W (2014) CHARMM-GUI membrane builder toward realistic biological membrane simulations. J Comput Chem 35(27):1997–2004

79. Xiang Z, Honig B (2001) Extending the accuracy limits of prediction for side-chain conformations. J Mol Biol 311(2):421–430

80. Xu J, Zhang Y (2010) How significant is a protein structure similarity with TM-score = 0.5? Bioinformatics 26:889–895

81. Xue Z, Xu D, Wang Y, Zhang Y (2013) ThreaDom: extracting protein domain boundary information from multiple threading alignments. Bioinformatics 29(13):i247–i256

82. Yang T, Zhou Y (2008) Ab-initio folding of terminal segments with secondary structures reveals the fine difference between two closely related all-atom statistical energy functions. Protein Sci 72:1212–1219

83. Zadeh LA (1994) Fuzzy logic, neural networks, and soft computing. Commun ACM 37 (3):77–84

84. Zemla A (2003) LGA—a method for finding 3D similarities in protein structures. Nucleic Acids Res 31(13):3370–3374

85. Zhang T, Faraggi E, Xue B, Dunker AK, Uversky VN, Zhou Y (2012) SPINE-D: accurate prediction of short and long disordered regions by a single neural-network-based method. J Biomol Struct Dyn 29(4):799–813

86. Zhang Y (2008) Progress and challenges in protein structure prediction. Curr Opin Struct Biol 18(3):342–348

87. Zhang Y (2010) I-TASSER: fully automated protein structure prediction in CASP8. Proteins: Struct, Funct, Bioinf 77(9):100–113

88. Zhang Y, Skolnick J (2005) The protein structure prediction problem could be solved using the current PDB library. Proc Nat Acad Sci USA 102(4):1029–1034

89. Zhang Y (2014) Interplay of I-TASSER and QUARK for template-based and ab initio protein structure prediction in CASP10. Proteins: Struct, Funct, Bioinf 82(2):175–187

90. Zheng W, Jesse E, Cheng J (2010) MULTICOM: a multi-level combination approach to protein structure prediction and its assessments in CASP8. Bioinformatics 26(7):882–888

91. Zhou H, Zhou Y (2005) SPEM: improving multiple sequence alignment with sequence profiles and predicted secondary structures. Bioinformatics 21(18):3615–3621

92. Zwanzig R, Szabo A, Bagchi B (1992) Levinthal's paradox. Proc Nat Acad Sci USA 89:20–22

Hybridizing Differential Evolution Variants Through Heterogeneous Mixing in a Distributed Framework

G. Jeyakumar and C. Shunmuga Velayutham

Abstract While hybridizing the complementary constituent soft computing techniques has displayed improved efficacy, the hybridization of complementary characteristics of different Differential Evolution (*DE*) variants (could as well be extended to evolutionary algorithms variants in general) through heterogeneous mixing in a distributed framework also holds a great potential. This chapter proposes to mix competitive *DE* variants with diverse characteristics in a distributed framework as against the typical distributed (homogeneous) Differential Evolution (*dDE*) algorithms found in *DE* literature. After an empirical analysis of 14 classical *DE* variants on 14 test functions, two heterogeneous *dDE* frameworks *dDE_HeM_best* and *dDE_HeM_worst* obtained by mixing best *DE* variants and worst *DE* variants, respectively, have been realized, implemented and tested on the benchmark optimization problems. The simulation results have validated the robustness of the heterogeneous mixing of best variants. The chapter also hybridized *DE* and dynamic *DE* variants in a distributed framework. The robustness of the resulting framework has been validated by benchmarking it against the state-of-the-art *DE* algorithms in the literature.

Keywords Differential evolution · Distributed differential evolution · Mixing of *DE* variants · Co-operative evolution · Heterogeneous mixing

G. Jeyakumar (✉) · C. Shunmuga Velayutham
Department of Computer Science and Engineering, Amrita School of Engineering,
Amrita Vishwa Vidyapeetham, Coimbatore, India
e-mail: g_jeyakumar@cb.amrita.edu

C. Shunmuga Velayutham
e-mail: cs_velayutham@cb.amrita.edu

© Springer India 2016
S. Bhattacharyya et al. (eds.), *Hybrid Soft Computing Approaches*,
Studies in Computational Intelligence 611,
DOI 10.1007/978-81-322-2544-7_4

107

1 Introduction

Differential Evolution (*DE*), proposed by Storn and Price [40], is a simple yet powerful stochastic real-parameter optimization algorithm. It has been proven a robust global optimizer and has been successfully applied to many global optimization problems with superior performance in both widely used benchmark functions and real-world applications. *DE* is a population-based stochastic global optimizer employing mutation, recombination and selection operators like other Evolutionary Algorithms (*EAs*).

However, *DE* has some characteristics that makes it different from other members of the *EA* family. The major differences are in the way the trial vectors are generated and in the selection mechanism employed to transit to the next generation. In fact, *DE* uses a *differential mutation* operation based on the distribution of parent solutions in the current population. It is coupled with recombination with a predetermined target vector to generate a trial vector. Then a one-to-one greedy selection between the trial vector and the target vector is carried out.

This conceptual and algorithmic simplicity, ease of implementation and hence experimentation and high convergence characteristics of *DE* has attracted, is still attracting many researchers who are working on its improvement. This consequently has resulted in multitude of *DE* variants.

Despite the active research on *DE* during the last decade, there are still many open problems. One such problem is the difficulty in choosing the right *DE* variant given an optimization problem. The efficacy of *DE* in solving an optimization problem depends on the selected strategy as well as control parameter values. In addition to that, different optimization problems (depending on modality and decomposability) require different mutation strategies. Few research attempts [36] have been carried out in the literature in this direction.

The multitude of *DE* variants, existing in the literature, display different search efficacies by virtue of their differential mutation-cum-crossover combination. This results in the problem of choosing the right *DE* variant while solving a diverse range of optimization problems. It is understandable that, to solve a diverse range of problems, a *DE* algorithmic framework should display diverse as well as complementary search characteristics. The typical practice in the *DE* literature is to tediously select appropriate *DE* variant to solve different problems. However, this chapter proposes to achieve such diverse as well as complementary search characteristics by mixing the search dynamics of different *DE* variants. Since a distributed framework facilitates exchange of search solutions thus promoting co-operative evolution, it becomes one natural choice for implementing the mixing of search characteristics of different *DE* variants.

In an effort towards this direction, this chapter proposes to mix competitive *DE* variants with diverse characteristics in a distributed framework such that they are allowed to evolve independently but are made to suitably exchange their search information among others. The chapter hypothesizes that such a co-operative evolution by heterogeneous mix of *DE* variants will enhance the efficacy of the system as a whole with robust optimization characteristics as compared to when the variants operate separately.

The remainder of the chapters is organized as follows. Section 2 describes the *DE* algorithm which is followed by a discussion on dynamic *DE* in Sect. 3. Section 4 details the empirical analysis of *DE* variants. Section 5 presents the theoretical insights about the performance of the variants. Section 6 discusses the mixing strategies followed in this chapter. In Sect. 7, the design of experiments is presented. Section 8 presents the detailed analysis on the simulation results. The idea of mixing *DE* and dynamic *DE* is presented in Sect. 9. The benchmarking results are presented in Sect. 10 and finally the last section concludes the work.

2   Differential Evolution (*DE*) Algorithm

DE aims at exploring the search space by sampling at multiple, randomly chosen *NP* D-dimensional parameter vectors (population of initial points), so-called individuals, which encode the candidate solution $x_{i,G} = \left\{x_{i,G}^1, \ldots, x_{i,G}^D\right\}, i = i, \ldots, NP$. The initial population should sufficiently cover the search space as much as possible, by uniformly randomizing individuals, for better exploration. After population initialization an iterative process is started and at each iteration (generation) a new population is produced until a stopping criterion is satisfied. Figure 1 depicts the algorithmic description of a typical *DE*.

At each generation, *DE* employs the *differential mutation* operation to produce a mutant vector $V_{i,G} = \left\{v_{i,G}^1, \ldots, v_{i,G}^D\right\}$, with respect to each individual $X_{i,G}$ the so called target vector, in the current population. The mutant vector is created using the weighted difference of parent solutions in the current population. A number of differential mutation strategies have been proposed in the literature that primarily differs in the way the mutant vector is created. Along with the strategies came a notation scheme to classify the various *DE* variants. The notation is defined by *DE/a/b/c* where *a* denotes the base vector or the vector to be perturbed (which can be a random vector, a best vector, or a sum of target vector and weighted vector difference between random and best vector/random vectors/best vectors); *b* denotes the number of vector differences used for perturbation (which can be one or two pairs); and *c* denotes the crossover scheme used (which can be *binomial* or *exponential*) between the mutant vector $V_{i,G}$ and the target vector $X_{i,G}$ to create a trial vector $U_{i,G} = \left\{u_{i,G}^1, \ldots, u_{i,G}^D\right\}$. The seven commonly used (classical) mutation strategies [30, 34, 35] are listed in Table 1.

In Table 1, $X_{best,G}$ denotes the best parent vector in the current generation, *F* and *K* commonly known as the *scaling factor* or *amplification* factor is a positive real number that controls the rate of evolution of the population. The indices $r_1^i, r_2^i, r_3^i, r_4^i, r_5^i, i \varepsilon 1, \ldots, NP$ are randomly generated anew for each mutant vector and are mutually exclusive $r_1^i \neq r_2^i \neq r_3^i \neq r_4^i \neq r_5^i \neq i$.

Initialize $P_G = \{X_{i,G},...,X_{NP,G}\}$ with $X_{i,G} = \{X^1_{i,G},..., X^D_{i,G}\}$ $i = 1,...,\ NP$; $G = 0$;

Compute $\{f(X_{i,G}),..., f(X_{NP,G})\}$

WHILE stopping criterion is not satisfied *DO*

FOR $i = 1\,to\,NP$

  /* Differential Mutation */

    •   Generate a mutant vector $V_{i,G} = \{v^1_{i,G},...,v^D_{i,G}\}$ for each target vector $X_{i,G}$ via one
         of the classical mutation strategies listed in Table 1.

  /* Crossover */

    •   Generate a trial vector $U_{i,G} = \{u^1_{i,G},...,u^D_{i,G}\}$ for each target vector $X_{i,G}$ by employing
         binomial crossover scheme via equation (1) or using exponential crossover
         scheme via equation (2).

  /* Selection (assuming a minimization problem) */

      IF $f(U_{i,G}) < f(X_{i,G})$*THEN*

        $X_{i,G+1} \leftarrow U_{i,G}$

      ELSE

        $X_{i,G+1} \leftarrow X_{i,G}$

      END IF

END FOR

$G = G+1$; Compute $\{f(X_{i,G}),..., f(X_{NP,G})\}$

END WHILE

Fig. 1 The general structure of a typical *DE* algorithm

Table 1 The commonly used differential mutation strategies

S. no.	Nomenclature	Mutation strategy
1	DE/rand/1	$V_{i,G} = X_{r^i_1,G} + F \cdot \left(X_{r^i_2,G} - X_{r^i_3,G}\right)$
2	DE/best/1	$V_{i,G} = X_{best,G} + F \cdot \left(X_{r^i_1,G} - X_{r^i_3,G}\right)$
3	DE/rand/2	$V_{i,G} = X_{r^i_1,G} + F \cdot \left(X_{r^i_2,G} - X_{r^i_3,G} + X_{r^i_4,G} - X_{r^i_5,G}\right)$
4	DE/best/2	$V_{i,G} = X_{best,G} + F \cdot \left(X_{r^i_2,G} - X_{r^i_3,G} + X_{r^i_4,G} - X_{r^i_5,G}\right)$
5	DE/current-to-rand/1	$V_{i,G} = X_{i,G} + K \cdot \left(X_{r^i_3,G} - X_{i,G}\right) + F \cdot \left(X_{r^i_1,G} - X_{r^i_2,G}\right)$
6	DE/current-to-best/1	$V_{i,G} = X_{i,G} + K \cdot \left(X_{best,G} - X_{i,G}\right) + F \cdot \left(X_{r^i_1,G} - X_{r^i_2,G}\right)$
7	DE/rand-to-best/1	$V_{i,G} = X_{i,G} + K \cdot \left(X_{best,G} - X_{r^i_3,G}\right) + F \cdot \left(X_{r^i_1,G} - X_{r^i_2,G}\right)$

After the differential mutation strategy, *DE* then uses a crossover operation in which the mutant vector $V_{i,G}$ mixes with target vector $X_{i,G}$ and generates a trial vector $U_{i,G}$ or offspring. The two frequently used crossover schemes are binomial (uniform) crossover and exponential crossover. The binomial crossover is defined as follows.

$$u_{i,G}^j = \begin{cases} u_{i,G}, & \text{if}(rand_j[0,1) \le C_r) \vee (J = J_{rand}) \\ x_{i,G}^j, & \text{otherwise} \end{cases} \tag{1}$$

where $C_r \varepsilon(0,1)$ (crossover probability) is a user-specified constant; $rand_j[0,1)$ is the jth evaluation of uniform random number generator; $j_{rand} \varepsilon 1, \ldots, D$ is a random parameter index, chosen once for each i to make sure that at least one parameter is always selected from the mutant vector $V_{i,G}$.

The exponential crossover is defined as follows

$$u_{i,G}^j = \begin{cases} u_{i,G}, & \text{for}\langle n\rangle_D, \langle n+1\rangle_D, \ldots, \langle n+L-1\rangle_D \\ x_{i,G}^j, & \text{for all other } j\varepsilon[1,D] \end{cases} \tag{2}$$

where the acute brackets $\langle\rangle_D$ denote modulo functions with modulus D, the starting index n is a randomly selected integer in the range $[1, D]$, and the integer L (which denotes the number of parameters that are going to be exchanged between the mutant and trial vectors) is drawn from the same range with the probability C_r.

After the mutation and crossover operations, a one-to-one knockout competition between the target vector $X_{i,G}$ and its corresponding trial vector $U_{i,G}$ based on the objective function values decides the survivor, among the two, for the next generation. The greedy selection scheme is defined as follows (assuming a minimization problem without the loss of generality)

$$X_{i,G+1} = \begin{cases} U_{i,G}, & \text{if}(f(U_{i,G})) \le (f(X_{i,G})) \\ x_{i,G}, & \text{otherwise.} \end{cases} \tag{3}$$

The above 3 steps of differential mutation, crossover, followed by selection marks the end of one *DE* generation. These steps are repeated generation after generation until a stopping criterion is satisfied. With seven commonly used mutation strategies and two crossover schemes, there are fourteen possible, so called variants viz. *DE/rand/1/bin, DE/rand/1/exp, DE/best/1/bin, DE/best/1/exp, DE/rand/2/bin, DE/rand/2/exp, DE/best/2/bin, DE/best/2/exp, DE/current-to-rand/1/bin, DE/current-to-rand/1/exp, DE/current-to-best/1/bin, DE/current-to-best/1/exp, DE/rand-to-best/1/bin,* and *DE/rand-to-best/1/exp.*

3 Dynamic Differential Evolution (*DDE*) Algorithm

Despite its high convergence characteristics and robustness, from the point of view of population update, *DE* is still static i.e., *DE* responds to the population progress after a time lag (i.e., one generation). The whole population in *DE* remains unchanged until it is replaced by a new population in the subsequent generation. Inevitably, it results in slower convergence. To alleviate this problem, dynamic versions of *DE* variants *DE/rand/1* and *DE/best/1* (called Dynamic Differential

Evolution) have been proposed in [38, 39]. *DDE* updates population after every one-to-one greedy selection between the trial vector and the target vector in every generation, thus responding to any improvement immediately.

As can be seen from the *DE* algorithm, in Fig. 1, the repeated cycles of differential mutation and crossover do not make use of any progress taking place in the current generation, consequently employing the optimal individuals from previous generation to produce trial vectors. Even if better fit individuals are generated in the current generation, they are reserved for use in the next generation. In contrast, the dynamic differential evolution, shown in Fig. 2, employs a dynamic evolution mechanism in place of the above said static evolution mechanism of classical *DE*.

The dynamic evolution mechanism in *DDE* updates both the current optimal individual with the new competitive individual (if better than the current optimal) and the non-optimal individuals dynamically. Consequently, the trial vectors are always generated using the newly updated population, and thus *DDE* always responds to any progress immediately. Owing to the dynamicity in population update, the creation of every *NP* trial vectors is considered as one generation of *DDE* in the current work as against the usual mutation-crossover-selection cycle as one generation in case of a typical *DE*. Except this significant difference, *DDE* inherits all the basic operations of *DE* and its three crucial control parameters viz. population size (*NP*), scaling factor (*F*), and crossover rate (*C_r*).

Initialize $P_G = \{X_{i,G}, ..., X_{NP,G}\}$ with $X_{i,G} = \{X_{i,G}^1, ..., X_{i,G}^D\}$, $i = 1, ..., NP$; $G = 0$;

Compute $\{f(X_{i,G}), ..., f(X_{NP,G})\}$

WHILE stopping criterion is not satisfied *DO*
   FOR $i = 1$ *to NP*
     /* Differential Mutation */
      • Generate a mutant vector $V_{i,G} = \{v_{i,G}^1, ..., v_{i,G}^D\}$ for the target vector $X_{i,G}$ via
       one of the classical mutation strategies listed in Table 1.
     /* Crossover */
      • Generate a trial vector $U_{i,G} = \{u_{i,G}^1, ..., u_{i,G}^D\}$ for the target vector $X_{i,G}$ by
       employing binomial crossover scheme via equation (1) or using exponential crossover scheme via equation (2).
    /* Selection (assuming a minimization problem) */
    /* *Dynamic Evolution Mechanism* */
      IF $f(U_{i,G}) < f(X_{i,G})$ *THEN*
        $X_{i,G+1} \leftarrow U_{i,G}$
      END IF
      IF $f(U_{i,G}) < f(X_{best,G})$ *THEN*
        $best \leftarrow i$
      END IF
 END FOR
 $G = G + 1$; Compute $\{f(X_{i,G}), ..., f(X_{NP,G})\}$
 END WHILE

Fig. 2 The algorithmic description of *DDE*

It is straight forward to extend the dynamic evolution mechanism to the seven commonly used differential mutation strategies, as listed in Table 1, and two crossover schemes (binomial and exponential), resulting in fourteen possible variants of *DDE*. Following the standard *DE* nomenclature used in the literature, the fourteen *DDE* variants can be written as follows: *DDE/rand/1/bin*, *DDE/rand/1/exp*, *DDE/best/1/bin*, *DDE/best/1/exp*, *DDE/rand/2/bin*, *DDE/rand/2/exp*, *DDE/best/2/bin*, *DDE/best/2/exp*, *DDE/current-to-rand/1/bin*, *DDE/current-to-rand/1/exp*, *DDE/current-to-best/1/bin*, *DDE/current-to-best/1/exp*, *DDE/rand-to-best/1/bin*, and *DDE/rand-to-best/1/exp*.

This work also proposes to mix competitive *DE* and *DDE* variant in a distributed framework. Towards this, the chapter has undertaken an extensive empirical analysis of *DE* variants to identify the most competitive variants.

4 Empirical Analysis of Differential Evolution Variants

The insight about the performance efficacies of the *DE* and *DDE* variants is crucial in identifying competitive variants to model efficient co-operative evolutionary framework. This necessitates a consistent design of experiments and careful choice of test functions for the empirical comparative performance analysis of the above said *DE* algorithms. This chapter investigates the performance of all the variants on different classes of unconstrained global optimization problems. Fourteen test functions, of dimensionality 30, grouped by the features—unimodal separable (f_1, f_2, f_4, f_6 and f_7), unimodal nonseparable (f_3), multimodal separable (f_8, f_9 and f_{14}), and multimodal nonseparable (f_5, f_{10}, f_{11}, f_{12} and f_{13})—have been chosen from [46, 30] for benchmarking the variants. The test functions are one of the well-known benchmarks in the literature to test *EAs* for global optimization. The details of the test functions f_1–f_{14} are presented below.

f_1—Sphere model

$$f_{sp}(x) = \sum_{n=1}^{n} x_i^2$$
$$- 100 \leq x_i \leq 100;$$
$$x^* = (0, 0, \ldots, 0);$$
$$f_{sp}(x^*) = 0;$$

f_2—Schwefel's Problem 2.22

$$f_{sch1}(x) = \sum_{i=1}^{n} |x_i| + \prod_{i=1}^{n} |x_i|$$

$-10 \leq x_i \leq 10;$

$x^* = (0, 0, \ldots, 0);$

$f_{sch1}(x^*) = 0;$

f_3—Schwefel's Problem 1.2

$$f_{sch2}(x) = \sum_{i=1}^{n} \left(\sum_{j=1}^{i} x_j \right)^2$$

$-100 \leq x_i \leq 100;$

$x^* = (0, 0, \ldots, 0);$

$f_{sch2}(x^*) = 0;$

f_4—Schwefel's Problem 2.21

$$f_{sch3}(x) = \max_{i}\{ |x_i|, \ 1 \leq i \leq n \}$$

$-100 \leq x_i \leq 100;$

$x^* = (0, 0, \ldots, 0);$

$f_{sch3}(x^*) = 0;$

f_5—Generalized Rosenbrock's Function

$$f_{gr}(x) = \sum_{i=1}^{n} \left| 100\left(x_{i+1} - x_i^2\right)^2 + (x_i - 1)^2 \right|$$

$-30 \leq x_i \leq 30;$

$x^* = (0, 0, \ldots, 0);$

$f_{gr}(x^*) = 0;$

f_6—Step Function

$$f_{st}(x) = \sum_{i=1}^{n} \left(\lfloor x_i + 0.5 \rfloor \right)^2$$

$-100 \leq x_i \leq 100;$

$x^* = (0, 0, \ldots, 0);$

$f_{st}(x^*) = 0;$

f_7—Quartic Function with Noise

$$f_{qf}(x) = \sum_{i=1}^{n} i x_i^4 + random[0, 1)$$

$$- 1.28 \leq x_i \leq 1.28;$$

$$x^* = (0, 0, \ldots, 0);$$

$$f_{qf}(x^*) = 0;$$

f_8—Generalized Schwefel's Problem 2.26

$$f_{ah4}(x) = \sum_{i=1}^{n} \left(x_i \sin\left(\sqrt{|x_i|} \right) \right)$$

$$- 500 \leq x_i \leq 500;$$

$$x^* = (420.9678, \ldots, 420.9678);$$

$$fsch4(x^*) = -12569.486618164879;$$

f_9—Generalized Restrigin's Function

$$f_{grf}(x) = \sum_{i=1}^{n} \left[x_i^2 - 10 \cos(2\pi x_i) + 10 \right]$$

$$- 5.12 \leq x_i \leq 5.12;$$

$$x^* = (0, 0, \ldots, 0);$$

$$f_{grf}(x^*) = 0;$$

f_{10}—Ackley's Function

$$f_{ack}(x) = 20 + e - 20 \exp\left(-0.2 \sqrt{\frac{1}{n} \sum_{i=1}^{n} x_i^2} \right) - \exp\left(\frac{1}{n} \sum_{i=1}^{n} \cos(2\pi x_i) \right)$$

$$- 5.12 \leq x_i \leq 5.12;$$

$$x^* = (0, 0, \ldots, 0);$$

$$f_{ack}(x^*) = 0;$$

f_{11}—Generalized Griewank's Function

$$f_{gri}(x) = \frac{1}{4000} \sum_{i=1}^{n} x_i^2 - \prod_{i=1}^{n} \cos\left(\frac{x_i}{\sqrt{i}} \right) + 1$$

$$- 600 \leq x_i \leq 600;$$

$$x^* = (0, 0, \ldots, 0);$$

$$f_{gri}(x^*) = 0;$$

f_{12}—Generalized Penalized functions

$$f_{gpf12}(x) = \frac{\pi}{n} \left\{ 10\sin^2(\pi y_1) + \sum_{i=1}^{n-1} (y_i - 1)^2 \left[1 + 10\sin^2(\pi y_{i+1}) \right] + (y_n - 1)^2 \right\} + \sum_{i=1}^{n} u(x_i, 10, 100, 4)$$

$-50 \leq x_i \leq 50;$

$x^* = (0, 0, \ldots, 0);$

$f_{gpf12}(x^*) = 0;$

f_{13}—Generalized Penalized functions

$$f_{gpf13}(x) = 0.1 \left\{ \sin^2(\pi 3x_1) + \sum_{i=1}^{n-1} (x_i - 1)^2 \left[1 + \sin^2(3\pi x_{i+1}) \right] + (x_n - 1)^2 \left[1 + (2\pi x_n) \right] \right\} + \sum_{i=1}^{n} u(x_i, 5, 100, 4)$$

$-50 \leq x_i \leq 50;$

$x^* = (0, 0, \ldots, 0);$

$f_{gpf13}(x^*) = 0;$

f_{14}—Bohachevsky Functions

$$f_{bf}(x) = x_i^2 + 2x_{i+1}^2 - 0.3\cos(3\pi x_i) - 0.4\cos(4\pi x_{i+1}) + 0.7$$

$-100 \leq x_i \leq 100;$

$x^* = (0, 0, \ldots, 0);$

$f_{bf}(x^*) = 0;$

$$\text{where } y_i = 1 + \frac{1}{4}(x_i + 1)$$

$$u(x_i, a, k, m) = \begin{cases} k(x_i - a)^m, & x_i \rangle a \\ 0, & -a \leq x_i \leq a \\ k(-x_i - a)^m, & x_i \langle -a \end{cases}$$

All the test functions considered for benchmarking have an optimum value at zero except f_8. In order to show the similar results, the description of f_8 was adjusted to have its optimum value at zero by subtracting the optimal value 12569.486618164879 [30].

A typical *DE* has three control parameters that must be set by the user viz. *NP* (population size), *F* (scaling factor), and C_r (crossover rate) [41]. The population size *NP* has been kept as 60 with the maximum number of generations as 3000 (consequently, the maximum number of function evaluations calculate to 180,000). The moderate population size and number of generations have been chosen to demonstrate the efficacy of *DE* variants in solving the chosen problems. The choice of moderate population size is also influenced by an empirical study of the impact of *NP* on *DE* variants, which suggested that a large population size affects the ability of the algorithm to find the correct search direction [19]. The variants will stop before the maximum number of generations only if the tolerance error (which has been fixed as an error value of 1×10^{-12}) with respect to the global optimum is obtained. Following (Mezura-Montes 2006, Personal Communication with Mezura-Montes), the range for the scaling factor, F, is defined as [0.3, 0.9] and the value for $F \in$ [0.3, 0.9] is generated anew at each generation for all the variants. This value of F, when generated, is used for K (scaling factor in *current-to-rand*, *current-to-best* and *rand-to-best* variants) as well. Each experiment was repeated 100 times (runs). In all the runs, the maximum number of generations has been kept as *GMax* = 3000 and the maximum number of function evaluations (*MaxFE = NP * GMax*) as constant to provide a fair performance comparison.

The C_r, crossover rate, parameter is tuned for each *DE* variant-test function combination. Eleven different values for the C_r viz. {0.0, 0.1, 0.2, 0.3, 0.4, 0.5, 0.6, 0.7, 0.8, 0.9, 1.0} have been tested for each *DE* variant-test function combination. For each combination of variant-test function-C_r value, 50 independent runs have been conducted. Based on the obtained results, a bootstrap test has been conducted in order to determine the confidence interval for the mean objective function value. The C_r value corresponding to the best confidence interval (i.e. 95 %) was chosen to be used in the subsequent experiments. Based on an empirical study, sample size for the bootstrap test has been fixed as 500. The study involved finding the confidence interval with different bootstrap sample sizes viz. 500, 1000, 2000, 4000, 5000, 10000, 20000, 30000, 40000, and 50000. However, no noticeable change in the best confidence interval has been observed while increasing the bootstrap sample size i.e., the best C_r value for the sample size of 500 continued to be the best with large sample sizes also.

As *EAs* are stochastic in nature, 100 independent runs are performed for each Variant (from *DE* and *DDE* suite of variants)-Function combination by initializing the population for every run with uniform random initialization within the search range. The empirical comparative performances of *DE* and *DDE* algorithms have been analyzed using Probability of Convergence ($P_c(\%)$). The probability of convergence [12] is measured as the mean percentage of number of successful runs out of total number of runs, $P_c = (nc/nt)\%$, where nc is the total number of successful runs made by each variant for all the functions considered and nt is the total number of runs. The probability of convergence for each variant is measured to find the variants with higher probability of convergence for each variant-function combination.

The identified 14 classical *DE* variants are analyzed by measuring their probability of convergence (P_c). The $P_c(\%)$ values measured for the 14 *DE* variants on the 14 benchmark problems are presented in Table 2, by way of an example. The analysis identified *DE/rand/1/bin*, *DE/rand/2/bin*, *DE/best/2/bin*, and *DE/rand-to-best/1/bin* as the most competitive variants. The analysis also identified few worst performing variants viz. *DE/rand/2/exp*, *DE/current-to-rand/1/exp*, *DE/current-to-best/1/exp*, and *DE/rand-to-best/1/exp*. The comparison of *DDE* variants with their counterpart *DE* variants based on $P_c(\%)$ is also presented in Table 2. Invariably, the *DDE* variants outperformed their *DE* counterparts. Interestingly, the most competitive variants are *DDE/rand/1/bin*, *DDE/rand/2/bin*, *DDE/best/2/bin*, and *DDE/rand-to-best/1/bin* (the same *DE* variants identified earlier).

Table 2 The $P_c(\%)$ measured for the 14 *DE* and *DDE* variants

S. no.	Variant	$P_c(\%)$		S. no.	Variant	$P_c(\%)$	
		DE	DDE			DE	DDE
1	DE/rand/1/bin	73.93	82.20	8	DE/best/2/exp	43.36	48.93
2	DE/rand/1/exp	48.50	55.79	9	DE/current-to-rand/1/bin	46.43	41.93
3	DE/best/1/bin	23.00	26.00	10	DE/current-to-rand/1/exp	0.21	0.00
4	DE/best/1/exp	10.21	16.07	11	DE/current-to-best/1/bin	46.14	47.43
5	DE/rand/2/bin	57.36	64.70	12	DE/current-to-best/1/exp	0.36	0.50
6	DE/rand/2/exp	29.00	33.50	13	DE/rand-to-best/1/bin	74.21	81.80
7	DE/best/2/bin	74.57	82.80	14	DE/rand-to-best/1/exp	48.93	55.00

After an extensive empirical comparative performance analysis of 14 classical *DE* variants and their 14 *DDE* counterparts, we identified */rand/1/bin, */rand/2/bin, */best/1/bin, and */rand-to-best/1/bin (* represents both *DE* and *DDE*) variants as the most competitive *DE/DDE* variants for the given test suite.

5   Theoretical Insight

Having identified four competitive *DE* variants and their *DDE* counterparts, this work attempts to analyze the behavior of only the identified competitive *DE* variants to gather insight for possible improvement and/or effective design of the variants to make them more robust and efficient. The behavioral analysis of the four identified *DE* variants has been attempted through deriving analytical relationship as well as empirical observation of the evolution of population variance as against the number of generation.

In fact a theoretical relationship between the expected population variance after mutation-crossover and the initial population variance has been derived for *DE/rand/1/bin* in the literature [47]. We extended the theoretical measure of population diversity, derived for *DE/rand/1/bin* to the three other identified variants viz. *DE/rand/2/bin*, *DE/best/2/bin*, and *DE/rand-to-best/1/bin* [20]. The extended theorem summarizes the computation as follows (original theorem adapted as is from [47]).

Let $x = \{x_i, \ldots, x_m\}$ be the current population, $Y = \{y_1, \ldots, y_m\}$ the intermediate population obtained after applying the mutation, and $Z = \{z_1, \ldots, z_m\}$ the population obtained by crossing over the populations x and Y. If F and K are the parameters of the mutation step and P_c is the parameter of the crossover step, then the expected population variance for *DE/rand/1/bin* [47] is $\left(2F^2 p_c + 1 - \frac{2p_c}{m} + \frac{p_c^2}{m}\right) \mathrm{Var}(x)$.

The derived expected population variance for *DE/rand/2/bin* is

$$\left(4F^2 p_c + 1 - \frac{2p_c}{m} + \frac{p_c^2}{m}\right) \text{Var}(x). \tag{4}$$

The derived expected population variance for *DE/best/2/bin* is

$$\left(4F^2 p_c + \frac{(1-p_c)^2}{m} + \frac{m-1}{m}(1-p_c)\right) \text{Var}(X)$$
$$+ \left(\frac{m-1}{m}\right) p_c(1-p_c)(\bar{x} - x_{best})^2. \tag{5}$$

The derived expected population variance for *DE/rand-to-best/1/bin* is

$$\left(\left(\frac{m-1}{m}\right)\left(1 + p_c\left(k^2 - 2k + \frac{m}{m-1}2F^2\right)\right) + \frac{(1-p_c)^2}{m}\right) \text{Var}(x). \tag{6}$$

The derived theoretical measures were validated with success using a simple experimental setup. The experimental study involved the sphere function-f_1 of dimension 30, with a population size of $NP = 50$, a fixed value of $p_c = 0.5$, and four different values of parameter F viz. 0.1, 0.12, 0.2, and 0.3 for *DE/rand/2/bin* and *DE/best/2/bin* (by way of an example). The empirical population variance for both variants has been computed by component-wise variance averaging for 100 independent runs. Table 3 displays the (theoretical) expected and empirical population variances for *DE/rand/2/bin* and *DE/best/2/bin*. For the sake of convenience, 30 generations have been considered.

Figure 3 displays the (theoretical) expected and empirical population variances obtained for *DE/rand/2/bin* and *DE/best/2/bin*. As can be seen from Fig. 3, the decreasing and increasing pattern of theoretical expected variance is matched by the empirical variance. However, there is a large difference between the theoretical and empirical variances, which may be attributed to the fact that the theoretical derivation ignores the restriction that the indices of chosen solution vectors should not be equal to the parent vector index.

The experimental study is then extended, to gain insight about the behavior, for all four *DE* variants using four benchmark functions (viz. $f_3, f_4, f_9,$ and f_{10}) from the test suite. Interestingly, Zaharie [47] observed that there is a critical value (F_c) for scaling factor F under which the variance decreases and over which it increases. For the assumed population size (i.e., 50) and p_c value (0.5), the F_c values for *DE/best/2/bin*, *DE/rand/2/bin*, and *DE/rand-to-best/1/bin* are calculated to be, respectively, 0.44, 0.087, and 0.67. Zaharie calculated the F_c value to be 0.12 for *DE/rand/1/bin*. Accordingly, the empirical and expected population variances

Table 3 Empirical and theoretical variance measured for *DE/rand/2/bin* and *DE/best/2/bin*

DE/rand/2/bin's empirical/theoretical variance					DE/best/2/bin's empirical/theoretical variance				
G	$F = 0.1$	$F = 0.12$	$F = 0.2$	$F = 0.3$	G	$F = 0.1$	$F = 0.12$	$F = 0.2$	$F = 0.3$
1	8.75/8.92	8.81/9.14	8.46/9.65	8.44/11.31	1	8.44/8.41	8.48/8.90	8.68/8.58	8.39/8.09
4	9.11/9.76	9.52/10.70	11.34/16.44	15.51/36.72	4	1.55/0.62	1.59/0.35	2.35/0.98	3.66/1.81
7	9.18/10.36	9.94/11.95	13.82/24.36	24.13/88.36	7	0.38/0.09	0.42/0.05	0.83/0.19	1.97/0.56
10	9.37/10.98	10.29/13.35	16.54/36.09	37.87/212.61	10	0.09/0.01	0.11/0.01	0.29/0.04	1.05/0.18
13	9.53/11.66	10.71/14.91	20.29/54.47	60.24/511.55	13	0.02/0.00	0.02/0.00	0.10/0.01	0.54/0.06
16	9.64/12.38	11.26/16.67	24.34/79.21	93.68/1230.85	16	0.01/0.00	0.01/0.00	0.04/0.00	0.29/0.02
19	9.74/13.13	11.66/18.61	29.54/117.36	145.21/2961.55	19	0.00/0.00	0.00/0.00	0.01/0.00	0.15/0.01
22	9.93/13.04	11.99/20.79	35.78/173.86	227.65/7125.80	22	0.00/0.00	0.00/0.00	0.01/0.00	0.09/0.00
25	10.11/14.79	12.57/23.23	43.36/257.59	354.89/17145.41	25	0.00/0.00	0.00/0.00	0.00/0.00	0.05/0.00
28	10.27/15.70	13.09/15.34	52.87/381.63	550.62/41253.64	28	0.00/0.00	0.00/0.00	0.00/0.00	0.03/0.00
30	10.39/16.33	13.33/27.93	59.18/495.97	749.29/74075.03	30	0.00/0.00	0.00/0.00	0.00/0.00	0.01/0.00

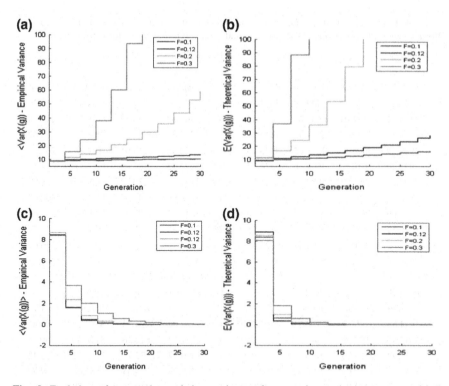

Fig. 3 Evolution of expected population variance after mutation and crossover—empirical evolution and theoretical expected evolution, respectively, in **a**, **b** for *DE/rand/2/bin* and **c**, **d** for *DE/best/2/bin*

for the four *DE* variants have been measured for two values of parameter *F* (one less than F_c and the other greater than F_c).

Tables 4, 5, 6 and 7 display the (theoretical) expected and empirical population variances of all four *DE* variants for functions f_3, f_4, f_9, and f_{10}, respectively. As can be seen from Tables, the *'pure'* rand variants' (*DE/rand/1/bin* and *DE/rand/2/bin*) runs display gradual loss of variance or gradual increase in population variance. This can be easily observed by comparing the variances at the beginning and at the end of runs in case of each variant. It is worth mentioning that while the variance decreases invariably for all four variants for the first *F* value, it increases invariably for the second *F* value as both *F* values are on either side of the critical value (F_c).

The best variants' (*DE/best/2/bin*) runs display a sharp loss of variance or a sharp increase in population variance. This greedy behavior is appreciable, in case of *DE/best/2/bin*, as the variant primarily exploits and focus on the best solution found so far. However, in case of *DE/rand-to-best/1/bin*, both decrease and increase in population variance are much more balanced as against those of rand and best variants. This behavior of *DE/rand-to-best/1/bin* is by virtue of the contribution of both random and best candidate solutions in the population.

Table 4 Empirical variance (*EmpVar*) and theoretical variance (*TheVar*) measured for *DE/rand/1/bin*, *DE/rand/2/bin*, *DE/best/2/bin*, and *DE/rand-to-best/1/bin* for the function—f_3

DE/rand/1/bin					DE/rand/2/bin				
G	F = 0.1		F = 0.2		G	F = 0.1		F = 0.2	
	EmpVar	TheVar	EmpVar	TheVar		EmpVar	TheVar	EmpVar	TheVar
1	3228.38	3229.2	3413.19	3401.59	1	3335.65	3277.33	3254.07	3230.3
4	3125.11	3181.01	3643.63	3663.15	4	3307.67	3255.75	3258.54	3241.94
7	3025.44	3133.53	3949.41	3944.81	7	3316.33	3234.31	3266.23	3253.62
10	2984.39	3086.76	4273.62	4248.13	10	3293.48	3213.01	3294.81	3265.35
13	2935.46	3040.69	4597.16	4574.77	13	3292.73	3191.85	3322.97	3277.12
16	2892.77	2995.31	4929.56	4926.52	16	3261.74	3170.83	3338.87	3288.93
19	2836.17	2950.6	5241.39	5305.33	19	3251.99	3149.95	3347.85	3300.79
22	2795.69	2906.57	5630.24	5713.26	22	3245.3	3129.21	3359.84	3312.68
25	2716.81	2863.19	6041.17	6152.55	25	3203.83	3108.6	3377.16	3324.62
28	2682.63	2820.45	6453.4	6625.63	28	3184.13	3088.13	3381.96	3336.61
30	2679.86	2792.32	6767	6961.05	30	3172.41	3074.55	3361.8	3344.62
DE/best/2/bin					DE/rand-to-best/1/bin				
G	F = 0.4		F = 0.5		G	F = 0.6		F = 0.7	
	EmpVar	TheVar	EmpVar	TheVar		EmpVar	TheVar	EmpVar	TheVar
1	3434.69	3128.32	4065.81	3777.05	1	3539.92	3224.95	3650.18	3294.41
4	3407.5	2577.66	6625.35	5271.03	4	3534.62	2622.56	5178.06	3590.46
7	3343.98	2123.93	10717.98	7355.93	7	3490.97	2132.7	7122.32	3913.12
10	3154.49	1750.06	17373.5	10265.5	10	3479.07	1734.34	9885.88	4264.77
13	2999.27	1442.01	27154.9	14325.93	13	3412.8	1410.38	13612.78	4648.03
16	2843.86	1188.18	42586.56	19992.42	16	3410.79	1146.94	18819.72	5065.72
19	2614.89	979.03	69145.88	27900.25	19	3327.77	932.7	26061.99	5520.96
22	2442.59	806.69	108194.19	38935.94	22	3224.32	758.48	36114.22	6017.1
25	2343.26	664.69	170622.72	54336.7	25	3250.18	616.81	48858.64	6557.82
28	2263.49	547.69	263904.66	75829.09	28	3240.14	501.6	67608.43	7147.14
30	2509.31	481.37	358122.97	94695.84	30	3258.95	437.01	83610.71	7569.16

In fact, an evolutionary algorithm converges to global optimum successfully if it achieves and maintains balance between exploration and exploitation of the search space. The domination of the exploitation process (as is obtained in *DE/best/2/bin*) results quicker loss of population diversity (variance) leading to a suboptimal solution. This situation is often called as premature convergence. The domination of the exploration process (as is observed in *rand* variants) results in slower convergence leading to what is called stagnation. Despite the fact that the above observations of exploitative *best* variant and explorative *rand* variants have been inferred from a small, 30 generations, window it very well matches with the observations in the literature [36].

Extending this observation further, the population variance behavior of *DE/rand-to-best/1/bin* suggests/hints that mixing of perturbation schemes may contribute to exploration–exploitation balance resulting in robust optimization characteristics.

Table 5 Empirical variance (*EmpVar*) and theoretical variance (*TheVar*) measured for *DE/rand/1/bin*, *DE/rand/2/bin*, *DE/best/2/bin*, and *DE/rand-to-best/1/bin* for the function—f_4

DE/rand/1/bin					DE/rand/2/bin				
G	F = 0.1		F = 0.2		G	F = 0.08		F = 0.09	
	EmpVar	TheVar	EmpVar	TheVar		EmpVar	TheVar	EmpVar	TheVar
1	3300.94	3284.09	3404.11	3472.47	1	3316.91	3275.90	3261.83	3215.74
4	3249.90	3235.07	3685.18	3739.47	4	3319.95	3254.32	3250.12	3227.33
7	3200.36	3186.79	3999.44	4027.00	7	3320.32	3232.89	3255.33	3238.97
10	3158.17	3139.22	4274.15	4336.63	10	3258.87	3211.60	3272.42	3250.64
13	3106.72	3092.37	4619.10	4670.08	13	3227.06	3190.45	3274.23	3262.36
16	3069.08	3046.22	4999.93	5029.17	16	3213.12	3169.44	3256.22	3274.12
19	3020.12	3000.75	5475.81	5415.86	19	3215.54	3148.57	3247.09	3285.92
22	2986.40	2955.97	5906.00	5832.29	22	3176.39	3127.83	3243.95	3297.76
25	2980.51	2911.85	6254.59	6280.74	25	3164.37	3107.24	3222.96	3309.65
28	2951.76	2868.39	6650.37	6763.67	28	3150.11	3086.77	3208.02	3321.58
30	2886.31	2839.78	6943.25	7106.08	30	3109.21	3073.21	3227.78	3329.55
DE/best/2/bin					DE/rand-to-best/1/bin				
G	F = 0.4		F = 0.5		G	F = 0.6		F = 0.7	
	EmpVar	TheVar	EmpVar	TheVar		EmpVar	TheVar	EmpVar	TheVar
1	3342.77	3208.27	3903.15	3744.55	1	3356.34	3119.07	3822.26	3489.10
4	3001.70	2643.53	5906.83	5225.67	4	3264.92	2536.46	5136.06	3802.65
7	2641.97	2178.20	8461.86	7292.64	7	3072.23	2062.68	6808.68	4144.37
10	2348.26	1794.78	12265.96	10177.18	10	2947.88	1677.39	9052.86	4516.81
13	2113.32	1478.86	17538.24	14202.67	13	2839.70	1364.08	11686.05	4922.71
16	1885.07	1218.54	25785.57	19820.40	16	2675.51	1109.28	15477.90	5365.09
19	1693.73	1004.05	35992.53	27660.19	19	2547.64	902.08	19962.22	5847.22
22	1470.83	827.31	53432.97	38600.92	22	2476.62	733.58	25997.04	6372.68
25	1274.92	681.68	75244.32	53869.17	25	2322.93	596.56	33739.90	6945.36
28	1140.57	561.69	109039.80	75176.64	28	2226.83	485.13	44163.33	7569.51
30	1040.30	493.67	137379.18	93881.05	30	2131.15	422.66	52464.43	8016.47

To summarize, the robustness and efficiency of *DE* algorithm can be achieved by maintaining a balance between exploration (crossover and mutation) and exploitation (selection) processes. The analytical representation as well as the empirical observation of evolution of population variances for the identified four *DE* variants reveals that *DE/rand-to-best/1/bin* displays a balance between exploration and exploitation. This may be attributed to the presence of two perturbation elements viz. *rand* (explorative in nature) and *best* (greedy in nature). However, this mixing of perturbation elements is intrinsic to the variant. On the contrary, mixing perturbation schemes extrinsic through an island-based distributed framework would facilitate a co-operative-balanced exploration and exploitation of the search space.

Consequently, this chapter proposes that extrinsic mixing of effective *DE* variants (*DE/rand/1/bin*, *DE/rand/2/bin*, *DE/best/2/bin* and *DE/rand-to-best/1/bin*) with diverse characteristics, such that they are allowed to evolve independently but are

Table 6 Empirical variance (*EmpVar*) and theoretical variance (*TheVar*) measured for *DE/rand/1/bin*, *DE/rand/2/bin*, *DE/best/2/bin*, and *DE/rand-to-best/1/bin* for the function—f_9

DE/rand/1/bin					DE/rand/2/bin				
G	F = 0.1		F = 0.2		G	F = 0.08		F = 0.09	
	EmpVar	TheVar	EmpVar	TheVar		EmpVar	TheVar	EmpVar	TheVar
1	8.66	8.62	8.74	8.88	1	8.47	8.42	8.55	8.62
4	8.41	8.49	9.50	9.56	4	8.42	8.37	8.44	8.65
7	8.26	8.37	10.14	10.30	7	8.41	8.31	8.49	8.68
10	8.07	8.24	10.87	11.09	10	8.38	8.26	8.47	8.72
13	7.90	8.12	11.77	11.94	13	8.27	8.20	8.43	8.75
16	7.75	7.99	12.81	12.86	16	8.26	8.15	8.41	8.78
19	7.63	7.88	13.88	13.85	19	8.22	8.10	8.36	8.81
22	7.56	7.76	14.83	14.91	22	8.07	8.04	8.45	8.84
25	7.48	7.64	16.16	16.06	25	8.06	7.99	8.48	8.87
28	7.37	7.53	17.28	17.30	28	8.06	7.94	8.53	8.91
30	7.35	7.45	18.05	18.17	30	8.01	7.90	8.62	8.93
DE/best/2/bin					DE/rand-to-best/1/bin				
G	F = 0.4		F = 0.5		G	F = 0.6		F = 0.7	
	EmpVar	TheVar	EmpVar	TheVar		EmpVar	TheVar	EmpVar	TheVar
1	8.81	8.26	8.84	9.88	1	8.65	8.21	9.60	8.97
4	8.25	6.81	13.62	13.79	4	8.25	6.68	12.83	9.78
7	7.61	5.61	19.90	19.24	7	7.89	5.43	17.01	10.66
10	6.88	4.62	28.08	26.85	10	7.48	4.42	22.13	11.61
13	6.24	3.81	39.39	37.47	13	7.27	3.59	28.60	12.66
16	5.56	3.14	54.52	52.29	16	6.98	2.92	36.46	13.80
19	5.01	2.59	76.08	72.98	19	6.72	2.38	47.17	15.04
22	4.66	2.13	107.24	101.84	22	6.43	1.93	58.39	16.39
25	4.20	1.76	151.78	142.12	25	6.18	1.57	74.92	17.86
28	3.76	1.45	202.44	198.34	28	5.92	1.28	94.43	19.46
30	3.58	1.27	248.29	247.68	30	5.70	1.11	111.41	20.61

made to suitably exchange their search information amongst others, will co-operatively enhance the efficacy of the system as a whole with robust optimization characteristics.

6   Mixing Strategies of *Differential Evolution* Variants

Given an optimization problem at hand, trial-and-error searching for the appropriate *DE* variant and tuning its associated parameter values is a general practice followed for successfully solving the problem. This is because different *DE* variants perform with different efficacies when solving different optimization problems. Rather than

Table 7 Empirical variance (*EmpVar*) and theoretical variance (*TheVar*) measured for DE/rand/1/bin, DE/rand/2/bin, DE/best/2/bin, and DE/rand-to-best/1/bin for the function—f_{10}

DE/rand/1/bin					DE/rand/2/bin			
G	F = 0.1		F = 0.2		G	F = 0.08		F = 0.09	
	EmpVar	TheVar	EmpVar	TheVar		EmpVar	TheVar	EmpVar	TheVar
1	335.37	336.50	344.37	345.31	1	345.58	349.58	344.20	348.79
4	338.60	331.48	375.17	371.86	4	343.23	347.28	346.48	350.04
7	336.84	326.53	401.58	400.46	7	340.05	344.99	346.68	351.30
10	329.50	321.66	428.09	431.25	10	335.03	342.72	346.78	352.57
13	322.98	316.85	464.38	464.41	13	331.98	340.46	347.69	353.84
16	315.86	312.13	506.78	500.11	16	331.16	338.22	348.77	355.12
19	311.82	307.47	537.42	538.57	19	329.48	335.99	352.34	356.40
22	303.31	302.88	578.40	579.98	22	334.23	333.78	354.31	357.68
25	297.65	298.36	625.74	624.58	25	327.97	331.58	358.02	358.97
28	294.45	293.90	669.74	672.60	28	320.43	329.40	360.73	360.26
30	290.80	290.97	706.59	706.65	30	318.09	327.95	360.09	361.13
DE/best/2/bin					DE/rand-to-best/1/bin			
G	F = 0.4		F = 0.5		G	F = 0.6		F = 0.7	
	EmpVar	TheVar	EmpVar	TheVar		EmpVar	TheVar	EmpVar	TheVar
1	337.24	326.06	408.31	385.69	1	336.47	316.50	375.12	353.84
4	301.96	268.66	605.03	538.25	4	322.11	257.38	503.42	385.64
7	268.24	221.37	982.36	751.14	7	307.65	209.31	673.36	420.30
10	238.95	182.40	1713.00	1048.25	10	288.91	170.21	903.38	458.07
13	214.05	150.30	3096.05	1462.88	13	270.72	138.42	1264.69	499.23
16	187.71	123.84	5714.53	2041.51	16	251.21	112.56	1932.89	544.10
19	167.07	102.04	10476.12	2849.00	19	243.13	91.54	2938.19	592.99
22	143.58	84.08	18828.17	3975.90	22	229.76	74.44	4436.88	646.28
25	122.54	69.28	33318.53	5548.53	25	213.99	60.53	6788.27	704.36
28	113.36	57.08	61272.86	7743.21	28	198.76	49.23	10117.51	767.66
30	107.33	50.17	92632.89	9669.77	30	187.93	42.89	13054.89	812.98

searching for appropriate variant or improving individual variants to solve the given optimization problem, in an effort towards alleviating trial-and-error search of *DE* variants, this chapter proposes to mix effective *DE* variants with diverse characteristics, each in an island, in a distributed frame work.

The *distributed heterogeneous mixed variants DE (dDE_HeM)* is different from a typical distributed *DE (dDE)* in that the islands are populated by different *DE* variants not the same variants. The proposed distributed mixed best variants (called *dDE_HeM_best)* constitute the identified competitive *DE* variants viz. (*DE/rand/1/bin, DE/rand/2/bin, DE/best/2/bin,* and *DE/rand-to-best/1/bin*) each in an island as subpopulations with each variant evolving independently but also exchanging information amongst others to co-operatively enhance the efficacy of the *dDE_HeM_best* as a whole. The operational scheme of *dDE_HeM_best* is shown in Fig. 4.

To understand the dynamics of mixing, this chapter also intends to mix *DE* variants with very poor performance. Based on the earlier empirical analyses, the *DE* variants with very poor performance include *DE/rand/2/exp*, *DE/current-to-rand/1/exp*, *DE/current-to-best/1/exp*, and *DE/rand-to-best/1/exp*. The mixing of these 4 variants each in an island will be called *dDE_HoM_Worst* for notational sake. The operational scheme of *dDE_HeM_Worst* is shown in Fig. 5. The algorithmic description for *dDE_HeM* is shown in Fig. 6. It is worth noting that this chapter restricts its focus to only 4 islands and best/worst variants combinations in each island. Relaxing these constraints, however, would result in huge number of possibilities of mixing the *DE* variants.

Although the very idea of co-operative evolution by employing genetic algorithms with different configurations in each island has been attempted in [17], the possibility of mixing various *DE* variants in island-based distributed framework has not yet been addressed, in literature. However, there has been very few research works attempted in this direction. A variant of *DE* called *SaDE* has been proposed in [36], which is essentially a serial *DE* which adapts different *DE* strategies in a pool of predetermined strategies at different phases of *DE* evolution. Mallipeddi

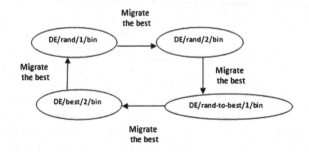

Fig. 4 The operational scheme of *dDE_HeM_best*

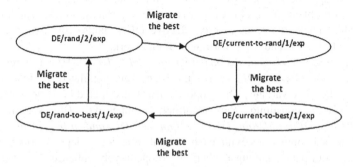

Fig. 5 The operational scheme of *dDE_HeM_worst*

Population Initialization

Initialize $P_G = \{X_{i,G},...,X_{NP,G}\}$ with $X_{i,G} = \{X_{i,G}^1,...,X_{i,G}^D\}, i = 1,...,NP; G = 0;$

Compute $\{f(X_{i,G}),..., f(X_{NP,G})\}$

Divide the population of size NP into ni subpopulations of size S_NP

Scatter the sub populations to the nodes in the cluster

Place different DE variants in all the four nodes. /* for dDE_HeM */

/* at each node */

WHILE stopping criterion is not satisfied DO

$FOR\ i = 1\ to\ S\_NP$

DO

      1.   Mutation Step

      2.   Crossover Step

      3.   Selection Step

END DO

END FOR

  DO (For every mf generation)

  1.   Send nm candidates, selected by sp, to next node as per mt

  2.   Receive the nm candidate from the previous node as per mt

  3.   Replace the candidates selected by rp by the received nm candidates

END DO

$$G = G + 1$$

Compute $\{f(X_{i,G}),..., f(X_{NP,G})\}$

END WHILE

Fig. 6 The general description of dDE_HeM algorithms

et al. [28] proposed a serial DE with ensemble of parameters and mutation strategies.

Interestingly, Mallipeddi and Suganthan [29] have proposed a DE with ensemble of population where fitness evaluation to each population is self-adapted. Weber et al. [42] proposed a Distributed DE with explorative–exploitative populations (DDE-$EEPF$) where a DE variant has been employed in one family for exploration and in other family of subpopulation for exploitation of the decision space. It is worth noting that almost all of the above existing work implements the distributed version with same variants on every island.

Researchers have also attempted to hybridize DE with other global search algorithms like particle swarm optimization (PSO) [23], ant colony systems [11], artificial immune systems (AIS) [10], bacterial foraging optimization ($BFOA$) [33], and simulated annealing (SA) [24].

The hybridization of DE with PSO, swarm differential evolution algorithm, was first proposed in [16]. Zhang and Xie [50] proposed $DEPSO$ as another hybrid algorithm comprising DE and PSO. In $DEPSO$, the PSO and DE algorithms are alternatively used to search in odd and even generations. Das et al. [7] proposed

PSO-DV called as particle swarm optimization with differentially perturbed velocity. In *PSO-DV*, the velocity-update operation of *PSO* is done by the *DE* operators. Continuing this trend various hybridization algorithms have been proposed in the literature [14, 21, 22, 27, 31, 32, 43]. The integration of fuzzy logic with *DE* for *DE's* parameter setting was attempted in [25, 26, 27].

Hybridizing *SA* with *DE*, to replace the selection logic of *DE* using *SA*, was attempted in [8]. Another hybrid version of *DE* and *SA*, called *SaDESA*, was proposed in [18]. Biswas et al. [3] proposed hybridization of *BFOA* with *DE*. Hybridization of ant colony optimizer with *DE* was studied in [51, 5, 6]. He and Han [15] proposed to mix *AIS* with *DE*.

7 Design of Experiments

The distributed implementation of *DE* variants has very well been conceived by researchers in the literature. Consequently, to serve as a frame of reference as well as comparison against the proposed distributed framework, distributed versions of 14 *DE* variants (called *dDE_HoM* to represent homogeneous composition) have all need to be implemented and analyzed.

Typical island-based *dDE* is characterized by the following parameters—number of islands *ni*, migration frequency *mf*, number of migrants *nm*, selection policy *sp* that decides what individuals migrate, replacement policy *rp* that determines which individual(s) the migrant(s) replace and the migration topology *mt* that decides which islands can send (or receive) the migrants.

Of the above parameters, the number of islands *ni* and number of migrants *nm* have been set as 4 and 1, respectively, for the sake of easier analysis. The migration frequency *mf* has been kept as 45 based on earlier empirical analysis. The best solution in each island migrates using a ring topology and replaces random solution (except the best solution) in the other islands.

The empirical comparative performances of *DE*, *dDE_HoM*, *dDE_HeM_best*, and *dDE_HeM_Worst* algorithms have been analyzed using the following performance metrics chosen from the *EA* literature.

- Mean Objective Function Value (*MOV*);
- Convergence Measure (C_m);
- Success Rate (*SR*);
- Probability of Convergence ($P_c(\%)$);
- Quality Measure (Q_m); and
- Success Performance (*SP*).

Mean Objective Function Value (*MOV*): The Mean Objective Function Value is measured as mean of the objective function values obtained by each variant for a particular function, in all the 100 independent runs.

Convergence Measure (C_m): The convergence measure is calculated as the mean percentage out of the total number of function evaluations required by each of the variant to reach its best objective function value, for all the independent runs [30]. The variants with good convergence speed will reach the global optimum with lesser number of function evaluations than the maximum (i.e., lower the mean percentage better the convergence measure), hence they will have the mean percentage less than 100.

Success Rate (*SR*): The success rate is calculated as the percentage of number of successful runs out of total runs for each function, as follows, $SR = (nc\_f/nt\_t)\%$, where nc_f is total number of successful runs made by a variant for a function and nt_t is the total number of runs, in this current work $nt\_t = 100$. A run is considered successful when the tolerance error is reached before the maximum number of generations.

Probability of Convergence ($P_c(\%)$): The probability of convergence [12] is measured as described already in Sect. 4.

Quality Measure (Q_m): Quality measure or simply *Q-measure* [12] is an empirical measure of an algorithm's convergence. It compares the *objective function convergence* of evolutionary algorithms. In the experiments, *Q-measure* combines convergence rate and probability of convergence as follows: $Q_m = C_s/P_c$, where $P_c(\%)$ is the probability of convergence and C_s is the convergence measure for successful runs. The convergence measure (C_s) is calculated as $C_s = \sum_{j=1}^{nc} FE_j/nc$, where $j = 1, ..., nc$ are successful trials and FE_j is the number of functions evaluations in *j*th trial. The probability of convergence ($P_c(\%)$) is calculated as described earlier. A best performing variant is expected to have higher probability of convergence and successful trials; lower number of function evaluations and consequently, lower Q_m measure.

Success Performance (*SP*): The success performance compares the overall performance of the variants on all functions by plotting empirical distribution of normalized success performance [13, 36]. It is akin to Q_m. The *SP* has been calculated as follows:

$$SP = \frac{\text{mean (\# function evaluations for successful runs)} * (\text{\# total runs})}{\text{\# successful runs}}. \quad (7)$$

A run is considered *successful* if the global optimum is reached with the given precision, before the maximum number of functions evaluations is reached. The success performances of all the variants on each benchmark function are calculated and are normalized by dividing them by the best *SP* on the respective function.

As *DE* is stochastic in nature (similar to all *EAs*), 100 independent runs were performed for each variant—function combinations and their average performances are reported. The population was initialized randomly for each run. In each run, the variants are allowed up to 3000 generations. In order to maintain uniformity among all the variants, a threshold value, for the mean objective function value, of 1×10^{-12} was fixed as the stopping criteria. The variants will stop their evolution either on reaching the threshold or at crossing the maxing number of generations.

8 Simulation Results and Analysis

The MOV, $P_c(\%)$, and Q_m values obtained for the classical DE and their corresponding distributed DE with homogenous mixing (dDE_HoM) are shown in Table 8. The results show the dDE_HoM variants have solved significantly more number of benchmarking functions than the classical DE variants, except for the variant $rand\text{-}to\text{-}best/1/bin$. It is observed from the $P_c(\%)$ values that, consistently in all the functions, the dDE_HoM variants have shown significant increase in its probability of convergence.

As the quality measure combines the convergence measure and probability of convergence as a single value, it is to be minimized by a best performing variant. The variants with higher probability of convergence are expected to have lower quality measure values. As it is evident from the results that the dDE_HoM variants are outperforming their counterpart classical DE variants by their quality measure values also. However, in the case of $rand/1/bin$ both the DE and dDE_HoM variants have obtained same quality measure value.

Figure 7 depicts the overall performances of all DE and dDE_HoM variants by plotting empirical distribution of normalized success performance.

In Fig. 7, the normalized success performance for all the fourteen variants are calculated, by dividing the SP values of each variant on each function by the best SP value for the respective function. The variant which solves all the considered benchmarking will have higher empirical distribution of 1, in the Y axis. Similarly, the variants which solve the function with lesser number of function evaluations will have small values of SP. The graph plots the $SP/SPbest$ value and the empirical distribution values in the X and Y axis, respectively. Naturally, the variants which solve more functions with lesser number of function evaluations are preferable. Such variants will reach the top in Y axis with lesser values in X axis.

Figure 7 shows that, while considering the success performance of the DE and dDE_HoM variants for all the fourteen functions, only the $dDE\_HoM/best/1bin$, $dDE\_HoM/best/2/bin$, $dDE\_HoM/rand/2/bin$, and $dDE\_HoM/rand\text{-}to\text{-}best/1/bin$ variants reaches the top of the graph. They are the preferable variants, because they produce at least one successful run for each of the benchmarking function. Conversely, none of the classical DE variants could reach the top of the graph. In case of the other variants also the dDE_HoM variants show significant outperformance, either by increased Y axis value or by decreased X axis value or by the both. Interestingly, the $DE/best/1/exp$, $DE/current\text{-}to\text{-}rand/1/exp$, and $DE/current\text{-}to\text{-}best/1/exp$ variants are not shown in the graph due to their bad performance. However, their corresponding dDE_HoM variants are present in the graph by virtue of their performance superior. Overall, on comparing all the DE and dDE_HoM variants, it is clearly evident that all the dDE_HoM outperforms their sequential counterparts in solving more functions with less number of function evaluations, except few cases in each category.

The MOV values measured for dDE_HeM_best are presented in Table 9, along with the MOV values of the constituent variants in classical DE and dDE_HoM. For

Table 8 The MOV, P_c(%), and Q_m values measured for DE and dDE_HoM variants for the functions f_1–f_{14}

Variant		MOV														P_c(%)	Q_m
		f_1	f_2	f_3	f_4	f_5	f_6	f_7	f_8	f_9	f_{10}	f_{11}	f_{12}	f_{13}	f_{14}		
rand/1/bin	DE	0.00	0.00	0.07	0.00	21.99	0.02	0.00	0.13	0.00	0.09	0.00	0.00	0.00	0.00	73.93	1.40E+03
	dDE_HoM	0.00	0.00	0.04	0.00	34.51	0.00	0.00	0.08	0.00	0.18	0.00	0.00	0.00	0.00	77.79	1.40E+03
rand/1/exp	DE	0.00	0.00	0.31	3.76	25.48	0.00	0.02	0.1	47.93	0.09	0.05	0.00	0.00	0.00	48.5	3.40E+03
	dDE_HoM	0.00	0.00	0.00	1.65	25.84	0.00	0.01	0.02	0.00	0.04	0.00	0.00	0.00	0.00	79.07	1.50E+03
best/1/bin	DE	457.25	0.14	13.27	1.96	585,900	437.25	0.09	0.00	4.33	3.58	3.72	15.78	973,097	12.93	23	7.80E+03
	dDE_HoM	0.00	0.00	0.00	0.06	31.1	0.00	0.01	0.00	0.32	0.00	0.00	0.00	0.19	0.00	77.57	1.20E+03
best/1/exp	DE	583.79	4.05	57.39	37.36	64,543.8	591.85	0.06	0.01	50.74	6.09	5.91	131,449	154,435	32.18	10.21	1.80E+04
	dDE_HoM	0.00	0.00	0.00	9.11	61.79	0.00	0.03	0.00	15.29	0.00	0.00	0.00	0.72	0.00	66.57	2.50E+03
rand/2/bin	DE	0.00	0.00	1.64	0.06	19.01	0.00	0.01	0.22	0.00	0.09	0.00	0.00	0.00	0.00	57.36	2.10E+03
	dDE_HoM	0.00	0.00	0.00	0.01	25.44	0.00	0.00	0.00	0.00	0.00	0.00	0.00	0.00	0.00	83.79	1.50E+03
rand/2/exp	DE	0.00	0.02	269.86	32.9	2741.32	0.00	0.05	0.27	101.38	0.01	0.21	0.00	0.01	0.01	29	6.90E+03
	dDE_HoM	0.00	0.00	0.02	4.53	75.47	0.00	0.01	0.00	0.07	0.00	0.00	0.00	0.00	0.00	76.07	2.00E+03
best/2/bin	DE	0.00	0.00	0.00	0.00	2.32	0.07	0.00	0.17	0.69	0.09	0.00	0.00	0.00	0.12	74.57	1.40E+03
	dDE_HoM	0.00	0.00	0.00	0.00	3.24	0.00	0.00	0.00	0.27	0.00	0.00	0.00	0.00	0.00	92.29	1.10E+03
best/2/exp	DE	0.00	0.00	0.00	0.05	1.12	0.39	0.01	0.08	80.63	0.83	0.03	0.14	0.00	2.53	43.36	3.30E+03
	dDE_HoM	0.00	0.00	0.00	0.61	6.04	0.00	0.01	0.00	12.15	0.00	0.00	0.00	0.00	0.00	73.93	1.90E+03
current-to-rand/1/bin	DE	0.00	0.02	3210.36	3.68	52.81	0.03	0.04	0.14	37.75	0.01	0.00	0.00	0.00	0.00	46.43	3.50E+03
	dDE_HoM	0.00	0.00	13.39	0.66	38.87	0.00	0.01	0.09	9.05	0.00	0.00	0.00	0.00	0.00	58.07	2.80E+03
current-to-rand/1/exp	DE	24.29	44.22	3110.9	57.52	199,243	43.07	0.27	0.12	235.14	13.83	1.21	10.89	24.11	18.35	0.21	8.40E+05
	dDE_HoM	0.00	0.37	7.31	13.67	100.93	0.00	0.02	0.02	87.62	1.04	0.05	0.06	0.03	1.05	36.21	4.50E+03
current-to-best/1/bin	DE	0.00	0.02	3444	3.71	56.91	0.00	0.04	0.19	37.04	0.01	0.00	0.00	0.00	0.00	46.14	3.60E+03
	dDE_HoM	0.00	0.00	14.58	0.63	41.08	0.00	0.01	0.3	9.16	0.00	0.00	0.00	0.00	0.00	57.79	2.90E+03
current-to-best/1/exp	DE	24.37	45.04	2972.62	56.67	119,686	41.95	0.26	0.1	232.8	13.69	1.21	10.37	23.04	18.21	0.36	5.00E+05
	dDE_HoM	0.00	0.28	7.61	12.21	182.89	0.00	0.02	0.04	91.48	0.74	0.03	0.05	0.02	1.34	34.43	4.60E+03
rand-to-best/1/bin	DE	0.00	0.00	0.07	0.00	17.37	0.00	0.00	0.22	0.00	0.09	0.00	0.00	0.00	0.00	74.21	1.40E+03
	dDE_HoM	0.00	0.00	0.04	0.00	22.57	0.00	0.02	0.18	0.00	0.32	0.00	0.00	0.00	0.00	77.5	1.30E+03
rand-to-best/1/exp	DE	0.00	0.00	0.2	3.38	24.54	0.00	0.01	0.12	48.09	0.09	0.05	0.00	0.00	0.00	48.93	3.90E+03
	dDE_HoM	0.00	0.00	0.00	1.51	20.73	0.00	0.01	0.02	5.82	0.00	0.00	0.00	0.00	0.03	72.79	1.60E+03

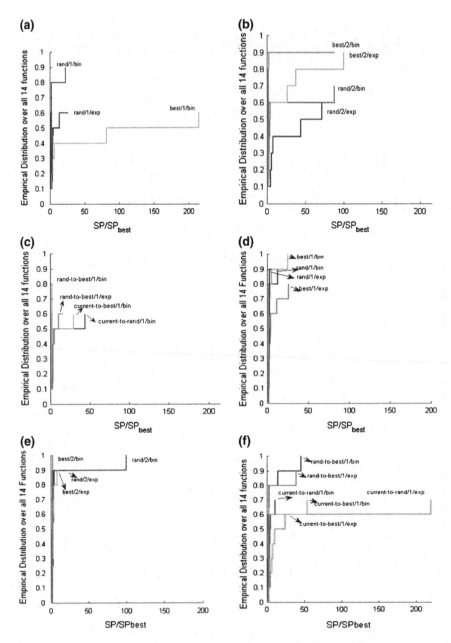

Fig. 7 Success performance for all the 14 functions **a–c** *DE* variants, **d–f** *dDE_HoM* variants

the functions f_1, f_2, f_{11}, f_{12}, and f_{13}, the *dDE_HeM_best* showed competitive performance as similar to other counterpart variants. In the remaining functions, the *dDE_HeM_best* algorithm has shown only marginal performance difference

Table 9 The *MOV* measured for *dDE_HeM_best* variant (along with their constituent classical *DE* and *dDE_HoM* variants)

Variant		f_1	f_2	f_3	f_4	f_5	f_6	f_7	f_8	f_9	f_{10}	f_{11}	f_{12}	f_{13}	f_{14}
rand/1/bin	*DE*	0.00	0.00	0.07	0.00	21.99	0.02	0.00	0.13	0.00	0.09	0.00	0.00	0.00	0.00
	dDE_HoM	0.00	0.00	0.04	0.00	34.51	0.00	0.00	0.08	0.00	0.18	0.00	0.00	0.00	0.00
rand/2/bin	*DE*	0.00	0.00	1.64	0.06	19.01	0.00	0.01	0.22	0.00	0.09	0.00	0.00	0.00	0.00
	dDE_HoM	0.00	0.00	0.00	0.01	25.44	0.00	0.00	0.00	0.00	0.00	0.00	0.00	0.00	0.00
best/2/bin	*DE*	0.00	0.00	0.00	0.00	2.32	0.07	0.00	0.17	0.69	0.09	0.00	0.00	0.00	0.12
	dDE_HoM	0.00	0.00	0.00	0.00	3.24	0.00	0.00	0.00	0.27	0.00	0.00	0.00	0.00	0.00
rand-to-best/1/bin	*DE*	0.00	0.00	0.07	0.00	17.37	0.00	0.00	0.22	0.00	0.09	0.00	0.00	0.00	0.00
	dDE_HoM	0.00	0.00	0.04	0.00	22.57	0.00	0.02	0.18	0.00	0.32	0.00	0.00	0.00	0.00
dDE_HeM_best		0.00	0.00	0.00	0.00	18.00	0.00	0.00	0.00	0.00	0.01	0.00	0.00	0.00	0.00

compared to other variants. In some cases of the functions, the *DE/best/2/bin* and its *dDE_HoM* have shown better performance than the *dDE_HeM_best*. While considering the *MOV*, due to the fact that all the considered algorithms are already the best variants of *DE* and their improved variants (i.e., *dDE_HoM*), the superiority of *dDE_HeM_best* is insignificant.

However, by the $P_c(\%)$, the *dDE_HeM_best* variant shows significant performance improvement by achieving the P_c value of 93.14 %. The *dDE_HoM* variant of *DE/best/2/bin* performs closely but next to *dDE_HeM_best* with the P_c value of 92.29 %.The $P_c(\%)$ values measured for the variants are presented in Table 10. The similar trend was observed in measuring the Q_m values (Table 11) also. The *dDE_HeM_best* variant has outperformed all the counterpart variants significantly by achieving least Q_m value of 1.02E+03. The *dDE_HoM* variant of *DE/best/2/bin* performs closer but next to *dDE_HeM_best* (with the Q_m value of 1.10E+03).

The empirical distribution of the success performance of the *dDE_HeM_best* variant is presented in Fig. 8c. Figure 8a, b depicts the graph for the best performing *DE* variants and their *dDE_HoM* variants, respectively. The graphs reiterate the efficacy of *dDE_HeM_best*. The *dDE_HeM_best* shows significant performance improvement by reaching the top of the graph faster than all the counterpart variants, except the *dDE_HoM* variant of *dDE/best/2/bin*.

Table 12 displays the *MOV* values obtained by *dDE_Hem_Worst* on the 14 test functions along with those obtained by the classical and distributed versions of constituent variants. Interestingly *dDE_HeM_Worst* is outperformed by all the variants considered. Exceptionally in some cases the *dDE_HeM_Worst* is outperforming some of the counterpart variants viz *current-to-rand/1/exp* and *current-to-best/1/exp*.

The $P_c(\%)$ values measured for all the variants are shown in Table 13. It is seen from the results that the *dDE_Hem_Worst* is outperformed by all the counterpart variants, except *DE/current-to-rand/1/exp* and *DE/current-to-best/1/exp*. The similar performance was shown by the *dDE_HeM_worst* variant in its quality measure (Table 14) value and success performance plot (Fig. 9) also.

As it is shown in Tables 12, 13 and 14, the *dDE_HoM* variant of all the worst performing variants outperforms their corresponding classical *DE* variants by all the performance measures (*MOV*, $P_c(\%)$, Q_m, and *SP*). However, when they are hybridized to form co-operative co-evolution, the *dDE_HeM_Worst* variant fails behind all the counterpart variants.

Thus, the experiments done on the worst performing variants show that the idea of mixing the worst performing variants in the distributed frame work does not bring any improvement in the performance against the constituent variants. The performance of mixing best performing variants and the performance of mixing worst performing variants is contradictory.

The contradictory behavior of the *dDE_HeM_best* and *dDE_HeM_worst* variants calls for an extensive empirical study which involves mixing of numerous varieties of *DE* variants with different optimization characteristics. We also admit that a clear theoretical insight about this contradictory performance is also essential to form a basis for the further research in this direction. However, it can be inferred

Table 10 The $P_c(\%)$ measured for dDE_HeM_best variant (along with their constituent classical DE and dDE_HoM variants)

Variant		f_1	f_2	f_3	f_4	f_5	f_6	f_7	f_8	f_9	f_{10}	f_{11}	f_{12}	f_{13}	f_{14}	nc	$P_c(\%)$
rand/1/bin	DE	100	100	73	100	0	98	60	4	100	0	100	100	100	100	1035	73.93
	dDE_HoM	100	100	8	100	0	100	88	11	100	82	100	100	100	100	1089	77.79
rand/2/bin	DE	100	100	0	0	0	100	2	1	100	0	100	100	100	100	803	57.36
	dDE_HoM	100	100	100	1	2	100	71	99	100	100	100	100	100	100	1173	83.79
best/2/bin	DE	100	100	100	100	38	95	75	1	47	0	100	99	100	89	1044	74.57
	dDE_HoM	100	100	100	100	22	100	96	100	74	100	100	100	100	100	1292	92.29
rand-to-best/1/bin	DE	100	100	10	0	0	100	0	6	0	0	69	100	100	100	685	48.93
	dDE_HoM	100	100	100	0	0	100	41	77	4	100	100	100	100	97	1019	72.79
dDE_HeM_best		100	100	100	100	27	100	77	99	99	100	100	100	100	100	1304	93.14

Table 11 The Q_m values measured for *dDE_HeM_best* variant (along with their constituent classic *DE* and *dDE_HoM* variants)

Variant		$Q_m = C/P_c$
rand/1/bin	*DE*	1.40E+03
	dDE_HoM	1.40E+03
rand/2/bin	*DE*	2.10E+03
	dDE_HoM	1.50E+03
best/2/bin	*DE*	1.40E+03
	dDE_HoM	1.10E+03
rand-to-best/1/bin	*DE*	1.40E+03
	dDE_HoM	1.30E+03
dDE_HeM_best		1.02E+03

that since mixing of *DE* variants is primarily a co-operative evolution unless one/more constituent variants co-operate (in terms of the quality of solution) the mixing fails to work.

Having demonstrated the robustness and efficacy of *dDE_HeM_best*, the next step could have been extending the idea of mixing to *DDE* variants in each island. However, the intention of this chapter is not to keep improving the performance alone but to understand and observe the dynamics of mixing between variants. Consequently, this chapter attempts one more level down of mixing i.e., mixing *DE* and *DDE* variants in a distributed framework. Thus, resulting in yet another class of algorithm called distributed heterogeneous mixing of *DE* and *DDE* variants (*dDE_HeM_DDE*).

The *dDE_HeM_DDE* is different from *dDE_HoM*, *dDE_HeM_best*, and *dDE_HeM_Worst* in that the islands are populated not only by different *DEs* but also by different *DDEs* as well. However, since we restrict the analysis using the island size of 4, only the variants *DE/best/2/bin* and *DDE/best/2/bin* have been considered for mixing (In fact any one of the 4 competitive *DE* variants and its dynamic version could have been considered as well with performance difference). This results in 4 *dDE_HeM_DDE* algorithms of which 3 are obtained by mixing *DE/best/2/bin* and *DDE/best/2/bin* in three proportions (i.e., 1:3, 2:2, and 3:1) and the fourth one is obtained by employing the dynamic versions of the four most competitive *DE* variants. Respectively, the four *dDE_HeM_DDE* algorithms will be called as *dDE_HeM_DEb2DDEb2/1:3*, *dDE_HeM_DEb2DDEb2/2:2*, *dDE_HeM_/DEb2DDEb2/3:1*, and *dDDE_HeM_best*. Table 15 lists the mixing schemes of four *dDE_HeM_DDE*. The algorithmic representation of *dDE_HeM_DEb2DDEb2/1:3*, by way of an example, is depicted in Fig. 10.

The *MOV* values obtained for all four *dDE_HeM_DDE* variants along with the *dDE_HeM_best* variant are shown in Table 15, and the $P_c(\%)$ values are shown in Table 16. As can be seen from Tables 8, and 16, with respect to *MOV*, in solving each function the *dDE_HeM_DDE* consistently retains/outperforms *dDE_HeM_best* variant in all the cases (Table 17). Table 18 compares the

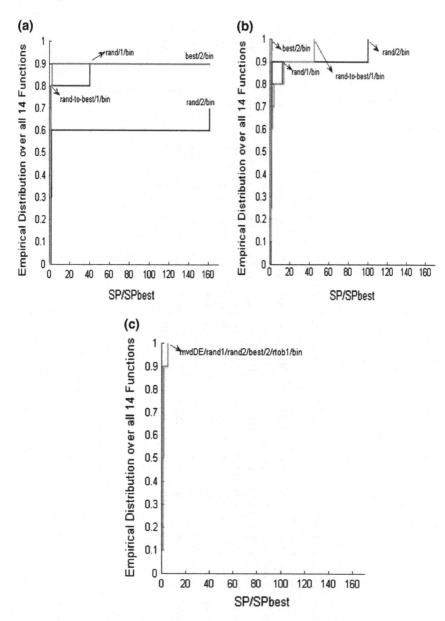

Fig. 8 Success performance for the chosen best four variants in **a** *DE*, **b** *dDE_HoM* and **c** *dDE_HeM_best*

number of function solved by *DE, DDE, dDE_HoM, dDE_HeM_best*, and four *dDE_HeM_DDE* variants. There has been an increase in performance albeit small, by *dDE_HeM_DDEs*, in terms of number of functions solved.

Table 12 The MOV measured for dDE_HeM_worst variant (along with their constituent classical DE and dDE_HoM variants)

Variant		f_1	f_2	f_3	f_4	f_5	f_6	f_7	f_8	f_9	f_{10}	f_{11}	f_{12}	f_{13}	f_{14}
rand/2/exp	DE	0.00	0.02	269.86	32.90	2741.32	0.00	0.05	0.27	101.38	0.01	0.21	0.00	0.01	0.01
	dDE_HoM	0.00	0.00	0.02	4.53	75.47	0.00	0.01	0.00	0.07	0.00	0.00	0.00	0.00	0.00
current-to-rand/1/exp	DE	24.29	44.22	3110.90	57.52	199243.32	43.07	0.27	0.12	235.14	13.83	1.21	10.89	24.11	18.35
	dDE_HoM	0.00	0.37	7.31	13.67	100.93	0.00	0.02	0.02	87.62	1.04	0.05	0.06	0.03	1.05
current-to-best/1/exp	DE	24.37	45.04	2972.62	56.67	119685.68	41.95	0.26	0.10	232.80	13.69	1.21	10.37	23.04	18.21
	dDE_HoM	0.00	0.28	7.61	12.21	182.89	0.00	0.02	0.04	91.48	0.74	0.03	0.05	0.02	1.34
rand-to-best/1/exp	DE	0.00	0.00	0.20	3.38	24.54	0.00	0.01	0.12	48.09	0.09	0.05	0.00	0.00	0.00
	dDE_HoM	0.00	0.00	0.00	1.51	20.73	0.00	0.01	0.02	5.82	0.00	0.00	0.00	0.00	0.03
dDE_HeM_Worst		0.00	51.79	0.10	65.53	46.93	25985.34	0.04	0.01	248.87	0.10	313.28	2.48E+08	4.99E+08	969.83

Table 13 The $P_c(\%)$ measured for dDE_HeM_worst variant (along with their constituent classical DE and dDE_HoM variants)

Variant		f_1	f_2	f_3	f_4	f_5	f_6	f_7	f_8	f_9	f_{10}	f_{11}	f_{12}	f_{13}	f_{14}	nc	$P_c(\%)$
rand2/exp	DE	61	0	0	0	0	100	0	2	0	64	3	100	50	26	406	29
	dDE_HoM	100	100	27	0	1	100	41	97	99	100	100	100	100	100	1065	76.07
current-to-rand/1/exp	DE	0	0	0	0	0	0	0	3	0	0	0	0	0	0	3	0.21
	dDE_HoM	94	9	0	0	0	100	1	71	0	1	32	77	75	47	507	36.21
current-to-best/1/exp	DE	0	0	0	0	0	0	0	5	0	0	0	0	0	0	5	0.36
	dDE_HoM	89	9	0	0	0	100	3	70	0	0	32	72	75	32	482	34.43
rand-to-best/1/exp	DE	100	100	10	0	0	100	0	6	0	0	69	100	100	100	685	48.93
	dDE_HoM	100	100	100	0	0	100	41	77	4	100	100	100	100	97	1019	72.79
dDE_HeM_Worst		100	0	8	0	1	0	1	82	0	91	0	0	0	0	283	20.21

Table 14 The Q_m values measure for *dDE_HeM_worst* variant (along with their constituent classical *DE* and *dDE_HoM* variants)	Variant		$Q_m = C/P_c$
	rand/2/exp	*DE*	6.90E+03
		dDE	2.00E+03
	current-to-rand/1/exp	*DE*	8.40E+05
		dDE	4.50E+03
	current-to-best/1/exp	*DE*	5.00E+05
		dDE	4.60E+03
	rand-to-best/1/exp	*DE*	3.90E+03
		dDE	1.60E+03
	dDE_HeM_Worst		6.83E+03

In terms of $P_c(\%)$, as can be seen in Tables 8, 17 and 19, all the *dDE_HeM_DDEs* display higher probability of convergence than that of *DE*, *DDE*, *dDE*, and *dDE_HeM_best* variants, except the cases where the *dDDE_HeM_best* stands next to *dDE/best/2/bin* and *dDDE/best/2/bin*. Similar to that, except *dDDE_HeM_best*, all the three *dDE_HeM_DDE* variants outperform the *dDE_HeM_best* algorithm, by their higher probability of convergence. This shows the effectiveness of mixing *DDE* variants with *DE* variants in the distributed frame work.

Table 20 presents the Q_m measures obtained for the four *dDE_HeM_DDEs*. The results show that interestingly, *dDE_HeM_DDEs* stand only next to *dDE_HeM_best*, but in a close margin.

The performance implication of the *dDE_HeM_DDE* and its robust optimization characteristics will be more evident when compared and benchmarked against the existing Differential Evolution algorithms in the literature.

9 Benchmarking *DDE_HeM_DDE* Against State-of-the-Art Serial Differential Evolution Algorithms

Despite the fact that it is only fair to compare a distributed Differential Evolution algorithm with its like lot, there have been works in the literature where distributed *DEs* have been compared with sequential algorithms. In a similar fashion, *dDE_HeM_DDE* has been compared with well-known sequential Differential Evolution algorithms viz. classification-based self-adaptive differential evolution (*p-ADE*) [1, 2], *DE* with Global and Local neighborhoods (*DEGL*) [9], adaptive *DE* with optional external archive (*JADE*) [48, 49], Self-adapting control parameters in *DE* (*jDE*) [4], and Self-adaptive differential evolution with neighborhood search and (*SaNSDE*) [45].

For the purpose of comparison between *dDE_HeM_DEb2DDEb2/1:3* and *p-ADE*, *DEGL*, *JADE*, *jDE* sequential algorithms 16 test problems (*p-ADE-f_1* to *p-ADE-f_{16}*) of dimensions 5, 10, and 30 as well as the simulation results of all

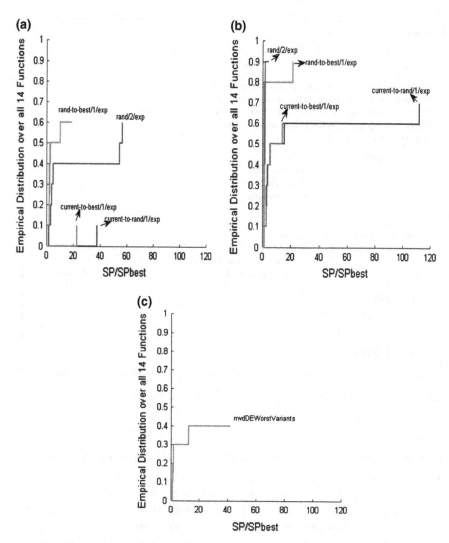

Fig. 9 Success performance for the chosen worst four variants in **a** *DE*, **b** *dDE_HoM* and **c** *dDE_HeM_worst*

Table 15 The four different mixed distribution schemes for *dDE_HeM_DDE*

S. no.	Variant name	Mixing proportions
1	*dDDE_HeM_best*	*DDE/rand/1/bin* in 1 node, *DDE/best/2/bin* in 1 node, *DDE/rand/2/bin* in 1 node and DDE/rand-to-best/1/bin in 1 node
2	*dDE_HeM_DEb2DDEb2/1:3*	*DE/best/2/bin* in 1 node, *DDE/best/2/bin* in 3 nodes
3	*dDE_HeM_DEb2DDEb2/2:2*	*DE/best/2/bin* in 2 nodes, *DDE/best/2/bin* in 2 nodes
4	*dDE_HeM_DEb2DDEb2/3:1*	*DE/best/2/bin* in 3 nodes, *DDE/best/2/bin* in 1 node

Population Initialization

Initialize $P_G = \{X_{i,G}, ..., X_{NP,G}\}$ with $X_{i,G} = \{X^1_{i,G}, ..., X^D_{i,G}\}, i = 1, ..., NP$; $G = 0$;

Compute $\{f(X_{i,G}), ..., f(X_{NP,G})\}$

Divide the population of size *NP* into *4* subpopulations of size *NP/4*

Scatter the sub populations to the nodes in the cluster

Place *DE/best/2/bin* variant in 1 node and *DDE/best/2/bin* in 3 nodes

WHILE stopping criterion is not satisfied DO

$FOR\ i = 1\ To\ NP\ /\ 4$

    DO

        1.   Mutation Step

        2.   Crossover Step

        3.   Selection Step

    END DO

END FOR

DO (For every *mf* generation)

        1.   Send *nm* candidates, selected by *sp*, to next node as per *mt*

        2.   Receive the *nm* candidate from the previous node as per *mt*

        3.   Replace the candidates selected by *rp* by the received *nm* candidates

END DO

$G = G + 1$

Compute $\{f(X_{i,G}), ..., f(X_{NP,G})\}$

END WHILE

Fig. 10 The algorithmic structure of *dDE_HeM_DEb2DDEb2/1:3* variant

Table 16 The *MOV* measured for *dDE_HeM_DDE* variants and *dDE_HeM_best* variant

S. no.	Variant	f_1	f_2	f_3	f_4	f_5	f_6	f_7
1	dDE_HeM_best	0.00	0.00	0.00	0.00	18.00	0.00	0.00
2	dDDE_HeM_best	0.00	0.00	0.00	0.00	11.69	0.00	0.00
3	dDE_HeM_DEb2DDEb2/1:3	0.00	0.00	0.00	0.00	2.19	0.00	0.00
4	dDE_HeM_DEb2DDEb2/2:2	0.00	0.00	0.00	0.00	2.61	0.00	0.00
5	dDE_HeM_DEb2DDEb2/3:1	0.00	0.00	0.00	0.00	4.01	0.00	0.00
S. no.	Variant	f_8	f_9	f_{10}	f_{11}	f_{12}	f_{13}	f_{14}
1	dDE_HeM_best	0.00	0.00	0.01	0.00	0.00	0.00	0.00
2	dDDE_HeM_best	0.00	0.00	0.00	0.00	0.00	0.00	0.00
3	dDE_HeM_DEb2DDEb2/1:3	0.00	0.01	0.00	0.00	0.00	0.00	0.00
4	dDE_HeM_DEb2DDEb2/2:2	0.00	0.00	0.00	0.00	0.00	0.00	0.00
5	dDE_HeM_DEb2DDEb2/3:1	0.00	0.01	0.00	0.00	0.00	0.00	0.00

Table 17 The $P_c(\%)$ values measured for *dDE_HeM_DDE* variants and *dDE_HeM_best* variant

Variant	f_1	f_2	f_3	f_4	f_5	f_6	f_7	f_8	f_9	f_{10}	f_{11}	f_{12}	f_{13}	f_{14}	nc	$P_c(\%)$
dDE_HeM_best	100	100	100	100	27	100	77	99	99	100	100	100	100	100	1304	93.14
dDDE_HeM_best	100	100	100	100	21	100	68	99	100	100	100	100	100	100	1288	92.00
dDE_HeM_DEb2DDEb2/1::3	100	100	100	100	39	100	100	100	93	100	100	100	100	100	1332	95.14
dDE_HeM_DEb2DDEb2/2:2	100	100	100	100	42	100	100	100	97	100	100	100	100	100	1339	95.64
dDE_HeM_DEb2DDEb2/3:1	100	100	100	100	32	100	100	100	94	100	100	100	100	100	1326	94.71

Table 18 The number of functions solved by the *DE, DDE, dDE*, and *dDE_HeM_DDE* variants

Variants	DE	DDE	dDE	dDE_HeM_best and dDE_HeM_DDEs	
				dDE_HeM_best	12
rand/1/bin	9	12	10	*dDDE_HeM_best*	13
rand/2/bin	8	9	12	*dDE_HeM_DEb2DDEb2/1:3*	12
best/2/bin	8	9	12	*dDE_HeM_DEb2DDEb2/2:2*	13
rand-to-best/1/bin	10	11	9	*dDE_HeM_DEb2DDEb2/3:1*	12

Table 19 The $P_c(\%)$ values measured for the *DE, DDE, dDE*, and *dDE_HeM_DDE* variants

Variants	DE	DDE	dDE	dDE_HeM_best and dDE_HeM_DDEs	
				dDE_HeM_best	93.14
rand/1/bin	73.93	82.21	77.79	*dDDE_HeM_best*	92.00
rand/2/bin	57.36	64.71	83.79	*dDE_HeM_DEb2DDEb2/1:3*	95.14
best/2/bin	74.57	82.79	92.29	*dDE_HeM_DEb2DDEb2/2:2*	95.64
rand-to-best/1/bin	74.21	81.79	77.50	*dDE_HeM_DEb2DDEb2/3:1*	94.71

sequential algorithms have been adopted from [1, 2]. The parameter setting for *dDE_HeM_DEb2DDEb2/1:3* has also been closely adopted from [1, 2]. The population size (*NP*) is set to be 100 for all the dimensions of 5, 10, and 30. The crossover rate C_r was tuned using bootstrap test, and the parameter *F* was generated anew at each generation from the range of [0.3, 0.9]. The *dmv/DEb2DDEb2/1:3* was run for 20,000, 150,000, and 500,000 fitness evaluations for dimensions 5, 10, and 30, respectively. The error tolerance is set to be = 1×10^{-6}. The mean of best-of-run objective function values for 30 independent runs of *dDE_HeM_DEb2DDEb2/1:3* along with *p-ADE, DEGL, JADE* and *jDE* is presented in Tables 21 and 22, for the dimensions *D* = 5, 10, and 30.

A close inspection of Tables 21 and 22 indicates that *dDE_HeM_DEb2DDEb2/1:3* outperforms or performs similar to other variants in most of the benchmark functions. *dDE_HeM_DEb2DDEb2/1:3* outperforms/performs similar to *p-ADE* in most of the functions for the dimension of 10 and for the other dimensions they are equally competitive.

Table 23 compares *dDE_HeM_DEb2DDEb2/1:3* with other variants by their probability of convergence ($P_c(\%)$) for *D* = 30. As can be seen in Table 18,

Table 20 The Q_m values measured for the $dmvD^2E$ and the best of dmvDE

S. no.	Variant	Q_m
1	*dDE_HeM_best*	1.02E+03
2	*dDDE_HeM_best*	1.11E+03
3	*dDE_HeM_DEb2DDEb2/1:3*	1.10E+03
4	*dDE_HeM_DEb2DDEb2/2:2*	1.10E+03
5	*dDE_HeM_DEb2DDEb2/3:1*	1.12E+03

Table 21 The MOV measured for *dDE_HeM_DEb2DDEb2/1:3*, *p-ADE*, *DEGL*, *JADE*, and *jDE* for *p-ADE-f₁* to *p-ADE-f₈*

Function	D	dDE_HeM_DEb2DDEb2/1:3	p-ADE	DEGL	JADE	jDE
p-ADE-f_1	5	3.5149E-198	0.0000E+00	4.3204E-57	1.8113E-39	1.3888E-39
	10	8.3187E-145	0.0000E+00	3.1643E-97	8.1522E-50	1.0755E-40
	30	1.0762E-62	6.6539E-299	3.6167E-40	4.8179E-22	1.4575E-17
p-ADE-f_2	5	0.0000E+00	8.8404E-72	1.5977E-59	5.0588E-41	3.0219E-17
	10	1.2047E-157	1.5627E-83	4.0419E-51	1.1317E-62	2.5071E-08
	30	1.0816E-12	6.5702E-52	5.3758E-21	1.9920E-02	1.1300E-03
p-ADE-f_3	5	0.0000E+00	2.9422E-205	3.0059E-61	3.2020E-49	2.4815E-37
	10	0.0000E+00	4.4082E-234	3.9845E-93	3.4547E-47	2.1021E-35
	30	1.1722E-134	1.6801E-302	3.7240E-136	4.2720E-28	5.4255E-28
p-ADE-f_4	5	0.0000E+00	6.2150E-55	1.2018E-32	7.4012E-24	6.5684E-14
	10	8.5193E-135	5.7593E-89	5.5981E-42	1.6956E-55	9.0628E-07
	30	6.9413E-04	2.7050E-51	4.5316E-10	5.3430E-01	1.4525E-01
p-ADE-f_5	5	2.5522E-05	3.4932E-07	1.3000E-03	1.6000E-03	3.7357E-04
	10	7.9783E-05	6.1547E-07	2.9303E-04	9.3776E-04	8.6951E-04
	30	1.0833E-03	5.5530E-07	8.3708E-04	7.7420E-03	7.4000E-03
p-ADE-f_6	5	1.1019E-04	5.3840E-01	0.0000E+00	2.7400E-04	7.6678E-18
	10	2.6577E-01	4.9213E+00	0.0000E+00	0.0000E+00	1.4369E-17
	30	5.3155E-01	2.5787E+01	1.2326E-30	9.8217E-03	1.5800E-02
p-ADE-f_7	5	1.3147E-03	0.0000E+00	9.0900E-02	2.2204E-16	7.4000E-06
	10	1.1503E-03	0.0000E+00	2.2100E-02	7.4000E-03	3.8940E-06
	30	0.0000E+00	0.0000E+00	1.4800E-02	2.0372E-41	4.4260E-40
p-ADE-f_8	5	4.4409E-16	-1.8285E-06	-1.8285E-06	-1.8285E-06	-1.8285E-06
	10	2.8126E-15	-1.8285E-06	-1.8285E-06	-1.8285E-06	-1.8285E-06
	30	3.9968E-15	-1.8285E-06	-1.8285E-06	-1.8285E-06	-1.8285E-06

Table 22 The MOV measured for dDE_HeM_DEb2DDEb2/1:3, p-ADE, DEGL, JADE, and jDE for p-ADE-f_9 to p-ADE-f_{16}

Function	D	dDE_HeM_DEb2DDEb2/1:3	p-ADE	DEGL	JADE	jDE
p-ADE-f_9	5	0.0000E+00	0.0000E+00	9.9500E-01	0.0000E+00	0.0000E+00
	10	0.0000E+00	0.0000E+00	4.9756E+00	0.0000E+00	0.0000E+00
	30	0.0000E+00	0.0000E+00	1.9899E+01	0.0000E+00	0.0000E+00
p-ADE-f_{10}	5	1.1734E-04	0.0000E+00	4.4240E+00	6.8048E+00	0.0000E+00
	10	1.3841E-04	1.2247E+01	2.5500E+01	2.6717E+01	1.7569E+01
	30	9.8631E-04	1.9800E+01	3.7752E+01	3.6916E+01	2.1900E+01
p-ADE-f_{11}	5	0.0000E+00	0.0000E+00	1.9852E+00	0.0000E+00	3.3300E-02
	10	0.0000E+00	0.0000E+00	6.0000E+00	0.0000E+00	0.0000E+00
	30	0.0000E+00	0.0000E+00	3.7000E+01	0.0000E+00	7.9769E-03
p-ADE-f_{12}	5	1.0408E-18	0.0000E+00	6.4277E-05	0.0000E+00	0.0000E+00
	10	4.7246E-10	0.0000E+00	2.6020E-07	0.0000E+00	7.2073E-04
	30	4.9860E-02	0.0000E+00	7.7920E-05	4.9267E-06	3.9525E-04
p-ADE-f_{13}	5	9.4233E-32	9.0000E-03	1.0391E-31	9.4233E-32	9.4233E-32
	10	4.7116E-32	4.0200E-02	4.7116E-32	4.7116E-32	8.9385E-32
	30	1.5705E-32	1.2800E-02	2.0100E-32	2.0100E-32	1.5705E-32
p-ADE-f_{14}	5	1.3498E-32	−9.3660E-02	−1.1504E+00	−1.1478E+00	−1.1496E+00
	10	1.3498E-32	1.5000E-02	−1.1437E+00	−1.1290E+00	−1.1382E+00
	30	1.3498E-32	3.4300E-02	−1.1504E+00	−1.1504E+00	−1.1504E+00
p-ADE-f_{15}	5	−1.4317E-05	2.1780E-01	7.9280E-01	9.0400E-02	1.4420E-01
	10	9.2903E-06	1.0380E+00	1.4154E+00	2.6490E-01	6.8780E+00
	30	−2.3713E-05	9.4417E-01	2.6707E+00	9.5610E-01	4.2835E+00
p-ADE-f_{16}	5	−2.9103E+02	3.8900E-02	2.3240E-01	7.4700E-02	4.8600E-02
	10	−2.9203E+02	8.8100E-02	4.9970E-01	1.6170E-01	2.7560E-01
	30	−2.8234E+02	8.4530E-01	1.1429E+01	1.9619E+00	4.1089E+00

Table 23 The $P_c(\%)$ values measured for $dDE\_HeM\_DEb2DDEb2/1:3$, p-ADE, $DEGL$, $JADE$, and jDE, for $D = 30$

Variant	f_1	f_2	f_3	f_4	f_5	f_6	f_7	f_8	f_9
$dDE\_HeM\_DEb2DDEb2/1:3$	30	30	30	0	0	26	30	30	30
p-ADE	30	30	30	30	27	0	30	30	30
$DEGL$	30	30	30	30	0	27	18	25.8	0
$JADE$	30	0	30	0	0	0	30	30	30
JDE	30	0	30	0	0	0	30	30	27

Variant	f_{10}	f_{11}	f_{12}	f_{13}	f_{14}	f_{15}	f_{16}	nc	$P_c(\%)$
$dDE\_HeM\_DEb2DDEb2/1:3$	0	30	12	30	30	0	0	308	64.17
p-ADE	0	30	30	0	0	0	0	297	61.88
$DEGL$	0	0	9	30	0	0	0	229.8	47.88
$JADE$	0	30	24	30	0	0	0	234	48.75
JDE	0	0	0	24	0	0	0	171	35.63

$dDE\_HeM\_DEb2DDEb2/1:3$ is emerging as the competitive variant by displaying increased probability of convergence.

Next, $dDE\_HeM\_DEb2DDEb2/1:3$ has been compared with $SaNSDE$, along with $SaDE$ [44] and $NSDE$ [37], by their MOV, on the test suite. The dimension has been set as 30 and population size as 60. The values for C_r and F are decided respectively by bootstrap test and in the range of [0.3, 0.9]. The maximum number of generation is set as 1500 for f_1–f_4, 5000 for f_5 and 1500 for f_6–f_{13}. The MOV measured for $dDE\_HeM\_DEb2DDEb2/1:3$, $SaNSDE$, $SaDE$, and $NSDE$ are presented in Table 24. The competitive performance of $dDE\_HeM\_DEb2DDEb2/1:3$ is once again evident.

Table 24 The MOV measured for $dDE\_HeM\_DEb2DDEb2/1:3$, $SaNSDE$, $SaDE$, and $NSDE$

Function	$dDE\_HeM\_DEb2DDEb2/1:3$	$SaNSDE$	$SaDE$	$NSDE$
f_1	4.40E-28	3.02E-23	7.49E-20	7.76E-16
f_2	1.69E-15	4.64E-11	6.22E-11	4.51E-10
f_3	2.42E+03	6.62E-22	1.12E-18	1.06E-14
f_4	7.15E-04	1.59E-03	2.96E-02	2.54E-02
f_5	5.96E-01	4.13E-30	2.10E+01	1.24E+01
f_6	0.00E+00	0.00E+00	0.00E+00	0.00E+00
f_7	3.69E-03	7.21E-03	7.58E-03	1.20E-02
f_8	2.44E-03	−1.26E+04	−1.26E+04	−1.26E+04
f_9	5.12E+01	1.84E-05	4.00E-08	7.97E-02
f_{10}	1.15E-14	2.36E-12	9.06E-11	6.72E-09
f_{11}	2.96E-04	0.00E+00	8.88E-18	4.68E-15
f_{12}	1.83E-27	5.94E-23	1.21E-19	5.63E-17
f_{13}	2.31E-27	3.12E-22	7.15E-19	5.52E-16

p-ADE utilizes a new mutation strategy *DE/rand-to-best/pbest* which employs both current and previous best solutions of each individual along with dynamic parameter adaption. *DEGL* employs local and global neighborhood of each population member along with self-adaptation weight factor to balance exploration and exploration. *JADE* employs optional archive of historical data thus providing information about progress direction thus diversifying the population and improving convergence performance. While *jDE* employs self-adaptive control parameters, *SaNSDE* is self-adaptive Differential Evolution with neighborhood search.

Unlike the above algorithms, *dDE_HeM_DDE* (and as a matter of fact *dDE_HeM_best* too) is a simple distributed Differential Evolution algorithm with a simple topology and migration topology. The competitive performance displayed by *dDE_HeM_DDE* (and *dDE_HeM_best)* may not simply be attributed to the distributed evolution but may largely be due to the simple but the effective idea of mixing competitive *DE* variants.

10 Conclusions

This chapter primarily focused on the idea of hybridizing different variants of *Differential Evolution* algorithm. The island-based distributed environment is used as frame work for hybridizing different *DE* variants. The *DE* variants with seven different mutation strategies and two different crossover schemes are chosen. To serve as a frame of reference, the study first discussed about same variant of *DE* in the island-based distributed frame work (*dDE_HoM*). Then the novel idea of hybridizing different variants of *DE* is implemented. Two pools of variants with different characteristics are chosen among the 14 classical *DE* variants; they are the best and worst performing variants of classical *DE*. This resulted two variants of *dDE* viz *dDE_HeM_best* and *dDE_HoM_worst*. While the former hybridizes the four best performing variants of *DE*, the later hybridizes the worst performing *DE* variants. The detailed empirical study among the variants (*DE, dDE_HoM, dDE_HeM_best*, and *dDE_HeM_Worst*) had shown that the *dDE_HoM* variants show significant performance improvement than their classical *DE* variants. The idea of hybridizing different variants of *DE* has shown contradictory performance in the cases of *dDE_HeM_best* and *dDE_HeM_worst*. The *dDE_HeM_best* variant resulted noteworthy performance improvement than its constituent counterpart *DE* and *dDE_HoM* variants. However, the *dDE_HeM_worst* variant fails behind all the constituent variants. The mixing of *DE* and *DDE* variants in a distributed framework (*dDE_HeM_DDE*) has also been attempted. The *dDE_HeM_DDE* algorithms have been benchmarked against five state-of-the-art sequential *DE* algorithms on a different set of benchmark problems. The simulation showed that *dDE_HeM_DDE* outperformed the compared algorithms in most of the cases displaying robust optimization characteristics.

Since the results reported in this chapter are based on hybridization of only few selected variants of *DE* and *DDE*, the *dDE_HeM_best*, *dDE_HeM_worst*, and *dDE_HeM_Worst* variants can be used as a prototype to study further extensively about numerous possibilities of mixing. This also raises a need for theoretical analysis of the performance of the hybridized variants.

References

1. Bi X, Xiao J (2010) p-ADE: self adaptive differential evolution with fast and reliable convergence performance. In: Proceedings of the 2nd international conference on industrial mechatronics and automation, pp 477–480
2. Bi X, Xiao J (2011) Classification-based self-adaptive differential evolution with fast and reliable convergence performance. Soft Comput—Fusion Found Methodol Appl 15(8):1581–1599 (Springer)
3. Biswas A et al (2007) A synergy of differential evolution and bacterial foraging algorithm for global optimization. Neural Netw World 17(6):607–626
4. Brest J et al (2006) Self adapting control parameters in differential evolution: a comparative study on numerical benchmark problems. IEEE Trans Evol Comput 10(6):646–657
5. Chiou JP, Wang FS (1999) Hybrid method of evolutionary algorithms for static and dynamic optimization problems with application to a fed-batch fermentation process. Comput Chem Eng 23:1277–1291
6. Chiou JP, Chang CF, Su CT (2004) Ant direction hybrid differential evolution for solving large capacitor placement problems. IEEE Transactions on Power Systems, vol 19. pp 1794–1800
7. Das S et al (2005) Improving particle swarm optimization with differentially perturbed velocity. In: Proceedings of the genetic and evolutionary computation conference, pp 177–184
8. Das S et al (2007) Annealed differential evolution. In: Proceedings of the IEEE congress on evolutionary computation, pp 1926–1933
9. Das S et al (2009) Differential evolution using a neighborhood-based mutation operator. IEEE Trans Evol Comput 13(3):526–533
10. Dasgupta D (ed) (1999) Artificial immune systems and their applications. Springer
11. Dorigo M, Gambardella LM (1997) Ant colony system: a cooperative learning approach to the traveling salesman problem. IEEE Trans Evol Comput 1(1):53–66
12. Feoktistov V (2006) Differential evolution in search of solutions. Optimization and its applications. Springer
13. Hansen N (2006).Compilation of results on the 2005 CEC benchmark function set. http://www.ntu.edu.sg/home/epnsugan/index_files/CEC-05/compareresults.pdf
14. Hao ZF et al (2007) A particle swarm optimization algorithm with differential evolution. In: Proceedings of the 6th international conference on machine learning and cybernetics, vol. 2, pp 1031–1035
15. He H, Han L (2007) A novel binary differential evolution algorithm based on artificial immune system. In: Proceedings of the IEEE congress on evolutionary computation, pp 2267–2272
16. Hendtlass T (2001) A combined swarm differential evolution algorithm for optimization problems. Lecture Notes in Computer Science, vol 2070. Springer, pp 11–18
17. Herrera F, Lozano M (2000) Gradual distributed real-coded genetic algorithms. IEEE Trans Evol Comput 4(1):43–63
18. Hu ZB et al (2008) Self-adaptive hybrid differential evolution with simulated annealing algorithm for numerical optimization. In: Proceedings of the IEEE congress on evolutionary computation, pp 1189–1194

19. Jeyakumar G, ShunmugaVelayutham C (2010) An empirical performance analysis of differential evolution variants on unconstrained global optimization problems. Int J Comput Inf Syst Ind Manage Appl 2:077–086
20. Jeyakumar G, ShunmugaVelayutham C (2010b) A comparative study on theoretical and empirical evolution of the population variance of the differential evolution variants. In: Lecture notes in computer science (LNCS-6457). Springer, pp 75–79
21. Kannan S et al (2004) Application of particle swarm optimization technique and its variants to generation expansion planning. Electric Power Syst Res 70(3):203–210
22. Kennedy J (2003) Bare bones particle swarms. In: Proceedings of the IEEE swarm intelligence symposium, pp 80–87
23. Kennedy J et al (2001) Swarm intelligence. The Morgan Kaufmann series in evolutionary computation. Academic Press, USA
24. Kirkpatrik S et al (1983) Optimization by simulated annealing. Sci J 220(4598):671–680
25. Liu J, Lampinen J (2002a) Adaptive parameter control of differential evolution. In: Proceedings of the 8th international mendel conference on soft computing, pp 19–26
26. Liu J, Lampinen J (2002b) A fuzzy adaptive differential evolution algorithm. In: Proceedings of the 17th IEEE region 10th international conference on computer, communications, control and power engineering, vol 1, pp 606–611
27. Liu J, Lampinen J (2005) A fuzzy adaptive differential evolution algorithm. Soft Comput—Fusion Found Methodol Appl 9(6):448–462 (Springer)
28. Mallipeddi R et al (2011) Differential evolution algorithm with ensemble of parameters and mutation strategies. Appl Soft Comput 11(2):1679–1696
29. Mallipeddi R, Suganthan PN (2009) Differential evolution algorithm with ensemble of populations for global numerical optimization. OPSEARCH 46(2):184–213
30. Mezura-Montes E et al (2006) A comparative study of differential evolution variants for global optimization. In: Proceedings of the genetic and evolutionary computation conference, pp 485–492
31. Moore PW, Venayagamoorthy GK (2006) Evolving digital circuit using hybrid particle swarm optimization and differential evolution. Int J Neural Syst 16(3):163–177
32. Omran MGH et al (2009) Bare bones differential evolution. Eur J Oper Res 196(1):128–139
33. Passino KM (2002) Biomimicry of bacterial foraging for distributed optimization and control. IEEE Control Syst Mag 52–67
34. Price K et al (2005) Differential evolution: a practical approach to global optimization. Springer
35. Price KV (1999) An introduction to differential evolution. In: Corne D, Dorigo M, Glover V (eds) New ideas in optimization. McGraw-Hill, pp 79–108
36. Qin AK et al (2009) Differential evolution algorithm with strategy adaptation for global numerical optimization. IEEE Trans Evol Comput 13(2):398–417
37. Qin AK, Suganthan PN (2005) Self-adaptive differential evolution algorithm for numerical optimization. In: Proceedings of the IEEE congress on evolutionary computation, pp 1785–1791
38. Qing A (2006) Dynamic differential evolution strategy and applications in electromagnetic inverse scattering problems. IEEE Trans Geosci Remote Sens 44(1):116–125
39. Qing A (2008) A study on base vector for differential evolution. In: Proceedings of the IEEE world congress on computational intelligence/2008 IEEE congress on evolutionary computation, pp 550–556
40. Storn R, Price K (1995) Differential evolution—a simple and efficient adaptive scheme for global optimization over continuous spaces. In: Technical report-95-012, ICSI
41. Tvrdik J (2006) Differential evolution: competitive setting of control parameters. In: Proceedings of the international multiconference on computer science and information technology, pp 207–213
42. Weber M et al (2009) Distributed differential evolution with explorative-exploitative population families. Genet Program Evolvable Mach 10(4):343–371

43. Xu X et al (2008) A novel differential evolution scheme combined with particle swarm intelligence. In: Proceedings of the IEEE congress on evolutionary computation, pp 1057–1062
44. Yang Z et al (2007) Making a difference to differential evolution. In: Michalewicz Z, Siarry P (eds) Advances in metaheuristics for hard optimization. Springer, pp 397–414
45. Yang Z et al (2008) Self-adaptive differential evolution with neighborhood search. In: Proceedings of the IEEE congress on evolutionary computation, pp 1110–1116
46. Yao D et al (2003) Fast evolutionary algorithms. In: Rozenberg G, Back T, Eiben A (eds) Advances in evolutionary computing: theory and applications. Springer, pp 45–94
47. Zaharie D (2001) On the explorative power of differential evolution algorithms. In: Proceeding of the 3rd international workshop on symbolic and numeric algorithms on scientific computing, SYNASC-2001
48. Zhang J, Sanderson AC (2007) JADE: self-Adaptive differential evolution with fast and reliable convergence performance. In: Proceedings of the IEEE congress on evolutionary computation, pp 2251–2258
49. Zhang J, Sanderson AC (2009) JADE: adaptive differential evolution with optional external archive. IEEE Trans Evol Comput 13(5):945–958
50. Zhang W-J, Xie X-F (2003) DEPSO: hybrid particle swarm with differential evolution operator. Proc IEEE Int Conf Syst Man Cybern 4:3816–3821
51. Zhang X et al (2008) DEACO: hybrid ant colony optimization with differential evolution. In: Proceedings of the IEEE congress on evolutionary computation, pp 921–927

Part II
Hybrid Soft Computing Approaches: Applications

Collaborative Simulated Annealing Genetic Algorithm for Geometric Optimization of Thermo-electric Coolers

Doan V.K. Khanh, Pandian M. Vasant, Irraivan Elamvazuthi and Vo N. Dieu

Abstract Thermo-electric Coolers (TECs) nowadays are applied in a wide range of thermal energy systems. This is due to its superior features where no refrigerant and dynamic parts are needed. TECs generate no electrical or acoustical noise and are environment friendly. Over the past decades, many researches were employed to improve the efficiency of TECs by enhancing the material parameters and design parameters. The material parameters are the most significant, but they are restricted by currently available materials and module fabricating technologies. Therefore, the main objective of TECs design is to determine a set of design parameters such as leg area, leg length, and the number of legs. Two elements that play an important role when considering the suitability of TECs in applications are rated of refrigeration (ROR) and coefficient of performance (COP). In this chapter, the technical issues of TECs were discussed. After that, a new method of optimizing the dimension of TECs using collaborative simulated annealing genetic algorithm (CSAGA) to maximize the rate of refrigeration (ROR) was proposed. Equality constraint and inequality constraint were taken into consideration. The results of optimization obtained by using CSAGA were validated by comparing with those obtained by using stand-alone genetic algorithm and simulated annealing optimi-

D.V.K. Khanh (✉)
Department of Fundamental and Applied Sciences, Universiti Teknologi PETRONAS,
Perak, Malaysia
e-mail: kimkhanh2906@gmail.com

P.M. Vasant
Department of Fundamental and Applied Sciences, UTP, Perak, Malaysia
e-mail: pvasant@gmail.com

I. Elamvazuthi
Department of Electrical & Electronic Engineering, UTP, Perak, Malaysia
e-mail: irraivan_elamvazuthi@petronas.com.my

V.N. Dieu
Department of Power Systems, HCMC University of Technology,
Ho Chi Minh City, Vietnam
e-mail: vndieu@gmail.com

© Springer India 2016
S. Bhattacharyya et al. (eds.), *Hybrid Soft Computing Approaches*,
Studies in Computational Intelligence 611,
DOI 10.1007/978-81-322-2544-7_5

155

zation technique. This work revealed that CSAGA was more robust and more reliable than stand-alone genetic algorithm and simulated annealing.

Keywords Thermo-electrics coolers · Thermal energy system · Rate of refrigeration · Coefficient of performance · Collaborative simulated annealing genetic algorithm · Geometric properties · Material properties · Genetic algorithm · Simulated annealing

1 Introduction

Mud-Pulse High-Temperature (MWD) is a system developed to perform drilling-related measurements down-hole and transmit information to the surface while drilling a well [1]. MWD systems can take several measurements like a natural gamma ray, directional survey, tool face, borehole pressure, temperature, vibration, shock, and torque. Maintaining optimal payload temperatures in a typical down-hole environment of 230 °C requires that the MWD cooling system is capable of pumping a significant load and requires a low thermal resistance path on the heat rejection (hot side). The application in the extreme environment of high temperature, high pressure, mechanical shock, and vibration requires the use of high-temperature TEC materials and assemblies. A typical High-Temperature MWD tool is shown in Fig. 1. Cooling of electronic components inside MWD housing is crucial for maintaining optimal operating conditions in the MWD. It has been identified that this can be accomplished using thin-film thermo-electric cooling devices.

TECs are solid-state cooling devices that use the Peltier effect through p-type and n-type semiconductor elements (unlike vapor-cycle-based refrigerators) [2]. These types of coolers are used to convert electrical energy into a temperature gradient. Thermo-electric coolers use no refrigerant and have no dynamic parts which make these devices highly reliable and require low maintenance. These coolers generate no electrical or acoustical noise and are ecologically clean. These coolers are compact in terms of size, light weight, and have high precision in temperature control. However, for this application, the most attractive feature of the

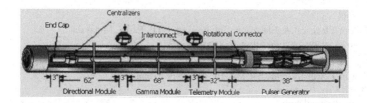

Fig. 1 High-temperature MWD tool

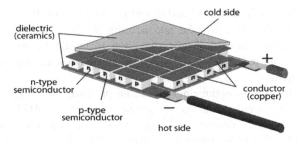

Fig. 2 Single-stage thermoelectric coolers (STECs) (Ferrotec)

Fig. 3 Two-stage thermo-electric coolers (TTECs) (Ferrotec)

TECs is that they have the capacity for cooling instruments such as MWDs under extreme physical conditions.

TECs can be a single-stage or multi-stages type (Figs. 2 and 3). The commercially available single-stage TECs (STECs) can produce a maximum temperature difference of about 60–70 K when hot side remains at room temperature [2]. Nevertheless, when a large temperature difference is required for some special applications, STECs will not be qualified. To enlarge the maximum temperature difference of TECs, we use two-stage TECs (TTECs) or multi-stages TEC. Thermo-electric module generally works with two heat sinks attached to its hot and cold sides in order to enhance heat transfer and system performance.

As mentioned previously, the application of TECs has been partitioned by their relatively low energy conversion efficiency and ability to dissipate only a limited amount of heat flux. Two parameters play a crucial role in characterization of TECs are the maximum rate of refrigeration (ROR) and the maximum coefficient of performance (COP). Thermo-electric coolers operate at about 5–10 % of Carnot cycle COP whereas compressor-based refrigerators normally operates at more than 30 %.

Several intelligent techniques that can be used for engineering design optimizations are discussed in [3]. However, one of the most effective and non-traditional methods used as an optimization technique for TECs is the Non-dominated Sorting Genetic Algorithm (NSGA-II) [4]. Similar sophisticated techniques in artificial intelligence, such as Simulated Annealing (SA) [5], other evolutionary algorithms (GA, Differential Evolution (DE) [6], and Particle Swarm Optimization (PSO) Poli et al. [7], can be used in their pure and hybrid form to enhance the effectiveness of the optimization of TECs. To take into account the large number of variables (physical, as well as geometrical properties) and create more performance in the problem, hybridized techniques such as hybrid simulated annealing genetic algorithm can be used.

This study focuses on optimizing the design parameters of single-stage TECs using collaborative simulated annealing genetic algorithm (CSAGA) optimization technique to create the maximum ROR under some defined constraints. The leg area, number of legs, and leg length of the thermo-electric module were optimized.

2 Optimization Matters in Designing TECs

The main drawback of thermo-electric coolers is the poor coefficient of performance and low ROR. They can be improved personally or simultaneously, from the parameters of the equation of TECs performance, we can group them into three categories which are specifications, material properties, and design parameter [8]. The specification is the operating temperature T_c and T_h, the required output voltage V, current I, and power output P. The specifications are usually provided by customers depending on the requirement of a particular application. The material parameters are restricted by currently materials and module fabricating technologies. Consequently, the main objective of the TEC design was to determine a set of design parameters which meet the required specifications or create the best performance at minimum cost.

3 Geometric Optimization

Table 1 lists some research in optimizing the geometric properties of TECs. In single-objective optimization problem, [8] combined a TEC model and a genetic algorithm to optimize the geometry and performance of the STECs. The geometric properties of STECs were considered as the search variables and were optimized

Table 1 Previous optimization techniques applied in optimizing performance of TECs

Type of optimizations	Type of TECs	Technique used	Author/year
SOP	STECs	GA	Cheng/2005
SOP	STECs	Conjugate-gradient method	Huang/2013
SOP	STECs	GA	Nain/2010
SOP	TTECs	GA	Cheng/2006
MOP	STECs	NSGA-II	Nain/2010
MOP	TTECs	TLBO-II	Rao/2013

simultaneously to reach the maximum ROR under the requirement of minimum COP, the confined volume of STECs, and the restriction on the maximum cost of the material. The optimal search used GA converged so rapidly (around 20 iterations).

Huang et al. [9] developed an optimization approach which integrates a complete multi-physics TEC model and a simplified conjugate-gradient method. Under a wide range of operating conditions of temperature difference and applied current, the geometric properties of STECs as search variables were optimized to reach the maximum ROR. The effects of applied current and temperature difference in the optimal geometry were also discussed.

For TTECs, [10] used GA for maximizing separately the ROR and COP. The author had considered the effect of thermal resistance and determined the optimum value of input current and number of legs for two different design configurations of TEC. The optimal search in this GA converges so rapidly with over 30 runs. These results were not different with those obtained from Xuan's work [11] and showed that GA had a robust behavior and effective search ability.

For multi-objective optimization problems (MOP), STECs will have a better design if we can find the optimal point of ROR and COP simultaneously. Nain et al. [12] used NSGA-II for multi-objective optimization of STECs. The value of geometric properties of STECs was optimized to achieve Pareto-optimal solutions at different values of thermal resistance. The authors point out the adverse effects of thermal resistance in obtaining the optimum value of cooling rate or COP.

For TTECs, [13] used modified teaching–learning-based optimization (TLBO) in optimizing the dimensional structure of TTECs. TLBO was based on the effect of the influence of a teacher on the output learners in a class. The algorithm mimics the teaching–learning ability of teacher and learners in a classroom; the teacher and learners are the two vital components of the algorithm. TLBO was modified and applied successfully to the multi-objective optimization of TTECs with a better performance than GA. The determination of the number of TE module in hot stage and cold stage as well as the supply current to the hot stage and the cold stage were considered as search variables. Two different configurations of TTECs, electrically separated and electrically connected in series, were investigated for the optimization.

4 Material Properties Optimization Matters

As shown in the above part, a good thermoelectric material should have high Seebeck coefficient, high electrical conductivity, and low thermal conductivity. However, since these three parameters are interrelated, following the Wiedenmann–Franz law, researches have to optimize these conflicting parameters to get the maximize ZT.

With the effectiveness of material properties on the performance of TEC, there have been conducted many research during the past ten years in finding a new material and structure for use in green, highly efficient cooling, and energy conversion system. Bismuth–Telluride (Bi_2Te_3) is one of the best thermo-electric materials with the highest value figure of merit [14]. Much effort has been made to raise ZT of bulk materials based on Bi_2Te_3 by doping or alloying other elements in various fabricating processes. However, ZT was not much more than one and are not sufficient to improve dramatically the cooling efficiency. The reason is due to the difficulty to increase the electrical conductivity or Seeback coefficient without increasing the thermal conductivity [15].

Recent advancements in improving ZT values include the work of Poudel et al., who achieved a peak ZT by 1.4 at 100 °C from a bismuth antimony Telluride (BiSbTe) p-type Nano crystalline bulk alloy [16]. This material is an alloy of Bi_2Te_3 and is made by hot pressing Nano powders that are ball-milled from crystalline ingots. ZT is about 1.2 at room temperature and peaks at about 1.4 at 100 °C, which makes these materials useful for microprocessor cooling applications.

5 Mathematical Modeling of Thermo-electric Coolers

Operation of TEC is based on the Peltier effect. TEC acts like a solid-state cooling device that can pump heat from one junction to the other junction when a DC current is applied. The energy balance equations at the hot junction and the cold junction for TEC can be described as in Eqs. 1, 2. ROR is the net rate of heat transfer in Watts. These equations show the completion between the Seebeck coefficient term, which is responsible for TEC cooling, and the parasitic effect of Joule heating and back heat conduction from the electrical resistance and thermal conductance terms, respectively. The heat flows $\alpha I T_h$ and $\alpha I T_c$ caused by the Peltier effect are absorbed at the cold junction and released from the hot junction, respectively. Joule heating $1/2 I^2$ ($\rho_r L/A + 2r_c/A$) due to the flow of electrical current through the material is generated both inside the TEC legs and at the contact surfaces between the TEC legs and the two substrates [8]. TEC is operated between temperatures T_c and T_h, so heat conduction $\kappa A(T_h - T_c)$ occurs through the TEC legs.

$$\text{ROR} = N\left[\alpha I T_c - \frac{1}{2}I^2\left(\rho_r\frac{L}{A} + \frac{2r_c}{A}\right) - \frac{kA(T_h - T_c)}{L}\right] \tag{1}$$

$$Q_h = N\left[\alpha I T_h + \frac{1}{2}I^2\left(\rho_r\frac{L}{A} + \frac{2r_c}{A}\right) - \frac{kA(T_h - T_c)}{L}\right]. \tag{2}$$

The input electrical power and coefficient of performance (COP) can be calculated using following relations (Eqs. 3–4):

$$P = Q_h - \text{ROR} \tag{3}$$

$$\text{COP} = \frac{\text{ROR}}{Q_h - \text{ROR}}. \tag{4}$$

α, ρ_r, k are Seebeck coefficient, electrical resistivity, and thermal conductivity of TE elements, respectively. They represent for thermo-electric material properties. A, L, N are geometric properties of TEC model.

COP is a common metric used to quantify the effectiveness of a heat engine. It is also important to quantify the amount of heat that a TEC can transfer and the maximum differential across the TEC. For an STEC, basically COP is between 0.3 and 0.7. The COP can be greater than 1.0 only when the module is pumping against the positive temperature gradient.

6 Relation Between COP and ROR

For a TEC with a specific geometry, ROR and COP are all dependent on its operating conditions which are the temperature difference ΔT and applied current. With a fixed ΔT, ROR and COP are first increased and then decreased as I is increased [9]. Unfortunately, with the same applied current, maximum ROR and maximum COP always cannot reach simultaneously. Similarly, with the same operating conditions, as the TEC geometry is varied, ROR and COP are all varied, but maybe cannot reach the maximums simultaneously [17].

7 Affections of Material Properties and Geometric Properties on TEC Performance

The performance of TEC (COP and ROR) strongly depends on thermo-electric materials. A good thermo-electric material should have a large Seebeck coefficient to get the greatest possible temperature difference per given amount of electrical potential (voltage), low electrical resistance to minimize the Joule heating [18], and low thermal conductivity to reduce the conduction from the hot side and back to the cold side. Pure metal has a low Seebeck coefficient, which leads to low thermal conductivity, whereas in insulators electrical resistivity is low which lead to higher Joule heating.

The performance evaluation index of thermo-electric materials is the figure of merit Z or dimensionless figure of merit ($ZT = \alpha^2 T/\rho K$), which combines the above properties. The increase in Z or ZT leads directly to the improvement in the cooling efficiency of Peltier modules.

The material properties are considered to be dependent on the average temperature of the cold side and hot side temperatures of each stage. Their values can be calculated from the following equations [8]:

$$\alpha_p = -\alpha_n = (-263,38 + 2.78T_{\text{ave}} - 0.00406T^2_{\text{ave}})10^{-6}$$
$$\alpha = \alpha_p - \alpha_n;$$

(5)

$$\rho_p = \rho_n = (22,39 - 0.13T_{\text{ave}} + 0.00030625T^2_{\text{ave}})10^{-6}$$
$$\rho_r = \rho_p + \rho_n;$$

(6)

$$\kappa_p = \kappa_p = 3.95 - 0.014T_{\text{ave}} + 0.00001875T^2_{\text{ave}}$$
$$\kappa = \kappa_n + \kappa_p.$$

(7)

From the Eqs. 1, 2, the geometric structure has remarkable effect on the TEC. The maximum ROR increases with the decrease of leg length until it reaches a maximum and then decreases with a further reduction in the thermo-element length [9]. The COP increases with an increase in thermo-element length. As the COP increases with the leg area, the ROR may decrease because the total available volume is limited. As the leg area is reduced, the ROR generally increases. A smaller leg area and a greater number of legs yield greater cooling capacity. When the leg length is below than this lower bound, the cooling capacity declines enormously [19]. Other elements have affection on the performance of TEC like the contact resistance, but it is very small in some calculation it can be neglected.

8 Meta-heuristic Optimization Algorithm

In the thermal energy sector, meta-heuristics have been used recently, to solve industrial problems as well as enhance current processes, equipment, and field operations. Table 2 lists some applications of meta-heuristic in thermal energy systems such as GA, SA, Particle Swarm Optimization, and Ant Colony Optimization.

Meta-heuristics are widely recognized as efficient approaches for many hard optimization problems. A meta-heuristic is an algorithm designed to solve approximately a wide range of hard optimization problems without having to deeply adapt to each problem. Almost all meta-heuristics share the following characteristic: they are nature-inspired (based on some principles from physic, biology or ethnology) [20]. A meta-heuristic will be successful on a given optimization problem if it can provide a balance between the exploration and exploitation [21]. Roughly speaking, the basic single-solution-based meta-heuristics are more exploitation oriented, whereas basic population-based meta-heuristics are more exploration oriented. Exploration (diversification) is needed to identify the parts of the search space with high-quality solutions. Exploitation (intensification)

Table 2 Recent application of intelligent strategies in thermal energy system

Author/year	Application	Technique
Xu and Wang [35]	Optimal thermal models of building envelope based on frequency domain	GA
Gozde and Taplamacioglu [36]	Automatic generation control application in a thermal power system	PSO
Kou et al. [37]	Optimum thermal design of micro-channel heat sinks	SA
Pezzini et al. [38]	Optimize energy efficiency	ACO, PSO, GA, ES, EP
Sharma [39]	Optimization of thermal performance of a smooth flat solar air heater	PSO
Eynard et al. [40]	Forecasting temperature and thermal power consumption	Wavelet and ANN

is important to intensify the search in some promising areas of the accumulated search experience [21]. The main differences between existing meta-heuristics concern the particular way in which they try to achieve this balance.

Categories of meta-heuristic are introduced in Fig. 4. SA is a point-based meta-heuristics which is normally started single initial solution and move away from it. GA is a population-based meta-heuristics which can deal with a set of solutions rather than with a single solution. This research mainly focuses on SA and GA meta-heuristic techniques.

9 Genetic Algorithm Optimization Technique

Genetic Algorithm (GA) is a method for solving both constrained and unconstrained optimization problems that are a random searching based on the mechanism of natural selection and survival of the fittest [3]. GA has repeatedly modified

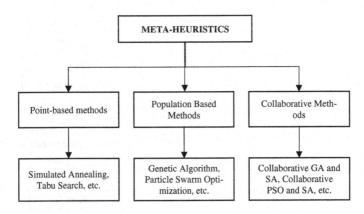

Fig. 4 Categories of meta-heuristic

a population of individual solutions. At each step, the genetic algorithm selects individuals at random from the current population to be parents and uses them to produce the children for the next generation. Over succeeding generations, the population "evolves" toward an optimal solution. The three most important phases involved in GA are selected, crossover, and mutation:

- *Selection*: Select the individuals, called parents that contribute to the population in the next generation.
- *Crossover*: Combine two parents to form children for the next generation.
- *Mutation*: Apply random changes to individual parents to form children.

GA can be used to solve a constrained optimization problem and can find a good local optimum solution [22]. GA is simple and quick to execute [23]. GA can be effectively applied in highly nonlinear problems [24] and solve a variety of optimization problems by searching a larger solution space [20]. However, GA requires determination of optimum controlling parameters such as crossover rate and mutation rate. Moreover, GA has a poor global search capability. Flow chart of GA algorithm is shown in (Fig. 5).

Fig. 5 Genetic algorithm flow chart

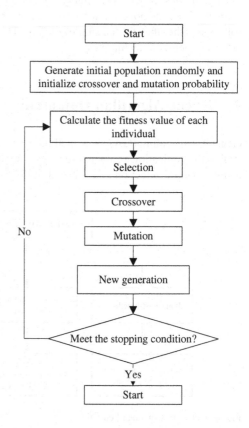

10 Simulated Annealing Optimization Technique

Simulated Annealing (SA) is a method for solving unconstrained and bound-constrained optimization problems [25]. The method models the physical process of heating a material and then slowly lowering the temperature to decrease defects, thus minimizing the system energy. The objective function of the problem similar to the energy of a material is then minimized, by introducing a fictitious temperature, which is a simple controllable parameter of the algorithm.

At each iteration of the SA algorithm, a new point is randomly generated. The distance of the new point from the current point, or the extent of the search, is based on a probability distribution with a scale proportional to the temperature. The algorithm not only accepts all new points that lower the objective, but also, with a certain probability, accepts points that raise the objective. By accepting points that raise the objective, the algorithm avoids being trapped in local minima, and is able to explore globally for more possible solutions. An annealing schedule is selected to systematically decrease the temperature as the algorithm proceeds. As the temperature decreases, the algorithm reduces the extent of its search to converge to a minimum.

In contrary to GA, SA has the ability to escape from local optima [26], flexibility, and ability to approach global optimality. SA can be applied to large-scale problems regardless of the conditions of differentiability, continuity, and convexity those are normally required in conventional optimization methods [27]. SA is easy to code even for complex systems and can deal with highly nonlinear models with many constraints. However, SA still suffers some disadvantages such as the difficulty in defining a good cooling schedule. Figure 6 presents the flow chart of SA.

Fig. 6 Simulated annealing flow chart

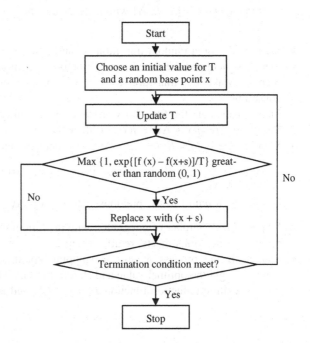

11 Introduction of Collaborative Simulated Annealing Genetic Algorithm

Traditional SA not only has strong global search ability in solving NP-hard problem, but also has defects such as premature and weak local search ability. GA has strong local search ability and no premature problem. Therefore, the combination of GA and SA can overcome the defects of each of the two methods, bring into play their respective advantages, and improve the solving efficiency. Flow chart of CSAGA is shown in Fig. 7. The algorithmic statement of CSAGA has been combined with STECs mathematical modeling and is described as follows:

- Step 1: Set the initial parameters of STECs model, CSAGA algorithm and create the initial point of design variables:

  - For STEC model, determine required initial parameters (T_h, T_c and I), set the boundary constraint of the design variables [A_{min}, L_{min}, N_{min}] and [A_{max}, L_{max}, N_{max}], nonlinear equality constraints as well. Consequently, the material properties of STEC are then calculated. Refer to "Parameter selection of STEC and CSAGA" for more details about choosing these parameters.
  - For SA, determine required parameters for the algorithm such as Initial annealing temperature T_o, Boltzmann annealing k_B, temperature reduction α, maximum number of iterations, stopping condition criteria: tolerance function value and final stopping temperature. Refer to "Parameter selection of STEC and CSAGA" for more details about choosing these parameters.
  - To implement the algorithm, first randomly initializing based point of design parameters $x = [A, L, N]$ within the boundary constraint [A_{min}, L_{min}, N_{min}] and [A_{max}, L_{max}, N_{max}].

- Step 2: Choose a random transition Δx and count the number of iterations.
- Step 3: Calculated the function value before transition $Q_{c(x)} = f(x)$.
- Step 4: Make the transition as $x = x + \Delta x$ within the range of boundary constraint.
- Step 5: Calculate the function value after transition $Q_{c(x+\Delta x)} = f(x + \Delta x)$.
- Step 6: If $\Delta f = f(x + \Delta x) - f(x) > 0$ then accept the state $x = x + \Delta x$.
- Step 7: Else If $\Delta f = f(x) - f(x + \Delta x) \leq 0$ then generate a random number (0,1).

  - If $e^{[f(x+\Delta x)-f(x)]/k_B \cdot T}$ is greater than random (0,1) then accept the state $x = x + \Delta x$.
  - Else then return to the previous state $x = x - \Delta x$.

- Step 8: Check number of integration with maximum number of iterations. If the number of iterations meets, return to step 2.
- Step 9: If the process meets the stopping condition, stop running the SA algorithm, get the optimal value and start to run GA. Otherwise, update T based on temperature reduction function $T_n = \alpha \cdot T_{n-1}$ and return to step 2.

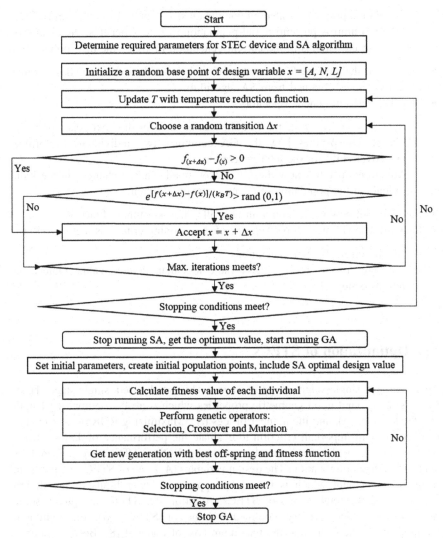

Fig. 7 Flow chart of collaborative simulated annealing genetic algorithm

- Step 10: GA optimization tool box in MAtlab is used. Set initial parameters of GA and create initial population points:
  - Population size, number of generations, crossover probability, mutation probability, and stopping condition values. Refer to "Parameter selection of STEC and CSAGA" for more details about choosing these parameters.
  - The initial population points [A, L, N] of STEC are created randomly at each iteration within the range of boundary constraint (upper bound and lower bound) by using Matlab random number generator. Number of initial

population points is limited by population size. Optimal value obtained from SA algorithm is put into initial population of GA. After this, the algorithm uses the individuals in the current generation to create the next population.

- Step 11: Calculate the fitness value of each individual that includes optimal design variables obtained from SA algorithm.
- Step 12: Perform genetic operators:

  - Selection: specify how the GA chooses parents for the next generation.
  - Crossover: specifies how the GA combines two individuals to form a crossover child for the next generation.
  - Mutation: specifies how the GA makes small random changes in the individuals in the population to create mutation children.

- Step 13: Get new generation with best offspring and fitness function.
- Step 14: The GA will stop when one of the stopping criteria is met as follows:

  - Maximum number of generations
  - Average change in the fitness value less than defined value. Otherwise, return to step 11.

12 Optimization of STECs

TECs can be single-stage of multi-stages. This work used single-stage TECs (STECs) and found good geometric properties which were the optimal leg length (L), the leg area (A), and the number of legs (N). Maximizing ROR or maximizing COP is an main important criterion to evaluate the performance of TECs. In this work, one objective function, namely maximizing ROR of STECs is considered for single-objective optimization. The design variables (A, L, N) of STECs are put in an inequality constraint which is bound by upper and lower limits of the design variables and the total area S of STECs. Additionally, STECs are put in some requirements which are the confined volume of STECs (S), the minimum requirement of the COP, and the maximum cost of material [8]. Because optimization of TECs geometry may cause the reduction in the COP, the COP is used as a constraint condition during the optimization in order to guarantee that the TECs with the optimal geometry have a relatively high COP [28]. Referring to Eqs. 1–4, the parameters T_c, T_h, and I are defined in the beginning of the calculation. The unknown term is material properties of TECs, which will be determined based on the Eqs. 5–7 with the values of T_h and T_c.

13   Test System Details

To verify the effectiveness of meta-heuristic optimization techniques, simulations tests were carried out on STECs model to find the optimal value of geometric properties. The purpose of the tests should be related to the ultimate goal of meta-heuristic methods: fast, high-quality solutions to important problems. Single-objective optimization is used under constraints. Parameter setting of STECs system, the proposed techniques CSAGA, and stand-alone optimization technique GA, SA are chosen. After testing the performance of the proposed technique CSAGA, better design parameters of STECs have been explored by running the system under various operating conditions such as various input currents, various cold side temperatures with constraint condition of COP.

14   Parameters Selection of STECs and CSAGA

Parameters of STECs are referred from Cheng's work [8]. Table 3 lists the parameters of STECs. STEC is placed in a confined volume with total area 100 mm^2 and a height of 1 mm. The objective function is to maximize the ROR. The temperatures of the cold side stage and the hot side stage are both fixed to 323 K. The effect of electrical resistance r_c is considered with the value 10^{-8} Ωm^2.

Parameter selection of CSAGA is shown in Table 4. Initial temperature ($T_o = 100$), the temperature is the control parameter in simulated annealing and it is decreased gradually as the algorithm proceeds [29]. Temperature reduction ($\alpha = 0.95$), temperature decrease is $T_n = \alpha \cdot T_{n-1}$. Experimentation is done with different alpha values: 0.70, 0.75, 0.85, 0.90, and 0.95. Boltzmann annealing ($k_B = 1$), k_B will be used in the Metropolis algorithm to calculate the acceptance probability of the points. Stopping criteria, the function tolerance is set as 10^{-6} and final stopping temperature is set as 10^{-10}. This value can be obtained as a function of minimum possible deterioration the objective function.

Table 3 Parameters setting of STECs	Group	Parameters setting	Specific values
	1	Objective function	Maximize ROR
	2	Variables	0.03 mm $< L <$ 1 mm 0.09 mm^2 $< A <$ 100 mm^2 $1 < N < 1000$
	3	Fixed parameters	$S = 100$ mm^2 $T_h = T_c = 323$ K $r_c = 10^{-8}$ Ωm^2
	4	Constraints availability	A.N < 100 mm^2 Maximum cost 385$ Required COP = 0.75; 1

Table 4 Parameters setting of collaborative simulated annealing genetic algorithm (CSAGA)

No.	Parameters setting	Specific values
1	Initial temperature	$T_o = 100$
2	Temperature reduction	$\alpha = 0.95$
3	Boltzmann annealing	$k_B = 1$
4	Stopping criteria of SA	Final stopping temperature 10^{-10} Function tolerance 10^{-6}
5	Population size	100
6	Fitness scaling function	Fitness scaling rank
7	Selection function	Selection tournament, 4
8	Crossover function	Crossover arithmetic
9	Crossover fraction	0.6
10	Mutation function	Mutation Adaptive Feasible
11	Stopping condition of GA	Function tolerance 10^{-6} Maximum number of generations 10,000

For GA, the population size, the maximum number of generations must be determined. Generation specifies the maximum number of iterations the genetic algorithm performs. Population specifies how many individuals are in each generation. Increasing the population size will increase the accuracy of GA [30] in finding a global optimum but cause the algorithm to run slowly. A population size of 100 for 10,000 generations is run. Furthermore, the selection operator, crossover operator, and the mutation operator are applied with probabilities, so as to construct a new set of solutions randomly. In selection process, option is selected by using selection function which has some options such as stochastic uniform, uniform or roulette, tournament. Tournament selection chooses each parent by choosing Tournament size four players at random and then choosing the best individual out of that set to be a parent. In crossover process, crossover arithmetic type with fraction 0.6 is chosen to create children that are weight arithmetic mean of two parents [31] so that children are always feasible with respect linear constraints and bounds. In mutation process, mutation adaptive feasible option is chosen which satisfies the constraints. Finally, for stopping criteria, the function tolerance was set at 10^{-6} or the algorithm will stop after getting over the maximum number of iterations 10,000. The optimization process terminated until one of these criteria is satisfied.

15 Robustness Test

Robustness of the meta-heuristic method is evaluated by measuring the less sensitive capability to different types of applied system as TEC's operating conditions (Barr et al. 1995), because the meta-heuristic method may not converge to an exact same solution at each random run. Therefore, their performances could not be judged by the results of a single run. Many trials should be done with some case studies of STEC's system to reach a useful conclusion about the robustness of the algorithm [32].

Based on Matlab 2013a platform, CSAGA was programmed and then run in 30 trials on a computer (CPU: Intel® Core™ i5-3470 CPU @ 3.2GHz 3.2 GHz; RAM: 4 GB DDR; OS: Windows 7) with three case studies as follows:

- Case study 1: Based on the setting of STECs model in Table 3, CSAGA is tested under one nonlinear inequality constraint which is confined volume of STECs $S = 100$ mm^2 (A.N < S).
- Case study 2: Based on the setting of STECs model in Table 3, CSAGA is tested under one nonlinear equality constraint which is a limitation of COP = 1.
- Case study 3: Based on the setting of STECs model in Table 3, CSAGA is tested under two above constraints which are nonlinear inequality constraint (A. N < S) and nonlinear equality constraint COP = 0.75.

To evaluate and compare the performance of the proposed technique, the stand-alone optimization techniques which are GA and SA were also run on this test in the same operating condition of STECs model. The best value, average value, lowest value, and standard deviation of 30 trials of each technique were collected.

After testing and collecting the data, CSAGA and GA can find the optimal dimension and satisfy the constraint in all three case studies; SA can solve the problem in case study 1 but get stuck in solving the problem with nonlinear equality constraint (case study 2). The comparison between maximum ROR obtained from the optimization technique with analytical results reveals that CSAGA, GA, and SA perform well and give exactly the same value as analysis results. Figures 8, 9 and 10 show the graphs of best fitness values which are maximum ROR and maximum COP after 30 trials. Table 5 shows the collected data of three case studies. From the figures, the line created by 30 trial runs of CSAGA is more stable than by SA and GA. As shown in Table 5, the range between maximum and minimum values of

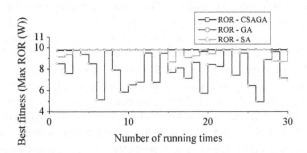

Fig. 8 Case study 1—run STECs system using CSAGA, GA, and SA under nonlinear inequality constraint A.N < 100 mm^2

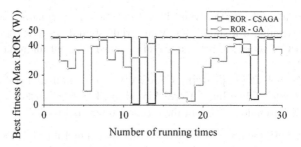

Fig. 9 Case study 2—run STECs system using CSAGA, GA, and SA under nonlinear equality constraint COP = 1

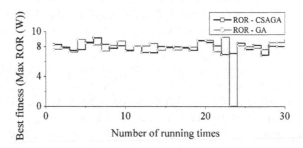

Fig. 10 Case study 3—run STECs system using CSAGA, GA, and SA under nonlinear inequality constraint A.N < 100 mm^2 and nonlinear equality constraint COP = 0.75

best fitness obtained by CSAGA approach is closer than by SA and GA approaches. In case study 1, the average value of maximum ROR obtained after 30 trial runs using CSAGA is 9.7874 W, 24.43 % higher than the obtained value by using GA (7.8654 W). In case study 2, the average value of maximum ROR obtained using CSAGA is 40.7097 W, increasing 38.83 % as compared to the GA approach (29.3253 W). In terms of standard deviation, CSAGA performs 0.0269 W for case study 1 and 0.5409 W for case study 2. The ranges of maximum value of ROR of CSAGA for case study 1 and case study 2 are (9.7874 ± 0.0269) and (7.916 ± 0.5409) W, respectively. As compared with other techniques, SA shows larger range (9.5893 ± 0.3681) W than CSAGA for case study 1; GA shows larger ranges (7.8654 ± 1.4389) and (7.8491 ± 1.5734) W than CSAGA for case study 1 and case study 2, respectively. A low standard deviation indicates that data points tend to be very close to the mean. These data demonstrate that the performance of CSAGA is more stable and more reliable when yields smaller range of maximum ROR than GA and SA. CSAGA has better robustness in solving optimization problem under constraints. CSAGA is more helpful for the designers to save their time in finding the optimal design parameters. By running the algorithm of CSAGA in only 1 time, the optimal design parameters of STECs can exactly better than other algorithms.

Table 5 Test results after running system for 30 trials

Technique used	CSAGA	SA	GA
Case study 1	Max ROR (W)	Max ROR (W)	Max ROR (W)
Standard deviation	0.0269	0.3681	1.4389
Average value of best fitness	9.7874	9.5893	7.8654
Minimum value of best fitness	9.6486	8.663	4.932
Maximum value of best fitness	9.7956	9.795	9.7798
Technique used	CSAGA	SA	GA
Case study 2	Max ROR (W)	Max ROR (W)	Max ROR (W)
Standard deviation	13.3064	SA cannot find optimal value	12.6172
Average value of best fitness	40.7097		29.3253
Minimum value of best fitness	0.6868		2.7407
Maximum value of best fitness	45.4319		45.1371
Technique used	CSAGA	SA	GA
Case study 3	Max ROR (W)	Max ROR (W)	Max ROR (W)
Standard deviation	0.5409	SA cannot find optimal value	1.5734
Average value of best fitness	7.916		7.8491
Minimum value of best fitness	6.907		0.0409
Maximum value of best fitness	9.1576		9.26

16 Computational Efficiency

Computational efficiency of all methods is compared based on the average CPU time taken to converge the solution. The CPU time taken by each solution is given in Table 6. SA can perform well in case study 1 with smallest time-consuming. GA takes longer computational time, especially in case study 3, about 44.97 s which is double as compared with CSAGA approach. CSAGA solves the problem for three cases with the same time-consuming. It demonstrates the stability of CSAGA in term of computational efficiency.

Table 6 Comparison of average execution times

Technique	CPU time (s)		
	Case study 1	Case study 2	Case study 3
CSAGA	27.24	28.78	20.35
GA	6.012	30.64	44.97
SA	14.06	–	–

17 Results

After evaluating and comparing the performance of STECs, results from optimizing the design of STECs were produced. Some case studies which were taken from previous works of [8] are as follows:

- Case study 1: STECs model is run with CSAGA under the two constraints which are the constraint of total area A.N < 100 mm² which is nonlinear inequality constraint and the maximum cost of material $385. The input current is varied from 0.1 to 8 A; hot side and cold side temperatures are set as 323 K ($T_h = T_c = 323$ K); the requirement of COP was neglected. Because COP is not taken into account, stand-alone SA can perform well and is run in the same condition, and results of Cheng's work using GA are taken for comparison.
- Case study 2: Same condition with case study 1 but cold side temperature is varied from 283 to 323 K; input current is set as 1 A.
- Case study 3: STECs model is run with CSAGA under various input currents, with three constraints which is the requirement of confined volume of STECs (S = 100 mm²), maximum cost of material $385, and the requirement of COP = 0.75 which is a non-linear equality constraint. In this case, SA cannot solve the problem with non-linear equality constraint; stand-alone GA is run in the same condition with CSAGA for comparison.

For case study 1, Table 7 presents the optimal design parameters of STECs model for various input currents from 0.1 to 8 A and the comparison is shown in Fig. 11. In Table 7, when maximum ROR is increased, leg area increases and number of legs decreases; leg length does not change and reach the lower bound by the limit (0.3 mm). As shown in Fig. 11, when the input current is larger than 0.5 A, the maximum values of ROR obtained by CSAGA and SA are approximately 7.42 W which seem unchanged. The table demonstrates that STECs using CSAGA

Table 7 Case study 1—collected data after running CSAGA and SA under various input currents

I (A)	CSAGA				SA			
	MaxROR (W)	N (unit)	A (mm²)	L (mm)	MaxROR (W)	N (unit)	A (mm²)	L (mm)
0.1	4.63	841.75	0.09	0.3	4.63	841.75	0.09	0.3
0.2	7.04	841.75	0.09	0.3	7.04	841.75	0.09	0.3
0.5	7.42	437.18	0.17	0.3	7.42	437.14	0.17	0.3
1	7.42	216.63	0.35	0.3	7.42	221.38	0.34	0.3
2	7.42	107.64	0.7	0.3	7.42	107.79	0.7	0.3
4	7.42	54.67	1.39	0.3	7.42	54.66	1.39	0.3
6	7.42	36.34	2.09	0.3	7.41	36.34	2.08	0.3
8	7.42	26.91	2.82	0.3	7.42	26.91	2.82	0.3

$T_h = T_c = 323$ K, maximum cost of the material was $385 and STECs were put in a confined volume 100 mm². The requirement of COP was ignored

Fig. 11 Case study 1—run STECs system under various input currents

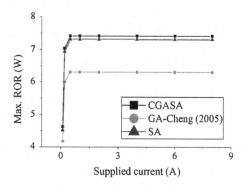

Table 8 Case study 2—collected data after running CSAGA and SA under various cold side temperatures

T_c (K)	CSAGA				SA			
	MaxROR (W)	N (unit)	A (mm^2)	L (mm)	MaxROR (W)	N (unit)	A (mm^2)	L (mm)
283	–	–	–	–	–	–	–	–
293	0.37	44.51	0.51	1	0.37	44.50	0.51	1
303	1.03	81.88	0.44	0.62	1.02	76.51	0.45	0.66
313	3.42	209.47	0.36	0.3	0.30	209.46	0.36	0.30
323	7.42	220.54	0.34	0.3	7.42	220.53	0.34	0.30
333	11.38	226.18	0.335	0.3	11.38	226.17	0.33	0.30

Supplied current for STECs was 1 A, maximum cost of the material was $385, and STECs were put in a confined volume 100 mm^2. The requirement of COP was ignored

can reach its maximum cooling capacity even under various input currents. Moreover, from Fig. 11, these obtained results are approximately 17.8 % higher than obtained results from [8] by using GA but same results obtained by stand-alone SA. Figure 11 demonstrates better performance of CSAGA compared to GA which was used in [8].

For case study 2, Table 8 presents the optimal design parameters of STECs model for various cold side temperatures from 283 to 333 K. In that table, the maximum value of ROR increases when the cold side temperature T_c is increased. The increasing of ROR is more obvious when T_c exceeds hot side temperature T_h. And as T_c is increased, the optimal leg area decreases and the number of TE couple increases. Leg length reaches the lower bound 0.3 mm to get the maximum ROR. Figure 12 plots the effects of the optimal dimension on maximizing ROR for various T_c. Maximum ROR increases with the cold side temperature and keep increasing if the value of cold side temperature is increased more than 333 K. The obtained results using CSAGA are slightly better than obtained results from Cheng's work (2005) but not quite different with the obtained results by using SA.

Fig. 12 Case study 2—run
STECs system under various
cold side temperatures

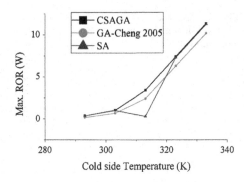

Table 9 Case study 3—collected data after running CSAGA and GA under various input currents

I (A)	CSAGA				GA			
	MaxROR (W)	N (unit)	A (mm^2)	L (mm)	MaxROR (W)	N (unit)	A (mm^2)	L (mm)
0.1	–	–	–	–	0.26	37.88	1.77	0.34
0.2	2.74	335.32	0.10	0.66	1.25	38.46	1.77	0.33
0.5	4.87	238.18	0.23	0.41	2.56	41.77	1.75	0.31
1	6.91	168.99	0.44	0.31	2.67	43.55	1.74	0.30
2	7.12	87.10	0.87	0.30	4.68	42.89	1.74	0.30
4	7.12	43.55	1.74	0.30	6.52	39.87	1.76	0.32
6	7.12	29.03	2.61	0.30	6.13	37.46	1.78	0.34
8	6.64	20.30	3.51	0.32	4.67	42.83	1.74	0.30

$T_h = T_c = 323$ K, maximum cost of the material was \$385, and STECs were put in a confined volume 100 mm^2. Requirement of COP = 0.75

For case study 3, Table 9 presents the optimal design parameters of STECs model for various input currents with the requirement of COP = 0.75. As discussed in the testing system part, SA cannot implement STECs model under nonlinear equality constraint. CSAGA has an opportunity to perform the higher advantage which is the capability to solve the problem with nonlinear equality constraint. Stand-alone GA is used in this case for comparison. In this case, the implementation will have some difficulties in finding an optimal value which must satisfy all the constraints (two nonlinear inequality constraints and one nonlinear equality constraint). For every values of input currents, CSAGA and GA were run in some limited time to collect the best value of maximum ROR. CSAGA often performs better than GA as shown in Fig. 13. In Table 9, the leg area must be increased to tolerate a higher input current.

Fig. 13 Case study 3—run
STECs system under various
cold side temperatures

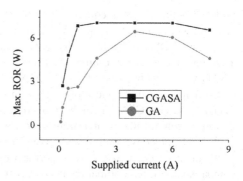

18 Future Research Direction

This work used STECs module for testing and applied optimization technique
which is CSAGA to obtain the optimal design. ROR and COP are two main
characteristics to evaluate the performance of TECs but they cannot get maximum
value simultaneously. Therefore, multi-objective optimization needs to be investigated to find the optimal dimension which can achieve better ROR and COP at the
same time. The enhancement of the research also depends on the required applications; other types of TECs such as two stages TECs or three stages TECs which
can achieve a larger temperature difference can be investigated. Other hybrid
meta-heuristic optimizations, such as hybrid differential evolution (DE) and simulated annealing, will be applied to the TECs system in order to look for more stable
and quality solutions for the objective function as well as the decision variables. DE
is a simple yet powerful algorithm that outperforms GA on many numerical
single-objective optimization problems [33]. DE explores the decision space more
efficiently than GA also when multiple objectives need to be optimized [34]. It
means that DE can achieve better results than GA also on numerical multi-objective
optimization problem.

19 Conclusion

In the present work, Collaborative Simulated Annealing and Genetic Algorithm was
used and was applied successfully in single-objective optimization of a single-stage
thermo-electric cooler module. ROR and COP are the two important criteria to
evaluate the performance of STEC. However, in this research, ROR is more focused
than COP and is an objective function. COP can be neglected or can be considered
as a satisfied condition (COP = 0.75, COP = 1) to obtain the optimal value of ROR.
CSAGA was tested and compared with stand-alone simulated annealing and
stand-alone genetic algorithm. Stand-alone SA can perform well with nonlinear
inequality constraint (A.N < S) but gets stuck in local optimum when solving the

problem with nonlinear equality constraint (COP = 0.75; 1). GA can solve the problem with nonlinear inequality constraint (A.N < S) and also nonlinear equality constraint (COP = 0.75; 1) but GA is less robust as compared to SA and CSAGA. Therefore, the obtained results of GA may not optimum and take more time to find. CSAGA is a combination of SA and GA. CSAGA can solve nonlinear inequality and equality constraint and more robust, and it consumes less computational time than SA and GA. CSAGA can be applied in diversity issues of TEC device, though it is more complicated in algorithm structure. In the future work, we would like to apply multi-objective optimization considering both important factors ROR and COP. In this case, CSAGA may perform even better than SA and GA and the comparison between optimal results of CSAGA approach and DE approach can be considerable.

Acknowledgments This research work was supported by Universiti Teknologi PETRONAS (UTP) under the Exploratory Research Grant Scheme-PCS-No. 0153AB-121 (ERGS) of Ministry of Higher Education Malaysia (MOHE). The authors would like to sincerely thank the Department of Fundamental and Applied Sciences (DFAS) and Centre of Graduate Studies (CGS) of UTP for their strong support in carrying out this research work.

Acronyms

Symbol	Description	Unit
A	Cross-sectional area of TEC legs	mm^2
L	Height of the confined volume	mm
N	Number of thermo-electric couple	–
S	Total area of STECs	mm^2
ROR	Rate of refrigeration	W
COP	Coefficient of performance	–
TE	Thermo-electric element	–
TECs	Thermo-electrics coolers	–
STECs	Single-stage thermo-electrics coolers	–
TTECs	Two-stage thermo-electrics coolers	–
MWD	Measurement while drilling	–
Z	Figure of merit	–
ZT	Dimensionless figure of merit	–
I	Supplied current to TECs	A
T_h	Hot side temperature	K
T_c	Cold side temperature	K
T_{ave}	Average of cold and hot side temperature	K
SOP	Single-objective optimization problem	–
MOP	Multi-objective optimization problem	–
CSAGA	Collaborative simulated annealing genetic algorithm	–

(continued)

(continued)

GA	Genetic algorithm	–
SA	Simulated annealing	–
PSO	Particle swarm cptimization	–
ACO	Ant olony ptimization	–
NSGA-II	Non-dominated sorting algorithm	–
TLBO	Teaching–learning-based optimization	–
TS	Tabu search	–
DE	Differential evolution	–
$\alpha_n = \alpha_p$	Seebeck coefficient of n-type and p-type thermo-electric element	V/K
$\rho_n = \rho_p$	Electrical resistivity of n-type and p-type of thermo-electric element	Ωm
$\kappa_n = \kappa_p$	Thermal conductivity of n-type and p-type of thermo-electric element	W/mK
r_c	Electrical contact resistance	Ωm^2

Key Terms and Definitions

Thermodynamics: is a branch of physics concerned with heat and temperature and their relation to energy and work. It defines macroscopic variables, such as internal energy, entropy, and pressure that partly describe a body of matter or radiation. It states that the behavior of those variables is subject to general constraints that are common to all materials, not the peculiar properties of particular materials. These general constraints are expressed in the four laws of thermodynamics. Thermodynamics describes the bulk behavior of the body, not the microscopic behaviors of the very large numbers of its microscopic constituents, such as molecules. Its laws are explained by statistical mechanics, in terms of the microscopic constituents. Thermodynamics apply to a wide variety of topics in science and engineering.

Thermo-electrics coolers: uses the Peltier effect to create a heat flux between the junctions of two different types of materials. A Peltier cooler, heater, or thermo-electric heat pump is a solid-state active heat pump which transfers heat from one side of the device to the other, with consumption of electrical energy, depending on the direction of the current. Such an instrument is also called a Peltier device, Peltier heat pump, solid-state refrigerator, or thermoelectric cooler (TEC). They can be used either for heating or for cooling (refrigeration), although in practice the main application is cooling. It can also be used as a temperature controller that either heats or cools.

Rate of refrigeration: ROR or can be called cooling rate is the rate at which heat loss occurs from the surface of an object.

Coefficient of performance: or COP of a heat pump is a ratio of heating or cooling provided to electrical energy consumed. Higher COPs equate to lower operating costs. The COP may exceed 1, because it is a ratio of output–loss, unlike the thermal efficiency ratio of output–Input energy. For complete systems, COP should include energy consumption of all auxiliaries. COP is highly dependent on operating conditions, especially absolute temperature and relative temperature between sink and system, and is often graphed or averaged against expected conditions.

Meta-heuristic: is a higher-level procedure or heuristic designed to find, generate, or select a lower-level procedure or heuristic (partial search algorithm) that may provide a sufficiently good solution to an optimization problem, especially with incomplete or imperfect information or limited computation capacity. Meta-heuristic may make few assumptions about the optimization problem being solved, and so they may be usable for a variety of problems.

Hybrid algorithm: is an algorithm that combines two or more other algorithms that solve the same problem, either choosing one (depending on the data), or switching between them over the course of the algorithm. This is generally done to combine desired features of each, so that the overall algorithm is better than the individual components. "Hybrid algorithm" does not refer to simply combining multiple algorithms to solve a different problem—many algorithms can be considered as combinations of simpler pieces—but only to combining algorithms that solve the same problem, but differ in other characteristics, notably performance.

Genetic Algorithm: is a search meta-heuristic that mimics the process of natural selection. This meta-heuristic was routinely used to generate useful solutions to optimization and search problems. Genetic algorithms belong to the larger class of evolutionary algorithms which generate solutions to optimization problems using techniques inspired by natural evolutions, such as inheritance, mutation, selection, and crossover.

Simulated Annealing: is a generic probabilistic meta-heuristic for the global optimization problem of locating a good approximation to the global optimum of a given function in a large search space. It is often used when the search space is discrete (e.g., all tours that visit a given set of cities). For certain problems, simulated annealing may be more efficient than exhaustive enumeration—provided that the goal is merely to find an acceptably good solution in a fixed amount of time, rather than the best possible solution.

Robust optimization: is a field of optimization theory that deals with optimization problems in which a certain measure of robustness is sought against uncertainty that can be represented as deterministic variability in the value of the parameters of the problem itself and/or its solution. There are a number of classification criteria for robust optimization problems/models. In particular, one can distinguish between problems dealing with local and global models of robustness, and between probabilistic and non-probabilistic models of robustness. Modern robust optimization deals primarily with non-probabilistic models of robustness that are worst-case oriented and as such usually deploy Wald's maximum models.

Nonlinear constraint optimization: is an important class of problems with a broad range of engineering, scientific, and operational applications. The form is

$$\text{Minimize} f(x) \text{ subject to } c(x) = 0 \text{ and } x \geq 0,$$

where the objective function, f: $R^n \to R$, and the constraint functions, $c : R^n \to R^m$, are twice continuously differentiable. We denote the multipliers corresponding to the equality constraints, $c(x) = 0$, by y and the multipliers of the inequality constraints, $x \geq 0$, by $z \geq 0$. An NCO may also have unbounded variables, upper bounds, or general range constraints of the form $l_i \leq c_i(x) \leq u_i$, which we omit for the sake of simplicity.

References

1. Scherbatskoy SA (1982) Systems, apparatus and methods for measuring while drilling: Google Patents
2. Goldsmid, HJ (2009) The thermoelectric and related effects. Introduction to thermoelectricity. Springer, Berlin, pp 1–6
3. Deb K (2001) Multi-objective optimization. Multi-objective optimization using evolutionary algorithms, pp 13–46
4. Deb K, Pratap A, Agarwal S, Meyarivan T (2002) A fast and elitist multiobjective genetic algorithm: NSGA-II. Evolut Comput. IEEE Trans 6(2):182–197
5. Van Laarhoven PJ, Aarts EH (1987) Simulated annealing. Springer, New York
6. Storn R, Price K (1995) Differential evolution-a simple and efficient adaptive scheme for global optimization over continuous spaces. ICSI Berkeley, Berkeley
7. Poli R, Kennedy J, Blackwell T (2007) Particle swarm optimization. Swarm Intell 1(1):33–57
8. Cheng Y-H, Lin W-K (2005) Geometric optimization of thermoelectric coolers in a confined volume using genetic algorithms. Appl Therm Eng 25(17–18):2983–2997. doi:10.1016/j.applthermaleng.2005.03.007
9. Huang Y-X, Wang X-D, Cheng C-H, Lin DT-W (2013) Geometry optimization of thermoelectric coolers using simplified conjugate-gradient method. Energy 59:689–697. doi:10.1016/j.energy.2013.06.069
10. Cheng Y-H, Shih C (2006) Maximizing the cooling capacity and COP of two-stage thermoelectric coolers through genetic algorithm. Appl Therm Eng 26(8–9):937–947. doi:10.1016/j.applthermaleng.2005.09.016
11. Xuan XC, Ng KC, Yap C, Chua HT (2002) Optimization of two-stage thermoelectric coolers with two design configurations. Energy Convers Manag 43(15):2041–2052. doi:10.1016/S0196-8904(01)00153-4
12. Nain PKS, Giri JM, Sharma S, Deb K (2010) Multi-objective performance optimization of thermo-electric coolers using dimensional structural parameters. In: Panigrahi B, Das, S, Suganthan P, Dash S (eds), Swarm, Evolutionary, and Memetic Computing, vol 6466. Springer, Berlin, pp 607–614
13. Venkata Rao R, Patel V (2013) Multi-objective optimization of two stage thermoelectric cooler using a modified teaching–learning-based optimization algorithm. Eng Appl Artif Intell 26(1):430–445. doi:10.1016/j.engappai.2012.02.016
14. Yamashita O, Sugihara S (2005) High-performance bismuth-telluride compounds with highly stable thermoelectric figure of merit. J Mater Sci 40(24):6439–6444. doi:10.1007/s10853-005-1712-6
15. Rodgers P (2008) Nanomaterials: Silicon goes thermoelectric. Nat Nano 3(2):76–76

16. Poudel B, Hao Q, Ma Y, Lan Y, Minnich A, Yu B, Vashaee D (2008) High-thermoelectric performance of nanostructured bismuth antimony telluride bulk alloys. Science 320 (5876):634–638
17. Goldsmid HJ (2009) Introduction to thermoelectricity, vol 121. Springer, Heidelberg
18. Yamashita O, Tomiyoshi S (2004) Effect of annealing on thermoelectric properties of bismuth telluride compounds doped with various additives. J Appl Phys 95(1):161–169. doi:10.1063/1.1630363
19. Rowe D, Min G (1996) Design theory of thermoelectric modules for electrical power generation. IEE Proc Sci Meas Technol 143(6):351–356
20. Boussaïd I, Lepagnot J, Siarry P (2013) A survey on optimization metaheuristics. Information sciences
21. Lee KY, El-Sharkawi MA (2008) Modern heuristic optimization techniques: theory and applications to power systems, vol 39. Wiley, New York
22. Geng H, Zhu H, Xing R, Wu T (2012) A novel hybrid evolutionary algorithm for solving multi-objective optimization problems. In: Huang, D-S, Jiang, C, Bevilacqua V, Figueroa J (eds), Intelligent computing technology, vol 7389. Springer, Berlin, pp 128–136
23. Vasant P (2010) Hybrid simulated annealing and genetic algorithms for industrial production management problems. Int J Comput Methods 7(2):279–297
24. Miettinen K (1999) Nonlinear multiobjective optimization, vol 12. Springer, New York
25. Blum C, Roli A (2008) Hybrid metaheuristics: an introduction. Hybrid metaheuristics. Springer, Heidelberg, pp 1–30
26. Blum C, Roli A, Sampels M (2008) Hybrid metaheuristics: an emerging approach to optimization, vol 114. Springer, Berlin
27. Shahsavari-Pour N, Ghasemishabankareh B (2013) A novel hybrid meta-heuristic algorithm for solving multi objective flexible job shop scheduling. J Manuf Syst 32(4):771–780. doi:10.1016/j.jmsy.2013.04.015
28. Zhao D, Tan G (2014) A review of thermoelectric cooling: materials, modeling and applications. Appl Therm Eng 66(1–2):15–24. doi:10.1016/j.applthermaleng.2014.01.074
29. Chen P-H, Shahandashti SM (2009) Hybrid of genetic algorithm and simulated annealing for multiple project scheduling with multiple resource constraints. Autom Constr 18(4):434–443. doi:10.1016/j.autcon.2008.10.007
30. Vasant P, Barsoum N (2009) Hybrid simulated annealing and genetic algorithms for industrial production management problems
31. Dingjun C, Chung-Yeol L, Cheol-Hoon P (2005, 16–16 Nov 2005) Hybrid genetic algorithm and simulated annealing (HGASA) in global function optimization. In: 17th IEEE international conference on paper presented at the tools with artificial intelligence, ICTAI 05, 2005
32. Hazra J, Sinha A (2009) Application of soft computing methods for economic dispatch in power systems. Int J Electr Electron Eng 2:538–543
33. Liu K, Du X, Kang L (2007) Differential evolution algorithm based on simulated annealing. In: Kang L, Liu Y, Zeng S (eds), Advances in computation and intelligence, vol 4683. Springer, Berlin, pp 120–126
34. Jing-Yu Y, Qing L, De-Min S (2006, 13–16 Aug 2006) A differential evolution with simulated annealing updating method. International Conference on paper presented at the machine learning and cybernetics, 2006
35. Xu X, Wang S (2007) Optimal simplified thermal models of building envelope based on frequency domain regression using genetic algorithm. Energy Build 39(5):525–536. doi:10.1016/j.enbuild.2006.06.010
36. Gozde H, Taplamacioglu MC (2011) Automatic generation control application with craziness based particle swarm optimization in a thermal power system. Int J Electr Power Energy Syst 33(1):8–16
37. Kou H-S, Lee J-J, Chen C-W (2008) Optimum thermal performance of microchannel heat sink by adjusting channel width and height. Int Commun Heat Mass Transfer 35(5):577–582

38. Pezzini P, Gomis-Bellmunt O, Sudrià-Andreu A (2011) Optimization techniques to improve energy efficiency in power systems. Renew Sustain Energy Rev 15(4):2028–2041
39. Sharma N (2012) A particle swarm optimization algorithm for optimization of thermal performance of a smooth flat plate solar air heater. Energy 38(1):406–413
40. Eynard J, Grieu S, Polit M (2011) Wavelet-based multi-resolution analysis and artificial neural networks for forecasting temperature and thermal power consumption. Eng Appl Artif Intell 24(3):501–516

Color Magnetic Resonance Brain Image Segmentation by ParaOptiMUSIG Activation Function: An Application

Sourav De and Siddhartha Bhattacharyya

Abstract Medical imaging is a technique to get images of the human body for medical science or clinical purposes. Segmentation of a medical image is a challenging task to isolate the suspicious region from the complex medical images. Genetic algorithms (GAs) are an effective tool to handle the problem of medical image segmentation. In this chapter, an application of color magnetic resonance (MR) brain image segmentation is presented by the parallel optimized multilevel sigmoidal (ParaOptiMUSIG) activation function with the parallel self-organizing neural network (PSONN) architecture. Not only confined within this approach, color MR brain image is also segmented by the NSGA-II-based ParaOptiMUSIG activation function to incorporate the multiple objective function-based scenario. These methods are compared with the process of color MR brain image segmentation by the MUSIG activation function with the PSONN architecture. All the methods are applied on a real-life color MR brain image and the quality of the segmented images are accessed by four standard objective functions. The comparison shows that the ParaOptiMUSIG activation function-based method and the NSGA-II-based ParaOptiMUSIG activation function-based method perform better than the MUSIG activation function-based method to segment the color MR brain image.

Keywords Segmentation · MLSONN architecture · PSONN architecture · MUSIG activation function · Evolution functions

S. De (✉)
Department of Information Technology, University Institute of Technology,
The University of Burdwan, Burdwan, West Bengal, India
e-mail: sourav.de79@gmail.com

S. Bhattacharyya
Department of Information Technology, RCC Institute of Information Technology,
Kolkata, West Bengal, India
e-mail: dr.siddhartha.bhattacharyya@gmail.com

© Springer India 2016
S. Bhattacharyya et al. (eds.), *Hybrid Soft Computing Approaches*,
Studies in Computational Intelligence 611,
DOI 10.1007/978-81-322-2544-7_6

1 Introduction

In any image processing application, image segmentation plays a vital and important role to express ourselves. The research and development in the field of digital image processing have progressed rapidly in the past decade due to the technological advancement in digital imaging, computer processors, and mass storage devices. Basically, it deals with the clustering of the pixels in an image into its constituent regions having same features such as intensity, shape, color, position, texture, and homogeneity. There are different types of modalities of medical images like magnetic resonance imaging (MRI), computed tomography (CT), X-rays, digital mammography, ultrasound, etc., to name a few [1, 2]. The segmentation of magnetic resonance (MR) brain image is a very important and critical stage in the field of biomedical image processing. These types of technologies are applied to analyze the normal and diseased anatomy like tumor identification, tissue and blood cell classification, multimodal registration, etc. [3]. In the earlier stage, segmentation is often needed for analyzing the medical images for computer-aided diagnosis (CAD) and therapy.

Different classical segmentation techniques are applied to segment the MRI brain images like edge detection and region growing [4, 5], thresholding [6] methods. The main objective of image segmentation using edge-based methods is to detect the boundaries of an image. But edge detection in a brain image may lead to an incorrect segmentation as different parts of the brain image are not clearly identifiable. Region growing techniques are not efficiently applied for brain image segmentation as the different regions of an image is not well defined. Image segmentation by thresholding techniques is totally dependent on image histogram and can be efficiently applied when the objects and the background of the image are clearly distinctive. As pixel distribution in brain image is not simple, thresholding techniques may not generate perfect segmentation for brain images.

Fuzzy C-means (FCM) [7], a well-known fuzzy clustering technique, tries to minimize an objective function-based on some criteria. In this method, a single data point may belong to more than one cluster specified by a membership grade. Brain MR images are segmented effectively by FCM in [8, 9]. But the drawback of this method is that it will find local optima when the cluster centroids are selected randomly. Different types of soft computing techniques like neural networks, genetic algorithms (GA), particle swarm optimization (PSO), etc., are applied to overcome this problem. The literature review on computer-aided diagnosis, medical image segmentation, edge detection, etc., by different neural network architectures are presented in [10]. In this article, the application of neural network in the field of medical image segmentation is illustrated with different types of examples. A new version of the self-organizing map (SOM) network, named moving average SOM (MA-SOM), is used to segment the medical images [11]. A cellular neural network (CNN) is proposed by Döhler et al. [12] to detect the hippocampal sclerosis in MRI. This network is a combination of cellular automata and artificial neural networks and creates an array of locally coupled nonlinear electrical circuits or cells to process a large amount of data in parallel and in real time. The histological grade, hormone status, and axillary lymphatic spread in breast cancer patients are predicted by a multivariate model [13]. In the multivariate methods, the variable reduction is done by principal component analysis (PCA) or partial least-squares regression-uninformative variable elimination (PLS-UVE), and modeled by PLS, probabilistic neural network (PNN), or cascade correlation neural network.

Genetic algorithms (GAs) [14] work as an effective tool in medical image segmentation [1]. A detailed survey of medical image segmentation using GAs is presented by Maulik [1]. Different techniques with different examples are presented to segment the medical images in this article. Zhao and Xie [2] also presented a good literature review on interactive segmentation techniques for medical images. A novel real-coded GA in connection with Simulated Binary Crossover (SBX)-based multilevel thresholding is applied to segment the T2 weighted MRI brain images [15]. In this article, the entropy of the test images is maximized to derive the optimum multilevel threshold. A hierarchical GA with a fuzzy learning-vector quantization network (HGALVQ) is proposed by Yeh and Fu [16] to segment multispectral human brain MRI. Lai and Chang [17] presented hierarchical evolutionary algorithms (HEA), mainly a variation of conventional GAs, for medical image segmentation. In this method, the exact number of classes is derived by applying the hierarchical chromosome structure and the image is segmented into those appropriate classes. Pavan et al. [18] presented an algorithm named automatic clustering using differential evolution (ACDE) to extract the shape of the tissues on different types of medical images automatically. The GA is applied to segment the two-dimensional slices of pelvic computed tomography (CT) images automatically [19]. In this method, the segmenting curve is applied as the fitness function in GA. A hybridized GA with seed region growing procedure is employed to segment the MRI images [20]. A good survey and application of brain MR images using PSO are presented in [21]. Nakib et al. [22] presented an MRI image segmentation method by two-dimensional exponential entropy (2DEE) and parameter-free PSO. A new idea of intracranial segmentation of MR brain image on the basis of pixel intensity values is proposed by the optimum boundary point detection (OBPD) method [23]. In this method, different soft computing techniques like, GA and PSO are applied in combination with FCM to solve the problem.

Pure color images can be effectively extracted from a noisy background using three independent and parallel self-organizing neural networks (TLSONN) from the parallel version of the SONN (PSONN) architecture [24]. The individual TLSONN of the PSONN architecture is employed to process the individual color components by using the generalized bilevel sigmoidal activation function with fixed and uniform thresholding. The standard backpropagation algorithm [25] is applied to adjust the network weights to gain stable solution. Bhattacharyya et al. [26] proposed a multilevel sigmoidal (MUSIG) activation function in connection with the PSONN architecture to segment true color images. This activation function is capable to map multilevel input information into multiple scales of gray. Since this activation function applies fixed and uniform thresholding parameters, it does not pay any heed to the heterogeneity of real-life images. The drawback of the MUSIG activation function is solved by De et al. [27]. The optimized MUSIG (OptiMUSIG) activation function is capable to segment multilevel grayscale images by incorporating the image heterogeneity in the MUSIG activation function. This activation function is applied in the multilevel self-organizing neural network (MLSONN) architecture [28] to segment the grayscale images. To segment the color images, the parallel version of the OptiMUSIG (ParaOptiMUSIG) activation function [29] is used in the PSONN architecture. In this method, the color images are segmented on the basis of single objective function. As these approaches are single evaluation criterion-based methods, the derived solutions may or may not ensure good results in respect of another objective criteria. To overcome from this type of problem, the solution may be the multi-objective optimization (MOO) [30]. In the real-world scenario, several competing constraints of different problems have to be optimized simultaneously to generate a set of alternative solutions instead of a single optimal set of final solutions. The resultant solutions are considered as the set of non-dominated or Pareto optimal set of solutions [30]. The optimal solutions have the equal priority as each solution is superior to these solutions when all constraints are considered. Some well-known and popular GA-based optimization techniques [30] are multi-objective genetic algorithm (MOGA), non-dominated sorting genetic algorithm (NSGA) [30], NSGA-II [31], strength Pareto evolutionary algorithm (SPEA) [32], SPEA 2 [33], etc. The drawbacks of the single objective-based ParaOptiMUSIG activation function for segmentation of color images are solved by introducing a NSGA-II-based ParaOptiMUSIG activation function [34]. In this method, the NSGA-II is applied to generate the optimized class levels and those class levels are used to design the ParaOptiMUSIG activation function. These class levels are generated after satisfying different objective criteria simultaneously.

Processing of the color MR brain image becomes a new field in the research arena. From the above discussion of the literature survey, we can notice that a very few research works are done to process the color MR brain image and it is also to be noted that it is hard to find the color MR brain images. We have tried to enlighten this field of research after getting one color MR brain image [35]. In this chapter, we have shown that the color MR brain medical image [35] can be segmented effectively to find out the different organs. For this purpose, we have applied two activation functions, the ParaOptiMUSIG activation function as well as the NSGA-II-based ParaOptiMUSIG activation function in connection with PSONN architecture. The pixel intensity of the test image is selected as the feature of the image data for the segmentation. The MUSIG activation function [26] based PSONN architecture [24] is also employed for the color MR brain image segmentation. At the initial stage, we have tried to show the application of color MR brain image segmentation using the ParaOptiMUSIG activation function-based approach and after that, the NSGA-II-based ParaOptiMUSIG activation is employed to segment the same color MR brain image. The image segmentation evaluation index like the standard measure of correlation coefficient (ρ) [26], entropy-based index [36] or quantitative-based index [37] are very much effective to assess the quality of the image segmentation results. A higher value of the correlation coefficient and lower quantitative value or entropy value signify better segmentations. The standard measure of correlation coefficient (ρ) [26] and three measures, viz., F due to Liu and Yang [38] and F' and Q due to Borsotti et al. [37] have been employed as the objective functions in the ParaOptiMUSIG activation function-based image segmentation method. In the NSGA-II method, three objective functions like the correlation coefficient (ρ) [26], F [38], and F' [37] are applied as the evaluation criteria. It is derived from the result that the color MR brain image segmentation by the ParaOptiMUSIG activation function as well as the NSGA-II-based ParaOptiMUSIG activation function performs better than the same image segmentation by the conventional MUSIG activation function.

The chapter commences with a discussion of the basic mathematical prerequisite on the multi-objective optimization using genetic algorithms and different image segmentation efficiency measures. A brief description of the PSONN architecture and its operation is given after that. The next section describes the mathematical representation and functionality of Parallel Optimized MUSIG (ParaOptiMUSIG) activation function and NSGA-II-based ParaOptiMUSIG activation function. The methodology for the color MR image segmentation by the ParaOptiMUSIG activation function as well as NSGA-II-based ParaOptiMUSIG activation function are also presented in the Methodology section. A comparative study of the results of segmentation of the test image by the aforesaid approaches vis-a-vis by the conventional MUSIG activation function are illustrated in the result section. The discussions and conclusion section concludes the chapter.

2 Mathematical Prerequisites

In this section, we will discuss the concept of multi-objective optimization using genetic algorithms (MOGA) in brief and different image segmentation quality evaluation metrics.

2.1 Multi-objective Optimization Using Genetic Algorithms

Most of the real-world problems involve simultaneous optimization of several conflicting and incomparable criteria/objectives and the single objective-based solutions cannot be considered in those cases. Usually, the process generates a set of alternative solutions instead of a single optimal solution. Each and every solution in that search space has the same priority, no one solution is superior to the other solution when all objectives are considered. There is no accepted definition to choose the best solution from a set of solutions. Hence, the concept of the multi-objective optimization is applied to solve multiple objectives with equal importance.

Mathematically, the definition of the multi-objective optimization is presented as [30]:

Determine the vector function

$$\bar{f}(\bar{x}) = \left[\bar{f_1}(\bar{x}), \bar{f_2}(\bar{x}), \ldots, \bar{f_k}(\bar{x})\right]^T \qquad (1)$$

subject to the m inequality constraints [30] $g_i(\bar{x}) \geq 0$, $i = 1, 2, \ldots, m$, and the p equality constraints [30], $h_i(\bar{x}) = 0, i = 1, 2, \ldots, p$ where k is the number of objective functions. The resultant solution $\bar{x}$ is a vector of n decision variables $\bar{x} = (x_1, x_2, \ldots, x_n,)^T$ where $x_i^L \leq \bar{x} \leq x_i^U$, $i = 1, 2, \ldots, n$. The lower bound and upper bound of each decision variable x_i are x_i^L and x_i^U, respectively. In case of a minimization problem, a decision vector $\overline{x^*}$ is determined as a Pareto optimal if and only if there is no $\bar{x}$ that dominates $\overline{x^*}$, i.e., there is no $\bar{x}$ such that $\forall i \in \{1, 2, \ldots, n\}$: $f_i(\bar{x}) \geq f(\overline{x^*})$ and $\exists j \in \{1, 2, \ldots, n\} : f_j(\bar{x}) > f_j(\overline{x^*})$. The Pareto optimal solutions are also noted as the non-dominated solutions.

According to different researchers, multi-objective search and optimization can be solved by evolutionary algorithms (EAs) efficiently than any other blind search algorithms. To solve the multi-objective optimization problems, evolutionary algorithms like genetic algorithms (GA), PSO, etc., are the better options as they process the set of solutions in parallel and the crossover is used to exploit the similarity of solutions. Some of the recently developed elitist multi-objective techniques are NSGA-II [31], SPEA [32], and SPEA 2 [33]. In this chapter, NSGA-II has been employed to generate the optimized class levels in the NSGA-II-based ParaOptiMUSIG activation function-based method.

2.2   Evaluation of Segmentation Efficiency

The segmentation efficiency of the existing segmentation algorithms is evaluated by several unsupervised subjective measures [39]. Among them, four different evaluation measures are discussed in the following subsections:

2.2.1   Correlation Coefficient (ρ)

The standard measure of correlation coefficient (ρ) [26] can be used to evaluate the quality of segmentation achieved and it is represented as [26]

$$\rho = \frac{\frac{1}{n^2} \sum\limits_{i=1}^{n} \sum\limits_{j=1}^{n} \left(R_{ij} - \overline{R}\right)\left(G_{ij} - \overline{G}\right)}{\sqrt{\frac{1}{n^2} \sum\limits_{i=1}^{n} \sum\limits_{j=1}^{n} \left(R_{ij} - \overline{R}\right)^2} \sqrt{\frac{1}{n^2} \sum\limits_{i=1}^{n} \sum\limits_{j=1}^{n} \left(G_{ij} - \overline{G}\right)^2}} \qquad (2)$$

where $R_{ij}, 1 \leq i,j \leq n$ and $G_{ij}, 1 \leq i,j \leq n$ are the original and segmented images respectively, each of dimensions $n \times n$, $\overline{R}$, and $\overline{G}$ are their respective mean intensity values. A higher value of ρ implies better quality of segmentation.

2.2.2 Empirical Goodness Measures

An overview of three different empirical goodness measures is illustrated in this subsection.

Let an image (Im) with an area S_{Im} be segmented into N number of regions. The area of region k is $S_k = |R_k|$ if R_k signifies the number of pixels in region k. The average value of the gray level intensity feature (τ) in region k is denoted as [38]

$$\overline{C_\tau}(R_k) = \frac{\sum\limits_{p \in R_k} C_\tau(p)}{S_k} \tag{3}$$

where $C_\tau(p)$ represents the value of τ for pixel p. The *squared color error* of region k is denoted as [38]

$$e_k^2 = \sum_{\tau \in (R,G,B)} \sum_{p \in R_k} \left(C_\tau(p) - \overline{C_\tau}(R_k)\right)^2 \tag{4}$$

Based on these notations three empirical measures (F, F', and Q) are discussed below.

1. *Segmentation efficiency measure* (F): The quantitative evaluation function F for image segmentation, proposed by Liu and Yang [38] is presented as

$$F(Im) = \sqrt{N} \sum_{k=1}^{N} \frac{e_k^2}{\sqrt{S_k}} \tag{5}$$

2. *Segmentation efficiency measure* (F'): The performance of Liu and Yang's method [38] is improved by Borsotti et al. [37]. They proposed another evaluation function, F', which is represented as [37]

$$F'(Im) = \frac{1}{1000 \times S_{Im}} \sqrt{\sum_{m=1}^{Maxarea} [N(m)]^{1+\frac{1}{m}} \sum_{k=1}^{N} \frac{e_k^2}{\sqrt{S_k}}} \tag{6}$$

where $N(m)$ is represented as number of regions in the segmented image of an area of m and *MaxArea* is used as the area of the largest region in the segmented image.

3. *Segmentation efficiency measure* (Q): The performance of F and F' is again improvised by another evaluation function Q and it is denoted as [37]

$$Q(Im) = \frac{1}{1000 \times S_{Im}} \sqrt{N} \sum_{k=1}^{N} \left[\frac{e_k^2}{1 + \log S_k} + \left(\frac{N(S_k)}{S_k} \right)^2 \right] \qquad (7)$$

It has been observed that lower values of F, F', and Q imply better segmentation in contrast to the correlation coefficient (ρ) where higher values dictate terms.

3 Parallel Self-Organizing Neural Network (PSONN) Architecture

The multilayer self-organizing neural network (MLSONN) [28] consists of an input layer, a hidden layer, and an output layer of neurons. It is a feedforward neural network architecture that operates in a self-supervised manner. The interested readers can get the detailed study of the architecture and operation of the MLSONN architecture in the article [28]. The miniature version of the generalized MLSONN architecture is the three-layer self-organizing neural network (TLSONN) architecture and it consists of a single hidden layer of neurons apart from the input and output layers.

The parallel version of the self-organizing neural network (PSONN) [24, 40] comprises of three independent single three-layer self-organizing neural network (TLSONN) architecture in addition to a source layer and a sink layer. Each TLSONN architecture of the PSONN architecture is applied for component-level processing of the multidimensional data and the source layer and the sink layer are used for the inputs to the network and for generating the final network output, respectively. The pure color objects can be extracted efficiently from a pure color image by the PSONN architecture [24, 40]. The functionality of the source layer is to disperse the primary color component information of true color images into the three parallel TLSONN architectures. However, the three PSONN architectures operate in a self-supervised mode on multiple shades of color component information [40]. The system errors are computed on the basis of the linear indices of fuzziness of the color component information, which have been obtained at the respective output layers. The system errors at the respective inter-layer interconnection weights are adjusted using the standard backpropagation algorithm [25]. This method of self-supervision proceeds until the system errors at the output layers of the three independent SONNs fall below some tolerable limits. The corresponding output layers of three independent SONN architectures produce the extracted color component outputs. At last, the extracted pure color output image is developed by mixing the segmented component outputs at the sink layer of the PSONN network architecture [24, 40].

Basically, the intensity features of the pixels of different regions play a vital role to segregate the object and the background regions in the input image. The multilevel input images, i.e., inputs which consist of different heterogeneous shades of image pixel intensity levels cannot be segmented by the PSONN [24, 40] network architecture. This is solely due to the fact that the constituent primitives/SONNs resort to the use of the standard bilevel sigmoidal activation functions [24, 40]. However, the problem of this network architecture can be solved by introducing a modified activation function instead of changing the architectural concept. The multilevel version of the generalized bilevel sigmoidal activation function is introduced as the new activation function of this network to overcome the problem. The main essence of the multilevel sigmoidal (MUSIG) activation [26] function consists of its ability to map inputs into multiple output levels and capable to segment true color images efficiently. The detailed review of the architecture and operational characteristics of the PSONN network architecture can be found in the literature [24, 40].

4 Parallel Optimized Multilevel Sigmoidal (ParaOptiMUSIG) Activation Function

The PSONN architecture is applied to segment the color components of the color gamut of individual pixel of the image into different class levels in parallel. For this purpose, the individual color component of the multidimensional color pixels are processed by the individual neural networks of the PSONN architecture parallelly and independently. The individual color component, such as, the red component of a color pixel can be taken as an individual grayscale object. The bilevel sigmoidal activation function which is applied in the PSONN architecture is used to segment the binary images. This activation function gives bipolar responses [0(low)/1(high)] corresponding to input information and disable to classify the grayscale object into multiple levels. To overcome this drawback, the multilevel sigmoidal (MUSIG) activation function [26] is employed to generate multilevel outputs corresponding to the multiple scales of the grayscale object and it is represented as [26]

$$f_{MUSIG}(x; \mu_\alpha, cl_\alpha) = \sum_{\alpha=1}^{K-1} \frac{1}{\mu_\alpha + e^{-\lambda[x-(\alpha-1)cl_\alpha-\theta]}} \tag{8}$$

where μ_α denotes the multilevel class responses. It is represented as [26] $\mu_\alpha = \frac{C_N}{cl_\alpha - cl_{\alpha-1}}$, where α denotes the grayscale object index ($1 \leq \alpha < K$) which is the total number of grayscale object classes. The transfer characteristics of the MUSIG activation function depends on the μ_α parameter. The cl_α and $cl_{\alpha-1}$ denote the grayscale contributions of the αth and (α-1)th classes, respectively. The maximum fuzzy membership of the grayscale contribution of neighborhood geometry is denoted by C_N. The threshold parameter (θ) in the MUSIG activation function is fixed and uniformly distributed. In this activation function, the characteristics of the underlying features of the color components are not considered and the class levels

(cl_α) are selected manually from the grayscale histograms of the input color components.

The optimized class boundaries are generated from the input image to overcome the drawback of the MUSIG activation function. The optimized version of the MUSIG activation (OptiMUSIG) [27] function is denoted as [27]

$$f_{\text{OptiMUSIG}} = \sum_{\alpha_{\text{opt}}=1}^{K-1} \frac{1}{\mu_{\alpha_{\text{opt}}} + e^{-\lambda\left[x-(\alpha-1)cl_{\alpha_{\text{opt}}}-\theta_{\text{var}}\right]}} \tag{9}$$

where $cl_{\alpha_{\text{opt}}}$ is the optimized grayscale contribution corresponding to optimized class boundaries. Based on the optimized class boundaries, the variable threshold,

$$\theta_{\text{var}} = cl_{\alpha_{\text{opt}}-1} + \frac{cl_{\alpha_{\text{opt}}} - cl_{\alpha_{\text{opt}}-1}}{2} \tag{10}$$

This OptiMUSIG activation function is efficient to segment the grayscale images and this activation function is applied to activate each individual SONN primitive of PSONN architecture with appropriate optimized parameter settings [24]. The parallel representation of the OptiMUSIG (ParaOptiMUSIG) activation function for color image is represented as [29]

$$f_{\text{ParaOptiMUSIG}} = \sum_{t \in \{R,G,B\}} f_{t_{\text{OptiMUSIG}}} \tag{11}$$

where $\{R, G, B\}$ represent the red, green, and blue layers, respectively, of the PSONN [26] architecture and $f_{t_{\text{OptiMUSIG}}}$ presents the OptiMUSIG activation function for one layer of the network. The collection of the OptiMUSIG functions of different layers is denoted by the $\sum$ sign.

The single objective-based ParaOptiMUSIG activation function executes efficiently to segment the color images. It has been observed that the single objective-based ParaOptiMUSIG activation function may not work efficiently with respect to another objective criterion. To get rid of this, the multi-objective criteria-based optimized class levels have to be generated and the NSGA-II-based ParaOptiMUSIG activation function $\left(\bar{f}_{\text{Para}_{\text{NSGA}}}(\bar{x})\right)$ is presented [34]. This is denoted as [34]

$$\bar{f}_{\text{Para}_{\text{NSGA}}}(\bar{x}) = \left[\bar{f}_{\text{Para}_{F_1}}(\bar{x}), \bar{f}_{\text{Para}_{F_2}}(\bar{x}), \ldots, \bar{f}_{\text{Para}_{Fn}}(\bar{x})\right]^T \tag{12}$$

subject to $g_{\text{Para}_{F_i}}(\bar{x}) \geq 0$, $i = 1, 2, \ldots, m$ and $h_{\text{Para}_{F_i}}(\bar{x}) = 0$, $i = 1, 2, \ldots, p$ where $f_{\text{Para}_{F_i}}(\bar{x}) \geq 0$, $i = 1, 2, \ldots, n$ are different objective functions in the NSGA-II-based ParaOptiMUSIG activation function. $g_{\text{Para}_{F_i}}(\bar{x}) \geq 0$, $i = 1, 2, \ldots, m$ and $h_{\text{Para}_{F_i}}(\bar{x}) = 0$, $i = 1, 2, \ldots, p$ are the inequality and equality constraints of this function, respectively.

5 Methodology of Segmentation

This chapter is intended to show the application of the ParaOptiMUSIG activation function using the color version of a MR brain image. For this purpose, two approaches are presented in this chapter. At first, the color MR brain image segmentation by the single objective function-based ParaOptiMUSIG activation function is narrated and after that, the methodology of the color MR brain image segmentation is presented with the NSGA-II-based ParaOptiMUSIG activation function. A combined pseudocode of both approaches will give a brief view of these algorithms and after that a detailed discussion is made.

1. Begin
2. Generate the optimized class levels for the color images.

    2.1 Input number of classes (K) and pixel intensity levels.

    2.2 Initialize each chromosome with K number of randomly selected class levels.

    2.3 Compute the fitnesses of each chromosome based on the fitness functions.

    2.4 Apply the genetic operators.

    2.5 Select a solution from the non-dominated set. /* *Not applicable for single objective-based method* */

3. Design the ParaOptiMUSIG activation function by the selected optimized class levels.
4. Input the color image pixel values to the source layer of the PSONN architecture and distribute the color component to three individual SONNs.
5. Segment the color component images by individual SONNs.
6. Fusion of individual segmented component outputs into a color image at the sink layer of the PSONN architecture.
7. End

5.1 *Generation of Optimized Class Levels for the Color Images*

The optimized class levels $(cl_{\alpha_{opt}})$ of the ParaOptiMUSIG activation function are generated in this phase. The optimized class levels can be generated by single objective-based method to design the ParaOptiMUSIG activation function and by NSGA-II algorithm to design the NSGA-II-based ParaOptiMUSIG activation function. The optimized class level generation method has gone through four intermediate steps for single objective-based method and five intermediate steps for NSGA-II-based method. The following subsections will illustrate these steps.

5.1.1 Input Phase

The pixel intensity levels of the color image and the number of classes (K) to be segmented are furnished as inputs to these optimization procedures. This step is common for both the approaches.

5.1.2 Chromosome Representation and Population Generation

The chromosomes are created with the randomly generated real numbers for both the processes and they are selected from the input color image information content. The individual elements in the chromosome are considered as the optimized class boundaries to segment the color image. Each pixel intensity of the color image information is differentiated into three color components, viz., red, green, and blue color components. Three different chromosome pools are originated for the three individual color components and individual chromosome pool is employed to generate the optimized class levels for the individual color component. A population size of 100 has been used for both the approaches.

5.1.3 Fitness Computation

In this phase for the single objective-based method, four segmentation evaluation criteria (ρ, F, F', Q), given in Eqs. 2, 5, 6 and 7 respectively, are employed as the fitness functions. These functions are used to evaluate the quality of the segmented images in this genetic algorithm-based optimization procedure. In the multi-objective-based approach, these segmentation evaluation functions are not applied individually to evaluate the quality of the segmented images. Three segmentation efficiency measures (ρ, F, F'), given in Eqs. 2, 5 and 6 respectively, are applied as the evaluation functions in the NSGA-II algorithm. These fitness functions are applied on three chromosome pools in cumulative fashion.

5.1.4 Genetic Operators

Three common genetic algorithm operators, i.e., selection, crossover, and mutation, are also applied in both the approaches.

In the selection phase of the single objective-based method, a proportionate fitness selection operator is applied to select the reproducing chromosomes. The selection probability of the ith chromosome is evaluated as [29]

$$p_i = \frac{f_i}{\sum_{j=1}^{n} f_j} \tag{13}$$

where f_i is the fitness value of the ith chromosome and n is the population size. The cumulative fitness P_i of each chromosome is evaluated by adding individual fitnesses in ascending order. Subsequently, the crossover and mutation operators are employed to evolve a new population. In this case, the crossover probability is equal to 0.8. A single point crossover operation is applied to generate the new pool of chromosomes. The mutation probability is taken as 0.01. The child chromosomes are propagated to form the new generation after mutation.

In the NSGA-II-based method, the crowded binary tournament selection operator is applied to select the fitter chromosomes. The crossover probability and the mutation probability are same like the single objective-based method. The non-dominated solutions among the parent and child populations are carried out for the next generation of NSGA-II. The near-Pareto-optimal strings furnish the desired solutions in the last generation.

5.1.5 Selecting a Solution from the Non-dominated Set

This step is not applied for the single objective-based segmentation process as any multi-objective algorithm generates a set of solutions instead of a single solution. A particular solution has to be selected from the non-dominated solutions those have rank one in the NSGA-II algorithm. It has been found from different research articles that Davies–Bouldin (DB) index [41], CDbw (Composed Density between and within clusters) [42] are some well-known cluster validity indices. It is not desirable that any cluster validity index can be relevant for all clustering algorithms. Zhang [39] showed in his article that the image segmentation results can be assessed by the image segmentation evaluation index like entropy-based index or quantitative-based index. A lower quantitative value or entropy value leads to better segmentation. The NSGA-II-based ParaOptiMUSIG activation function is generated from the selected chromosomes on the basis of the Q index [37], given in Eq. 7.

5.2 ParaOptiMUSIG Activation Function Design

In the single objective-based approach, the optimized class boundaries $(cl_{\alpha_{opt}})$ which have been obtained from the previous phase are applied to design the ParaOptiMUSIG activation function. The NSGA-II-based ParaOptiMUSIG activation function is generated by the optimized class boundaries $(cl_{\alpha_{opt}})$ which have been selected in the Pareto optimal set. The optimized class boundaries for individual color component in the selected chromosomes are applied to generate the individual OptiMUSIG activation function for that color component, viz., the class boundaries for the red component is employed to generate OptiMUSIG activation function for red and so on. The ParaOptiMUSIG function is rendered by collection of the individual OptiMUSIG functions generated for the individual color components using Eq. 11 and the NSGA-II-based ParaOptiMUSIG activation function is derived using Eq. 12.

5.3 Input the Color Image Pixel Values to the Source Layer of the PSONN Architecture and Distribute the Color Component to Three Individual SONNs

In this phase, the source layer of the PSONN architecture is fed with the pixel intensity levels of the color image. The number of neurons in the source layer of the PSONN architecture and the number of the pixels in the processed image are the same. The pixel intensity levels of the input color image are differentiated into three individual primary color components and the three individual three-layer component SONNs are fed with these independent primary color components, viz., the red component is applied to one SONN, the green component to another SONN, and the remaining SONN takes the blue component information at their respective input layers. The fixed interconnections of the respective SONNs with the source layer are responsible for this scenario.

5.4 Segmentation of Color Component Images by Individual SONNs

The individual color components of the color images are segmented by applying the corresponding SONN architecture channelized by the projected ParaOptiMUSIG activation function at the individual primitives/neurons. The neurons of the different layers of individual three-layer SONN architecture yield different input color component level responses, depending on the number of transition lobes of the ParaOptiMUSIG activation function. The subnormal linear index of fuzziness [26] is applied to decide the system errors at the corresponding output layers as the network has no a priori knowledge about the outputs. The standard backpropagation algorithm [25] is employed to adjust the interconnection weights between the different layers to minimize the errors. The respective output layers of the independent SONNs generate the final color component images after attaining stabilization in the corresponding networks.

5.5 Fusion of Individual Segmented Component Outputs into a Color Image at the Sink Layer of the PSONN Architecture

The segmented outputs derived at the three output layers of the three independent three-layer SONN architectures are fused at the sink layer of the PSONN architecture to get the segmented color image. The number of segments is a combination of the number of transition lobes of the designed ParaOptiMUSIG activation functions employed during component-level segmentation.

6 Experimental Results and Analysis

Medical image segmentation with the single objective-based ParaOptiMUSIG activation function and NSGA-II-based ParaOptiMUSIG activation function are demonstrated with one color MR brain image [35] of dimensions 170 × 170 using the self-supervised PSONN neural network architecture. The original image is shown in Fig. 1. This section is divided into two parts. In the first part, the color MR brain image is segmented with the single objective-based ParaOptiMUSIG activation function and the NSGA-II-based ParaOptiMUSIG activation function is applied to segment the same color image in the second part of this section.

Fig. 1 Color magnetic resonance (MR) brain image of dimensions 170 × 170

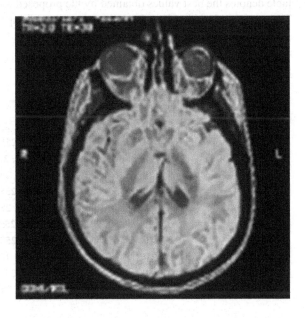

6.1 Segmentation Using ParaOptiMUSIG Activation Function

Experimental results of color MR brain image segmentation using single objective-based ParaOptiMUSIG activation function in connection with the PSONN architecture is presented in this section. The ParaOptiMUSIG activation function has been prepared with a fixed slope, $\lambda = \{2, 4\}$ for $K = \{6, 8\}$ classes. The fixed slope $\lambda = 4$ in combination with 8 classes are applied to derive and present the experimental results in this section of the chapter. Results are also reported for the segmentation of the color MR brain image with the self-supervised PSONN architecture guided by the MUSIG activation function with same number of class responses and with heuristic class levels. Four evaluation functions (ρ, F, F' and Q) have been applied to demonstrate the quantitative measures of the efficiency of the ParaOptiMUSIG and the conventional MUSIG activation functions for $K = 8$ in this section. The genetic algorithm-based optimization procedure generates the optimized sets of class boundaries $cl_{\alpha_{opt}}$ on the basis of four evaluation functions (ρ, F, F' and Q) for different number of classes. They are tabulated in Tables 1, 2, 3, and 4 for the test image. In the first column of these tables, the evaluation functions (EF_{op}) which are used as the fitness functions in this approach are shown and the optimized set of class boundaries are accounted in the third column of the tables. Two set of results per evaluation function of the test image are presented. The quality measures η [graded on a scale of **1** (best) to **2** (worst)] obtained by the segmentation of the test images based on the corresponding set of optimized class boundaries are shown in the last columns of the tables. All the evaluation function values are reported in normalized form in this chapter. The **boldfaced** result in each table denotes the best values obtained by the proposed approach for easy reckoning. The genetic algorithm-based quality measures are applied to compare with the

Table 1 Optimized and fixed class boundaries and evaluated segmentation quality measures, ρ for 8 classes of color MR brain image

EF	Set	Color levels	η
ρ_{op}	1	$R = \{0, 57, 69, 145, 152, 182, 234, 255\}$	0.9702 (2)
		$G = \{0, 42, 68, 101, 108, 161, 229, 255\}$	
		$B = \{0, 18, 44, 74, 110, 139, 191, 255\}$	
	2	$R = \{0, 88, 139, 142, 154, 222, 227, 255\}$	**0.9728 (1)**
		$G = \{0, 55, 66, 102, 189, 197, 213, 255\}$	
		$B = \{0, 15, 33, 82, 117, 159, 222, 255\}$	
ρ_{fx}	1	$R = \{0, 45, 70, 100, 120, 150, 180, 255\}$	**0.9371 (1)**
		$G = \{0, 30, 50, 90, 110, 180, 200, 255\}$	
		$B = \{0, 43, 65, 86, 106, 140, 175, 255\}$	
	2	$R = \{0, 35, 90, 100, 115, 130, 140, 255\}$	0.9135 (2)
		$G = \{0, 5, 74, 130, 155, 175, 200, 255\}$	
		$B = \{0, 50, 80, 120, 150, 175, 215, 255\}$	

Table 2 Optimized and fixed class boundaries and evaluated segmentation quality measures, F for 8 classes of color MR brain image

EF	Set	Color levels	η
F_{op}	1	$R = \{0, 8, 37, 104, 150, 203, 245, 255\}$	**0.655 (1)**
		$G = \{0, 20, 43, 69, 132, 227, 233, 255\}$	
		$B = \{0, 103, 136, 147, 185, 222, 230, 255\}$	
	2	$R = \{0, 6, 193, 228, 244, 250, 252, 255\}$	0.689 (2)
		$G = \{0, 25, 43, 75, 87, 126, 149, 255\}$	
		$B = \{0, 14, 133, 237, 244, 249, 252, 255\}$	
F_{fx}	1	$R = \{0, 20, 50, 100, 150, 200, 245, 255\}$	**0.862 (1)**
		$G = \{0, 20, 50, 100, 130, 210, 230, 255\}$	
		$B = \{0, 80, 120, 140, 180, 220, 230, 255\}$	
	2	$R = \{0, 25, 55, 105, 155, 205, 245, 255\}$	1.000 (2)
		$G = \{0, 30, 60, 110, 135, 215, 225, 255\}$	
		$B = \{0, 70, 115, 145, 175, 220, 235, 255\}$	

Table 3 Optimized and fixed class boundaries and evaluated segmentation quality measures, F' for 8 classes of color MR brain image

EF	Set	Color Levels	η
F'_{op}	1	$R = \{0, 13, 23, 83, 190, 210, 236, 255\}$	**0.466 (1)**
		$G = \{0, 13, 26, 123, 180, 191, 212, 255\}$	
		$B = \{0, 9, 67, 107, 149, 189, 227, 255\}$	
	2	$R = \{0, 12, 32, 58, 88, 100, 158, 255\}$	0.562 (2)
		$G = \{0, 16, 28, 81, 94, 181, 192, 255\}$	
		$B = \{0, 38, 77, 115, 152, 203, 209, 255\}$	
F'_{fx}	1	$R = \{0, 60, 90, 120, 150, 180, 250, 255\}$	**0.991 (1)**
		$G = \{0, 30, 80, 140, 170, 185, 210, 255\}$	
		$B = \{0, 50, 130, 160, 186, 195, 210, 255\}$	
	2	$R = \{0, 55, 85, 115, 145, 180, 225, 255\}$	1.000 (2)
		$G = \{0, 35, 80, 135, 165, 185, 215, 255\}$	
		$B = \{0, 50, 130, 160, 186, 195, 225, 255\}$	

quality measures which have been derived with the heuristically selected class boundary-based conventional MUSIG activation function.

The ParaOptiMUSIG activation function-based PSONN architecture is applied to generate the segmented multilevel test images for $K = 8$ and corresponding to the best segmentation quality measures (ρ, F, F' and Q) derived, as shown in Figs. 2, 3, 4 and 5.

The segmented test images obtained with the PSONN architecture characterized by the conventional MUSIG activation applying fixed class responses for $K = 8$ and with four segmentation quality measures (ρ, F, F' and Q) achieved, as shown in Figs. 6, 7, 8, and 9.

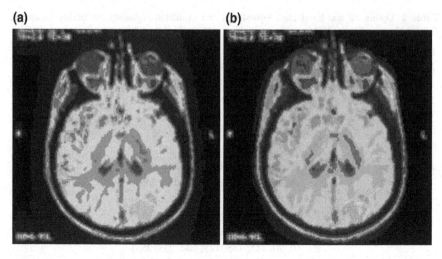

Fig. 2 8-class segmented 170 × 170 color MR brain image with the optimized class levels pertaining to **a** set 1 **b** set 2 of Table 1 for the quality measure ρ with ParaOptiMUSIG activation function

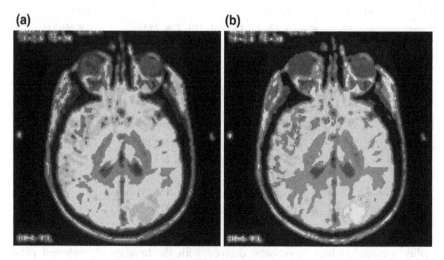

Fig. 3 8-class segmented 170 × 170 color MR brain image with the optimized class levels pertaining to **a** set 1 **b** set 2 of Table 2 for the quality measure F with ParaOptiMUSIG activation function

It is quite clear from the derived results in Tables 1, 2, 3, and 4 that the ParaOptiMUSIG activation function overwhelms its conventional MUSIG counterpart in respect of the segmentation quality of the images. The ρ values reported in Table 1 shows that the values derived by the genetic algorithm-based segmentation method are larger than the same derived by the heuristically generated segmentation

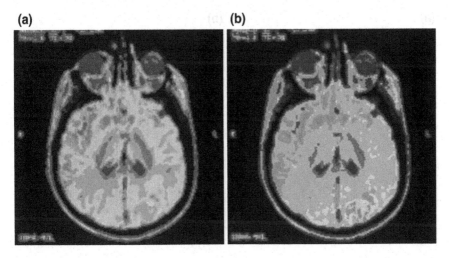

Fig. 4 8-class segmented 170 × 170 color MR brain image with the optimized class levels pertaining to **a** set 1 **b** set 2 of Table 3 for the quality measure F' with ParaOptiMUSIG activation function

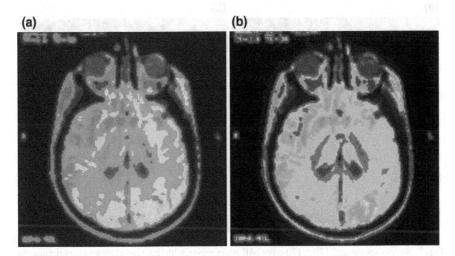

Fig. 5 8-class segmented 170 × 170 color MR brain image with the optimized class levels pertaining to **a** set 1 **b** set 2 of Table 4 for the quality measure Q with ParaOptiMUSIG activation function

method. Similarly, the ParaOptiMUSIG activation function-based segmentation method generates lower F, F', and Q values in Tables 2, 3 and 4, respectively, than the conventional MUSIG activation function-based segmentation method. It is also be evident that the ParaOptiMUSIG function incorporates the image heterogeneity as it can handle a wide variety of image intensity distribution prevalent in real life.

(a)   **(b)**

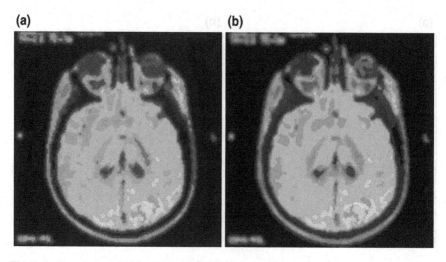

Fig. 6 8-class segmented 170 × 170 color MR brain image with the fixed class levels pertaining to **a** set 1 **b** set 2 of Table 1 for the quality measure ρ with MUSIG activation function

(a)   **(b)**

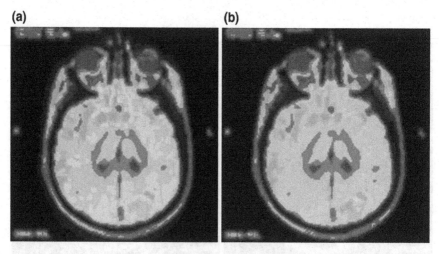

Fig. 7 8-class segmented 170 × 170 color MR brain image with the fixed class levels pertaining to **a** set 1 **b** set 2 of Table 2 for the quality measure F with MUSIG activation function

The segmented outputs generated by the ParaOptiMUSIG activation function-based method are more qualitative than the segmented outputs generated by the MUSIG activation function-based method. Different organs of the ParaOptiMUSIG activation function-based segmented images are more prominent than the MUSIG activation function-based segmented images.

(a) **(b)**

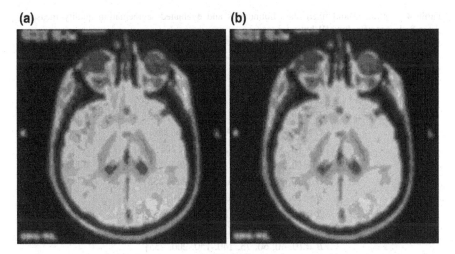

Fig. 8 8-class segmented 170 × 170 color MR brain image with the fixed class levels pertaining to **a** set 1 **b** set 2 of Table 3 for the quality measure F' with MUSIG activation function

(a) **(b)**

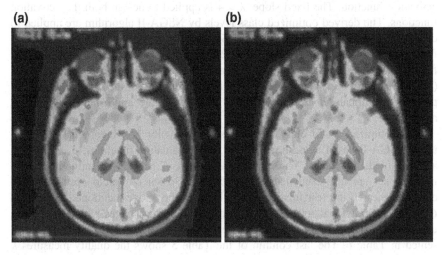

Fig. 9 8-class segmented 170 × 170 color MR brain image with the fixed class levels pertaining to **a** set 1 **b** set 2 of Table 4 for the quality measure Q with MUSIG activation function

6.2 Segmentation Using NSGA-II-Based ParaOptiMUSIG Activation Function

This section tries to show the color MR brain image segmentation using NSGA-II-based ParaOptiMUSIG activation function. Also in this case, we have conducted experiments with $K = \{6, 8\}$ classes and reported for 8 classes for the

Table 4 Optimized and fixed class boundaries and evaluated segmentation quality measures, Q for 8 classes of color MR brain image

EF	Set	Color levels	η
Q_{op}	1	$R = \{0, 47, 93, 109, 129, 178, 183, 255\}$	0.139 (2)
		$G = \{0, 32, 43, 97, 110, 207, 218, 255\}$	
		$B = \{0, 52, 137, 140, 152, 229, 252, 255\}$	
	2	$\mathbf{R = \{0, 18, 38, 51, 158, 214, 234, 255\}}$	**0.114 (1)**
		$\mathbf{G = \{0, 40, 62, 88, 96, 172, 176, 255\}}$	
		$\mathbf{B = \{0, 31, 120, 127, 174, 195, 203, 255\}}$	
Q_{fx}	1	$\mathbf{R = \{0, 30, 60, 90, 140, 180, 205, 255\}}$	**0.535 (1)**
		$\mathbf{G = \{0, 20, 50, 80, 135, 175, 215, 255\}}$	
		$\mathbf{B = \{0, 35, 55, 75, 125, 155, 210, 255\}}$	
	2	$R = \{0, 45, 65, 85, 135, 175, 200, 255\}$	1.000 (2)
		$G = \{0, 10, 50, 90, 130, 170, 210, 255\}$	
		$B = \{0, 40, 60, 75, 140, 150, 200, 255\}$	

NSGA-II-based ParaOptiMUSIG activation function and the conventional MUSIG activation function. The fixed slope, $\lambda = 4$ is applied to design both the activation functions. The derived optimized class levels by NSGA-II algorithm are applied to design the NSGA-II-based ParaOptiMUSIG activation function. Three evaluation functions ρ, F, and F' (presented in Eqs. 2, 5 and 6, respectively) are applied as the fitness functions in the NSGA-II algorithm. Another empirical evaluation function (Q) (presented in Eq. 7) has been applied for the selection of the chromosome from the Pareto optimal set and the selected class levels are applied in the PSONN architecture.

The optimized set of class boundaries for the color MR brain image derived using NSGA-II-based algorithm with the three evaluation functions (ρ, F, and F') are presented in Table 5. Three fitness values (ρ, F, and F') of individual chromosome set are also tabulated in that table. Two set of Pareto optimal set of chromosomes of the test image are tabulated in the above-mentioned table. The Q values of individual set of class boundaries for color images which have been used for the gradation system for that particular set of class boundaries are presented in Table 6. The last column of the Table 5 shows the quality measures κ [graded on a scale of 1 (best) onwards] obtained by the segmentation of the test image based on the corresponding set of optimized class boundaries. The first four good Q-valued chromosomes are indicated in that table for easy reckoning. In this chapter, all evaluation function values are reported in normalized form.

The heuristically selected class boundaries are employed to generate the conventional MUSIG activation function and those class boundaries are depicted in Table 7. The same evaluation functions are applied to evaluate the individual set of class boundaries and the results are reported in that table. After the segmentation process using the MUSIG activation function, the quality of the segmented images

Table 5 NSGA-II-based set of optimized class boundaries and corresponding evaluated segmentation quality measures for 8 classes of color MR brain image

Set no.		Class level	ρ	F	F'	κ
1	(i)	$R = \{2, 31, 71, 119, 135, 193, 213, 254\}$; $G = \{3, 43, 69, 112, 143, 182, 208, 236\}$; $B = \{0, 17, 52, 92, 128, 166, 206, 254\}$	0.9926	0.7877	0.6538	1
	(ii)	$R = \{2, 31, 75, 117, 135, 191, 213, 254\}$; $G = \{3, 45, 71, 116, 143, 180, 208, 236\}$; $B = \{0, 17, 52, 92, 126, 166, 206, 254\}$	0.9926	0.8027	0.6764	4
	(iii)	$R = \{2, 55, 63, 121, 137, 191, 213, 254\}$; $G = \{3, 44, 71, 112, 143, 180, 208, 236\}$; $B = \{0, 18, 52, 92, 128, 166, 206, 254\}$	0.9923	0.8454	0.8706	
	(iv)	$R = \{2, 55, 63, 121, 135, 191, 211, 254\}$; $G = \{3, 44, 71, 112, 143, 178, 208, 236\}$; $B = \{0, 18, 52, 92, 126, 166, 206, 254\}$	0.9915	0.9393	0.9430	
	(v)	$R = \{2, 55, 63, 123, 135, 191, 211, 254\}$; $G = \{3, 44, 73, 114, 143, 178, 208, 236\}$; $B = \{0, 18, 54, 93, 126, 166, 206, 254\}$	0.9906	0.9520	1.0000	
	(vi)	$R = \{2, 31, 73, 115, 135, 191, 213, 254\}$; $G = \{3, 43, 71, 116, 143, 180, 208, 236\}$; $B = \{0, 17, 52, 92, 126, 166, 206, 254\}$	0.9908	0.8191	0.6978	3
	(vii)	$R = \{2, 31, 71, 115, 135, 191, 213, 254\}$; $G = \{3, 43, 69, 116, 141, 180, 208, 236\}$; $B = \{0, 17, 48, 88, 126, 166, 206, 254\}$	0.9908	0.8265	0.6878	2
2	(i)	$R = \{2, 39, 63, 112, 138, 182, 196, 254\}$; $G = \{3, 46, 61, 113, 122, 169, 179, 236\}$; $B = \{0, 17, 52, 94, 131, 171, 213, 254\}$	0.9926	0.8069	0.6901	
	(ii)	$R = \{2, 39, 63, 110, 138, 182, 198, 254\}$; $G = \{3, 46, 61, 113, 122, 173, 179, 236\}$; $B = \{0, 17, 56, 94, 131, 175, 213, 254\}$	0.9924	0.7480	0.6455	2
	(iii)	$R = \{2, 41, 59, 112, 138, 182, 194, 254\}$; $G = \{3, 46, 59, 111, 120, 173, 179, 236\}$; $B = \{0, 17, 48, 92, 131, 171, 213, 254\}$	0.9915	0.8574	0.7681	
	(iv)	$R = \{2, 39, 63, 104, 134, 182, 194, 254\}$; $G = \{3, 46, 59, 113, 120, 173, 179, 236\}$; $B = \{0, 17, 52, 94, 131, 171, 213, 254\}$	0.9912	0.8002	0.7066	
	(v)	$R = \{2, 39, 63, 112, 132, 182, 198, 254\}$; $G = \{3, 17, 48, 92, 131, 171, 213, 236\}$; $B = \{0, 17, 56, 94, 131, 175, 213, 254\}$	0.9918	0.7847	0.6252	1
	(vi)	$R = \{2, 41, 59, 104, 134, 182, 194, 254\}$; $G = \{3, 46, 59, 113, 120, 173, 179, 236\}$; $B = \{0, 17, 48, 92, 131, 171, 213, 254\}$	0.9918	0.8384	0.7312	
	(vii)	$R = \{2, 39, 61, 112, 132, 182, 198, 254\}$; $G = \{3, 46, 61, 113, 122, 173, 177, 236\}$; $B = \{0, 17, 52, 94, 131, 175, 213, 254\}$	0.9915	0.8001	0.7216	4
	(viii)	$R = \{2, 39, 65, 112, 132, 182, 198, 254\}$; $G = \{3, 46, 59, 113, 124, 173, 177, 236\}$; $B = \{0, 17, 56, 94, 131, 175, 213, 254\}$	0.9919	0.7854	0.6624	3
	(ix)	$R = \{2, 39, 63, 112, 140, 182, 198, 254\}$; $G = \{3, 46, 61, 111, 122, 173, 177, 236\}$; $B = \{0, 17, 52, 94, 131, 173, 213, 254\}$	0.9921	0.8088	0.7703	

Table 6 The evaluated segmentation quality measure Q of each optimized class boundaries of Table 5

Set	No	(i)	(ii)	(iii)	(iv)	(v)	(vi)	(vii)
1	Q	0.8230	0.8477	0.8769	0.8856	0.8902	0.8382	0.8338
2	No	(i)	(ii)	(iii)	(iv)	(v)	(vi)	(vii)
	Q	0.9025	0.8893	0.9072	0.9001	0.8889	0.9028	0.8934
	No	(viii)	(ix)					
	Q	0.8905	0.9005					

Table 7 Fixed class boundaries and corresponding evaluated segmentation quality measures for 8 classes of color MR brain image

Set no.	Class level	ρ	F	F'
1	$R = \{2, 30, 60, 100, 135, 195, 215, 254\};$ $G = \{3, 30, 60, 100, 135, 195, 215, 236\};$ $B = \{0, 30, 60, 100, 135, 195, 215, 254\}$	0.9887	0.7895	0.5849
2	$R = \{2, 35, 70, 95, 130, 190, 218, 254\};$ $G = \{3, 35, 70, 95, 130, 190, 218, 236\};$ $B = \{0, 35, 70, 95, 130, 190, 218, 254\}$	0.9867	1.0000	0.8805
3	$R = \{2, 25, 65, 110, 145, 200, 220, 254\};$ $G = \{25, 65, 110, 145, 200, 220, 236\};$ $B = \{0, 25, 65, 110, 145, 200, 220, 254\}$	0.9902	0.6877	0.4739
4	$R = \{2, 28, 55, 105, 140, 210, 225, 254\};$ $G = \{3, 28, 55, 105, 140, 210, 225, 236\};$ $B = \{0, 28, 55, 105, 140, 210, 225, 254\}$	0.9868	0.9102	0.9032

are evaluated using the Q evaluation function and the results are tabulated in Table 8.

It is quite clear from Tables 6 and 8 that the fitness values derived by the NSGA-II-based ParaOptiMUSIG activation function for color image are better than those obtained by the conventional MUSIG activation function. In most of the cases, the Q value in Table 6 is lower than the same value in Table 8 for the test image. From the earlier discussion, it is to be mentioned that the lower value of Q signifies the better segmentation. It is evident from those tables that the image segmentation done by the proposed method generates the lower Q values than the image segmentation done by the heuristically selected class levels. In another way, the ρ, F, and F' values of Table 5 derived by the NSGA-II-based ParaOptiMUSIG activation function for the test image are better than those values, reported in Table 7 by the conventional MUSIG activation function for the same image.

On the basis of the evaluation metric, Q value, first four better results of each set from Table 6 are used in the PSONN network to segment the color MR brain image. The segmented color test image derived with the PSONN architecture using the NSGA-II-based ParaOptiMUSIG activation function for $K = 8$ is shown in Figs. 10 and 11.

Table 8 The evaluated segmentation quality measure Q of each fixed class boundaries of Table 7

Image	Set no	Q
MRI	1	0.9054
	2	1.0000
	3	0.7627
	4	0.8158

Now, the quality of the segmented images by the MUSIG activation function-based method is not good than the segmented images by the NSGA-II-based ParaOptiMUSIG activation function. The components of the

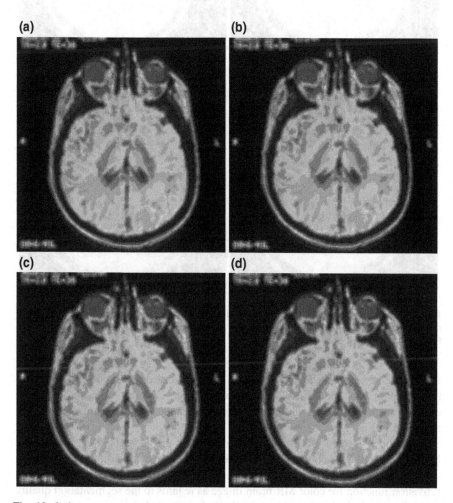

Fig. 10 8-class segmented 170 × 170 color MR brain image with the optimized class levels referring to **a–d** set 1 of Table 5 for first four better quality measure Q with NSGA-II-based ParaOptiMUSIG activation function

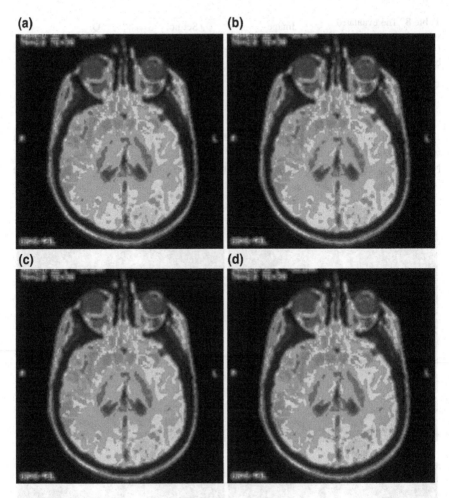

Fig. 11 8-class segmented 170 × 170 color MR brain image with the optimized class levels referring to **a–d** set 2 of Table 5 for first four better quality measure Q with NSGA-II-based ParaOptiMUSIG activation function

images are segmented effectively by the NSGA-II-based ParaOptiMUSIG activation function than the MUSIG activation function. Different organs of the segmented image in Figs. 10 and 11 are more prominent than the segmented images presented in Fig. 12.

From the above analysis, the image segmentation method using the conventional MUSIG activation function is outperformed by the image segmentation method by the ParaOptiMUSIG activation function and the NSGA-II-based ParaOptiMUSIG activation function for color MR brain image as regards to the segmentation quality of the images. It is also be evident that both the approaches incorporate the image

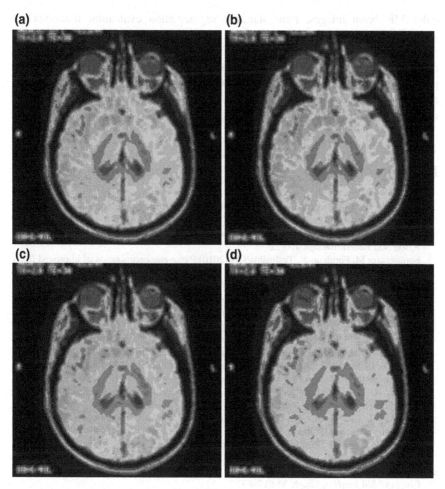

Fig. 12 8-class segmented 170 × 170 color MR brain image with the fixed class levels referring to **a–d** of Table 7 with MUSIG activation function

heterogeneity as it can handle a wide variety of image intensity distribution prevalent in real life.

7 Discussions and Conclusion

In this chapter, a comparison-based color MR brain image segmentation is presented. Three standard activation functions, viz., the ParaOptiMUSIG activation function, the NSGA-II-based ParaOptiMUSIG activation, and the MUSIG activation function, with the PSONN architecture are applied to segment the real-life color MR brain images. Four standard segmentation evaluation functions are applied in all cases. The ParaOptiMUSIG activation function-based segmentation method outperforms the standard MUSIG activation function-based segmentation method in all respect. The NSGA-II-based ParaOptiMUSIG activation function performs better in quantitatively and qualitatively than the conventional MUSIG activation function with fixed and heuristic class levels for the segmentation of real-life color MR brain image.

References

1. Maulik U (2009) Medical image segmentation using genetic algorithms. IEEE Trans Inf Technol Biomed 13(2):166–173
2. Zhao F, Xie X (2013) An overview of interactive medical image segmentation. Br Mach Vis Assoc Soc Pattern Recogn 7:1–22
3. Forouzanfar M, Forghani N, Teshnehlab M (2010) Parameter optimization of improved fuzzy c-means clustering algorithm for brain MR image segmentation. Eng Appl Artif Intell 23 (2):160–168
4. Tang H, Wu EX, Ma QY, Gallagher D, Perera GM, Zhuang T (2000) MRI brain image segmentation by multi-resolution edge detection and region selection. Comput Med Imaging Graph 24(6):349–357
5. Pohle R, Toennies KD (2001) Segmentation of medical images using adaptive region growing. Proc SPIE 4322, Medical Imaging 2001: Image Processing 1337. doi:10.1117/12.431013
6. Maitra V, Chatterjee A (2008) A novel technique for multilevel optimal magnetic resonance brain image thresholding using bacterial foraging. Measurement 41(10):1124–1134
7. Bezdek JC (1981) Pattern recognition with fuzzy objective function algorithms. Kluwer Academic Publishers, New York
8. Ji Z, Liu J, Cao G, Sun Q, Chen Q (2014) Robust spatially constrained fuzzy c-means algorithm for brain MR image segmentation. Pattern Recogn 47(7):2454–2466
9. Hung WL, Yang MS, Chen DH (2006) Parameter selection for suppressed fuzzy c-means with an application to MRI segmentation. Pattern Recogn Lett 27(5):424–438
10. Jiang J, Trundle P, Ren J (2010) Medical image analysis with artificial neural networks. Comput Med Imaging Graph 34:617–631
11. Torbati N, Ayatollahi A, Kermani A (2014) An efficient neural network based method for medical image segmentation. Comput Biol Med 44:76–87
12. Döhler F, Mormann F, Weber B, Elger CE, Lehnertz K (2008) A cellular neural network based method for classification of magnetic resonance images: towards an automated detection of hippocampal sclerosis. J Neurosci Methods 170(2):324–331
13. Bathen TF, Jensen LR, Sitter B, Fjösne HE, Halgunset J, Axelson DE, Gribbestad IS, Lundgren S (2007) MR-determined metabolic phenotype of breast cancer in prediction of lymphatic spread, grade, and hormone status. Breast Cancer Res Treat 104(2):181–189
14. Mitchell M (1996) An introduction to genetic algorithms. MIT Press, Cambridge, MA
15. Manikandan S, Ramar K, Willjuice Iruthayarajan M, Srinivasagan KG (2014) Multilevel thresholding for segmentation of medical brain images using real coded genetic algorithm. Measurement 47:558–568
16. Yeh JY, Fu JC (2008) A hierarchical genetic algorithm for segmentation of multi-spectral human-brain MRI. Expert Syst Appl 34(2):1285–1295
17. Lai CC, Chang CY (2007) A hierarchical evolutionary algorithm for automatic medical image segmentation. Expert Syst Appl 36(1):248–259

18. Pavan KK, Srinivas VS, SriKrishna A, Reddy BE (2012) Automatic tissue segmentation in medical images using differential evolution. J Appl Sci 12:587–592
19. Ghosh P, Mitchell M (2006) Segmentation of medical images using a genetic algorithm. In: Proceedings of the 8th annual conference on Genetic and evolutionary computation, pp 1171–1178
20. Zanaty EA, Ghiduk AS (2013) A novel approach based on genetic algorithms and region growing for magnetic resonance image (MRI) segmentation. Comput Sci Inf Syst 10(3):1319–1342
21. Forghani N, Forouzanfar M, Eftekhari A, Mohammad-Moradi S, Teshnehlab M (2009) Application of particle swarm optimization in accurate segmentation of brain MR images. In: Lazinica A (ed) Particle Swarm Optimization, InTech, pp 203–222. ISBN: 978-953-7619-48-0
22. Nakib A, Cooren Y, Oulhadj H, Siarry P (2008) Magnetic resonance image segmentation based on two-dimensional exponential entropy and a parameter free PSO. Artificial Intelligence, Springer-Verlag, Berlin, Heidelberg, pp 50–61
23. Agrawal S, Panda R, Dora L (2014) A study on fuzzy clustering for magnetic resonance brain image segmentation using soft computing approaches. Appl Soft Comput 24:522–533
24. Bhattacharyya S, Dutta P, Maulik U, Nandi PK (2007) Multilevel activation functions for true color image segmentation using a self supervised parallel self organizing neural network (PSONN) architecture: a comparative study. Int J Comput Sci 2(1):09–21
25. Haykin S (1994) Neural networks: a comprehensive foundation. Macmillan College Publishing Co., New York
26. Bhattacharyya S, Dutta P, Maulik U (2008) Self organizing neural network (SONN) based gray scale object extractor with a multilevel sigmoidal (MUSIG) activation function. Found Comput Decis Sci 33(2):131–165
27. De S, Bhattacharyya S, Dutta P (2010) Efficient grey-level image segmentation using an optimised MUSIG (OptiMUSIG) activation function. Int J Parallel Emergent Distrib Syst 26 (1):1–39
28. Ghosh A, Pal NR, Pal SK (1993) Self-organization for object extraction using a multilayer neural network and fuzziness measures. IEEE Trans Fuzzy Syst 1(1):54–68
29. De S, Bhattacharyya S, Chakraborty S (2012) Color image segmentation using parallel OptiMUSIG activation function. Appl Soft Comput 12:3228–3236
30. Deb K (2001) Multi-objective optimization using evolutionary algorithms. John Wiley & Sons Ltd, England
31. Deb K, Pratap A, Agarwal S, Meyarivan T (2002) A fast and elitist multiobjective genetic algorithm: NSGA-II. IEEE Trans Evol Comput 6(2):182–197
32. Zitzler E, Thiele L (1998) An evolutionary algorithm for multiobjective optimization: the strength pareto approach. Tech. Rep. 43, Gloriastrasse 35, CH-8092 Zurich, Switzerland
33. Zitzler E, Laumanns M, Thiele L (2001) SPEA2: Improving the strength pareto evolutionary algorithm. Tech. Rep. 103, Gloriastrasse 35, CH-8092 Zurich, Switzerland
34. De S, Bhattacharyya S, Chakraborty S (2013) Color image segmentation by NSGA-II based ParaOptiMUSIG activation function. International conference on machine intelligence research and advancement (ICMIRA-2013), pp 105–109
35. http://commons.wikimedia.org/wiki/File:Brain_MRI.jpg
36. Zhang J, Liu Q, Chen Z (2005) A Medical image segmentation method based on SOM and wavelet transforms. J Commun Comput 2(5):46–50
37. Borsotti M, Campadelli P, Schettini R (1998) Quantitative evaluation of color image segmentation results. Pattern Recogn Lett 19(8):741–747
38. Liu J, Yang YH (1994) Multi-resolution color image segmentation. IEEE Trans Pattern Anal Mach Intell 16(7):689–700
39. Zhang Y (1996) A survey on evaluation methods for image segmentation. Pattern Recogn 29 (8):1335–1346
40. Bhattacharyya S, Dasgupta K (2003) Color object extraction from a noisy background using parallel multilayer self-organizing neural networks. In: Proceedings of CSI-YITPA(E) 2003, pp 32–36

41. Davies DL, Bouldin DW (1979) A cluster separation measure. IEEE Trans Pattern Recogn Mach Intell 1(2):224–227
42. Halkidi M, Vazirgiannis M (2002) Clustering validity assessment using multi representatives. In: Proceedings of the hellenic conference on artificial intelligence (SETN'02), Thessaloniki, Greece

Convergence Analysis of Backpropagation Algorithm for Designing an Intelligent System for Sensing Manhole Gases

Varun Kumar Ojha, Paramartha Dutta, Atal Chaudhuri
and Hiranmay Saha

Abstract Human fatalities are reported due to excessive proportional presence of hazardous gas components in manhole, such as hydrogen sulphide, ammonia, methane, carbon dioxide, nitrogen oxide, carbon monoxide, etc. Hence, predetermination of these gases is imperative. A neural network (NN)-based intelligent sensory system was proposed for avoiding such fatalities. Backpropagation (BP) was applied for supervised training of the proposed neural network model. A gas sensor array consists of many sensor elements was employed for sensing manhole gases. Sensors in the sensor array were only responsible for sensing their target gas components. Therefore, the presence of multiple gases results in cross sensitivity that is a crucial issue to this problem. It is viewed as a pattern recognition and noise reduction problem. Various performance parameters and complexity of the problem influences NN training. In this chapter, performance of BP algorithm on such real-life application was comprehensively studied, compared, and contrasted with several hybrid intelligent approache, both in theoretical and in statistical senses.

Keywords Gas detection · Backpropagation · Neural network · Pattern recognition · Parameter tuning · Complexity analysis

V.K. Ojha (✉) · A. Chaudhuri
Department of Computer Science and Engineering, Jadavpur University, Kolkata, India
e-mail: varun.kumar.ojha@vsb.cz

A. Chaudhuri
e-mail: atalc23@gmail.com

V.K. Ojha
IT4Innovations, VŠB Technical University of Ostrava, Ostrava, Czech Republic

P. Dutta
Department of Computer and System Sciences, Visva-Bharati University, Santiniketan, India
e-mail: paramartha.dutta@gmail.com

H. Saha
Centre of Excellence for Green Energy and Sensors System, Indian Institute
of Engineering Science and Technology, Howrah, India
e-mail: sahahiran@gmail.com

© Springer India 2016
S. Bhattacharyya et al. (eds.), *Hybrid Soft Computing Approaches*,
Studies in Computational Intelligence 611,
DOI 10.1007/978-81-322-2544-7_7

215

1 Introduction

Computational Intelligence (CI) offers solutions to a wide range of real-life problems. In this chapter, we have resorted to using a CI approach to offer a design of an intelligent sensory system (ISS) for detecting manhole gas mixture. The manhole gas mixture detection problem was treated as a pattern recognition/noise reduction problem. For the past decades, neural network (NN) has been proven as a powerful tool for machine learning application in various domains. In this chapter, we have used backpropagation (BP) NN technique for designing an ISS. The primary aim of this chapter is to provide a comprehensive performance study of BP algorithm.

Decomposition of wastage and sewage into sewer pipeline leads to toxic gaseous mixture formation, known as a manhole gas mixture that usually has toxic gases such as hydrogen sulphide (H_2S), ammonia (NH_3), methane (CH_4), carbon dioxide (CO_2), nitrogen oxides (NO_x), etc., [1–3]. Often, human fatalities occur due to excessive proportion of the mentioned toxic gases. The persons, who have the responsibilities for the maintenance and cleaning of sewer pipeline are in need of a compact instrument that may able predetermine safeness of manholes. In the recent years, due to the toxic gas exposures, several instances of deaths were reported, including municipality labourers and pedestrians [4–8]. We have investigated the commercially available gas sensor tools. We found that the commercially available gas detectors are insufficient in sensing all the aforementioned gases and the cross sensitive in their response is a basic problem associated with these sensor units.

Rest of the chapter is organized as follows: Sect. 2.1 provides a brief literature survey followed by a concise report on our research contribution in Sect. 2.2. Readers may find discussion on design issues of the proposed intelligent sensory system (ISS) in Sect. 2.3, a discussion on data sample preparation in Sects. 2.5 and the crucial issue of the cross sensitivity in Sect. 2.4. Section 2.6 provides a brief discussion on NN configuration, training pattern and BP algorithm. Performance analysis of BP algorithm on such real application provided in Sect. 3 is central subject of this chapter. Finally, Sections 4 and 5 provide results and conclusion, respectively.

2 Mechanism

This section provides a detailed discussion on the various materials and methods acquired in the design and the development of the ISS. We thoroughly explains design issues of an intelligent gas detection system, data collection process and data preparation technique.

2.1 A Brief Literature Survey

In the past few years, several research work have been reported on the development and design of electronic nose (E-NOSE) and gas detection system. After an detailed literature examination, we may appreciate the effort by Li et al. [9] for his contribution in developing a NN-based mixed gas (NO_x, and CO) measurement system. On the other hand, Sirvastava et al. [10, 11] have proposed a design of intelligent E-NOSE system using BP and Neuro-genetic approach. A pattern recognition technique, based on the wallet transformation for gas mixture analysis using single tin-oxide sensor was presented by Llobet [12]. Liu et al. [13] addressed a genetic-NN algorithm to recognize patterns of mixed gases (a mixture of three component gases) using infrared gas sensor. Tsirigotis et al. [14] illustrated a NN-based recognition system for CO and NH_3 gases using metallic oxide gas sensor array (GSA). Lee et al. [15] illustrated uses of micro-GSA combined with NN for recognizing combustible leakage gases. Ambard et al. [16] have demonstrated use of NN for gas discrimination using a tin-oxide GSA for the gases H_2, CO and CH_4. In [17], authors have illustrated a NN-based technique for developing a gas sensory system for sensing gases in a dynamic environment. Pan et al. [18] have shown several applications of E-NOSE. Wongchoosuka et al. [19] have proposed an E-NOSE detection system based on carbon nanotube-SnO_2 gas sensors for detecting methanol. Zhang et al. [20] developed a knowledge-based genetic algorithm for detecting mixed gas in mines. Won and Koo [21] proposed a system for estimation of hazardous gas release rate using optical sensor and NN-based technique. A comprehensive study of the above-mentioned articles lead us to the following conclusion: (i) mostly, BP and NN-based techniques were used for gas detection problem in the respective application areas; (ii) mainly, two or three gas mixture detection was addressed and that too, the gases were those, whose sensors were not cross sensitive at high extent; and (iii) cross sensitivity was not addressed firmly. In the design of manhole gas mixture detection system, cross sensitivity due to the presence of several toxic gases is a vital issue. Present article firmly addresses this issue. Ojha et al. [22–24] have presented several approaches towards potential solution to manhole gas detection issue.

2.2 Present Approach and Contribution

In this chapter, manhole gas detection problem was treated as a pattern recognition/noise reduction problem, where a NN regressor was modelled and trained in supervised mode using BP algorithm. A semiconductor-based GSA containing distinct semiconductor-type gas sensors was used to sense the presence of gases according to their concentration in manhole gas mixture. Sensed values were cross sensitive as multiple gases exists in manhole gas mixture. Cross sensitivity was occurred because the gas sensors are sensitive towards non-target gases too. Our aim was to train a NN regressor such that cross sensitivity effect can be minimized. The developed ISS would help persons to be watchful against the presence of toxic gases before entering into manholes and thereby avoiding fatality. Various parameters of BP algorithm were tuned to extract-out best possible result. Performance of BP on various parameters tuning are comprehensively reported. In addition, performance analysis of BP algorithm against various hybrid intelligent approaches such as conjugate gradient, neuro-genetic (NN trained using genetic algorithm), and neuro-swarm (NN trained using particle swarm optimization algorithm) is reported, both in theoretical and statistical senses.

2.3 Basic Design of Intelligent Sensory System

The design illustrated in this chapter consisted of three constituent units: input, intelligent and output. Input unit consists of gas suction-motor chamber, GSA and data acquisition-cum data-preprocessor block. Intelligent unit receives data from input unit and after performing its computation, it sends result to output unit. Output unit presents systems' output in user-friendly form. Gas mixture sample is allowed to pass over semiconductor-based GSA. Preprocessing block receives sensed data values from GSA and make sure that the received data values are normalized before feeding it to NN. Later, output unit perform de-normalization of network response to generates alarm, if any of the toxic gas components exceeds their safety limit. For training of NN, several data samples were prepared. Block diagram shown in Fig. 1 is a lucid representation of above discussion.

2.4 Semiconductor-Based Gas Sensor Array (GSA) and Cross Sensitivity Issue

Metal oxide semiconductor (MOS) gas sensors were used for GSA. We used N distinct sensor element representing n gases. MOS sensors are resistance-type electrical sensors, where response was change in circuit resistance proportional to gas concentration, given as $\Delta R_s / R_0$, where ΔR_s is change in MOS sensor resistance and R_0 is base resistance value [15, 19]. A typical arrangement of GSA is shown in Fig. 2. The circuitry shown in Fig. 2 was developed in our laboratory. In Ghosh et al. [25, 26], we elaborately discussed sensor array and its working principles. Although gas sensor elements were supposed to detect only their target gases, they showed sensitivity towards other gases too. Hence, sensor array response always involved cross-sensitivity effect [9]. This indicates that sensors' responses were noisy. If we concentrate on the first and second rows in

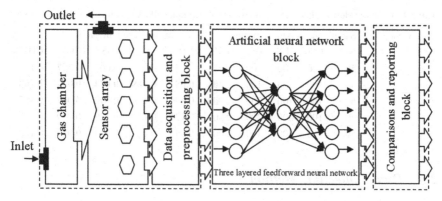

Fig. 1 Intelligent sensory system for manhole gas detection

Fig. 2 Gas sensor array with data acquisition system for sensing concentration of manhole gases

Table 1, we may observe the inherent cross-sensitivity effect in responses of sensors. The first and second samples in Table 1 indicate that changes in concentration of only methane gas resulted in change in response of all the other sensors, including the sensor earmarked for methane gas. It is indicated that the prepared data sample was containing cross-sensitive effect. We may observed that cross-sensitivity effect was not random, rather followed some characteristics and patterns. Hence, in operating (real-world) environment, raw sensor responses of GSA may not be used for predicting concentration of manhole gases. Therefore, to predict/forecast level of concentration of manhole gases, we proposed to use ISS, equipped with pattern recognition/noise reduction techniques that will help to filter-out noise induced on sensors due to cross sensitivity.

Table 1 Data sample for intelligent sensory system (samples illustrated for example purpose)

No. of Sample	Sample gas mixture (in ppm)					Sensor response ($\Delta R_s/R_0$)				
	NH$_3$	CO	H$_2$S	CO$_2$	CH$_4$	NH$_3$	CO	H$_2$S	CO$_2$	CH$_4$
1	50	100	100	100	**2000**	0.053	0.096	0.065	0.037	0.121
2	50	100	100	100	**5000**	**0.081**	**0.108**	**0.074**	**0.044**	**0.263**
3	50	100	100	200	2000	0.096	0.119	0.092	0.067	0.125
4	50	100	200	200	5000	0.121	0.130	0.129	0.079	0.274
5	50	100	200	400	2000	0.145	0.153	0.139	0.086	0.123

2.5 Data Collection Method

Data sample for experiment and NN training was prepared in following steps: In the first step, we collected information about safety limits of the component gases of manhole gas mixture. After that, distinct concentration values (level) around the safety limits of each manhole gas were recognized. Several gas mixture samples were prepared by mixing gas components in different combinations of their concentration level. As an example, if we have five gases and three recognized concentration levels of each gas, then, we may mix them in 243 different combinations. Hence, we obtained 243 gas mixture samples. When these samples were allowed one by one to pass over semiconductor-based GSA, as a result, we got data table shown in Table 1.

2.6 Neural Network Approach

We already mentioned in Sect. 2 that raw sensor response may not accurately represent real-world scenario. Therefore, we were inclined to use NN technique in order to reduce noise, so that prediction can achieve high accuracy with lowest possible error.

2.6.1 Multi-layer Perceptron

NN, "a massively parallel distributed processor that has a natural propensity for storing experiential knowledge and making it available for subsequent use" [27], trained using BP algorithm may offer a solution to aforementioned problem. NN shown in Fig. 3 contains 5 input nodes, n hidden nodes with l layers and 5 output nodes leading to a $5 - n \ldots n - 5$ network configuration. Nodes in input layer as well as in output layer indicate that our system was developed for detecting 5 gases from manhole gaseous mixture. A detailed discussion on NN performance based on network configuration is provided in Sect. 3.1.2.

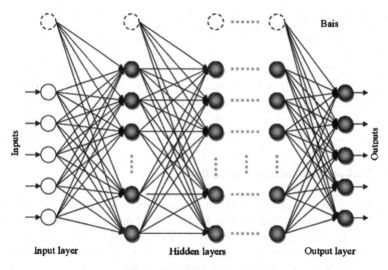

Fig. 3 NN architecture for five input manhole gas components

2.6.2 Training Pattern

We acquired supervised mode of learning for training of NN. Hence, the training pattern constituted of input vector and target vector. We mentioned above that normalized sensor responses were given as input to NN. Hence, input vector I consisted of normalized values of sensor responses. In given data sample, input vector was a five-element vector, where each element represents a gas of gas mixture. Input vector I can be represented as follows:

$$I = [i_1, \ i_2, \ i_3, \ i_4, \ i_5]^T. \tag{1}$$

The system output was presented in terms of concentration of gases. Hence, target vector T was prepared using values of gas mixture sample. In given data sample, target vector was a five-element vector, where each element represents a gas in gas mixture. Target vector T can be represented as follows:

$$T = [t_1, t_2, t_3, t_4, t_5]^T. \tag{2}$$

A training set containing input vector and target vector can be represented as per Table 2.

Table 2 Training set for neural network

No. of pattern	Input vector I					Target vector T				
	i_1	i_2	i_3	i_4	i_5	t_1	t_2	t_3	t_4	t_5
(I_1, T_1)	0.19	0.35	0.23	0.13	0.44	0.01	0.02	0.02	0.02	0.4
(I_2, T_2)	0.29	0.39	0.27	0.16	0.96	0.01	0.02	0.02	0.02	1
(I_3, T_3)	0.35	0.43	0.33	0.24	0.45	0.01	0.02	0.02	0.04	0.4
(I_4, T_4)	0.44	0.47	0.47	0.28	1	0.01	0.02	0.04	0.04	1
(I_5, T_5)	0.52	0.55	0.51	0.31	0.45	0.01	0.02	0.04	0.08	0.4

2.6.3 Backpropagation (BP) Algorithm

Let us have a glimpse of BP algorithm as described by Rumelhart [28]. BP algorithm is a form of supervised learning for training of multi-layer NNs, also known as generalized delta rule [28], where error data from output layer are backpropagated to earlier ones, allowing incoming weights to be updated [27, 28]. Synaptic weight matrix W can be updated using delta rule is as:

$$W(n + 1) = W(n) + \Delta W(n + 1), \tag{3}$$

where n indicates nth epoch training and $\Delta W(n + 1)$ is computed as:

$$\Delta W(n + 1) = \eta g(n + 1) + m \cdot g(n), \tag{4}$$

where η is learning rate, β is momentum factor, and $g(n)$ is gradient of nth epoch computed as:

$$g(n) = \delta_j(n)y_i(n). \tag{5}$$

Local gradient $\delta_j(n)$ is computed both at output layer and hidden layer as follows:

$$\begin{aligned}\delta_j(n) &= -e_j(n)\varphi\big(v_j(n)\big) \text{ for output layer} \\ &= \varphi\big(v_j(n)\Sigma_k\delta_k(n)w_{kj}(n)\big) \text{ for hidden layer.}\end{aligned} \tag{6}$$

The algorithm terminates either sum of squared error (SSE) reached to an acceptable minimum, or algorithm completes its maximum allowed iterations. SSE measures performance of BP algorithm [29], computed as:

$$\text{SSE} = \frac{1}{2}\sum_p \big(O_{pi} - t_{pi}\big)^2 \text{ for all } p \text{ and } i, \tag{7}$$

where O_{pi} and t_{pi} are actual and desired outputs, respectively, realized at output layer, p is total number of input pattern vectors and i is number of nodes in output layer. A flow diagram shown in Fig. 4 clearly illustrates aforementioned BP algorithm.

Fig. 4 Schematic of BP
algorithm flow diagram

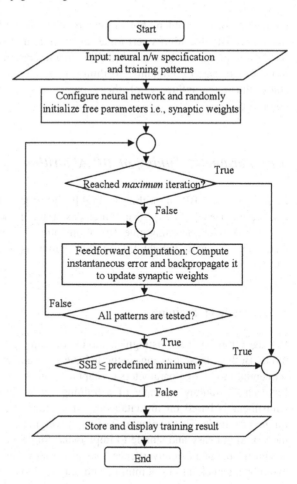

3 Performance Study Based on Various Parameters

We implement BP algorithm using JAVA programming language. Training of NN
model was provided by using data sample prepared as per method indicted in
Table 2. Thereafter, performance of BP algorithm was observed. An algorithm used
for training of NN for any particular application is said to be efficient if and only if,
SSE or mean square error (MSE) induced on NN for given training set can be
reduced to an acceptable minimum. BP algorithm is robust and popular algorithm
used for training of multi-layer perceptrons (MLPs). Performance analysis presented in this section aims to offer an insight of the strengths and weaknesses of BP
algorithm used for the mentioned application problem. Performance of BP algorithm depends on adequate choice of various parameters used in algorithm and
complexity of problem for which the algorithm used. We may not control the

complexity of the problem, but we may regulate the underlying parameters in order to enhance BP algorithm's performance. Even though BP algorithm is widely used for NN training, its several controlling parameters are one of the reasons that motivated research community to think of the alternatives of BP algorithm. Our study illustrates influence of these parameters over the performance of BP algorithm.

3.1   Parameter Tuning of BP Algorithm

Performance of BP algorithm is highly influenced by various heuristics and parameters used [27, 28]. These includes training mode, network configuration (network model), learning rate (η), momentum factor (β), initialization of free parameters (synaptic weights), i.e. initial search space (χ), volume of training set, and terminating criteria.

3.1.1   Mode of Training

BP algorithm has two distinguish modes of training: sequential mode and batch mode. Both of these training methods have their advantages and disadvantages. In sequential mode, free parameters (synaptic weights) of NN are made to be adjusted for each of training example of a training set. In other words, synaptic weight adjustment is based on instantaneous error induced on NN for each instance of training example. This particular fashion of weight adjustment makes the sequential mode training easy and simple to implement. Since the sequential mode training is stochastic in nature, convergence may not follow smooth trajectory, but it may avoid being stuck into local minima and may lead to a global optimum if one exists. On contrary, in batch mode, free parameters are updated once in an epoch. An epoch of training indicates the adjustment of free parameters of a NN that takes place once for entire training set. In other words, training of NN is based on SSE induced on NN. Hence, gradient computation in this particular fashion of training is more accurate. Therefore, the convergence may be slow, but may follow smooth trajectory. Figures 5 and 6 indicate performance of BP algorithm based on these two distinct modes of NN training. Figure 5 indicates that the convergence using batch mode training method was slower than that of sequential mode training. From Figs. 5 and 6, we may be observed that sequential mode training method may not follow smooth trajectory, but it may lead to global optimum, whereas, it is evident that in batch mode training method, convergence was following smooth trajectory.

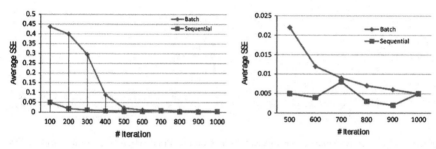

Fig. 5 Convergence: sequence versus batch mode

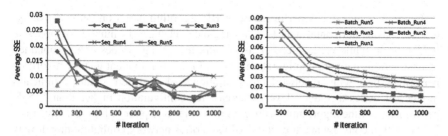

Fig. 6 Convergence trajectory in (*left*) sequence mode and (*right*) batch mode

3.1.2 Network Configuration

Multi-layer perceptrons are used for solving nonlinear problems. Basic MLP configuration consists of one input layer, one output layer with one or more hidden layer (s). Each layer may consist of one or more processing unit (neurons). The primary task in devising problems in NN is the selection of appropriate network model. Finding an optimum NN configuration is also known as structural training of NN. Network selection process starts with the selection of most basic three-layer architecture. Three-layer NN consists of a input layer, a hidden layer and a output layer. Number of neurons at input and output layers depends on problem itself. The gas detection problem was essentially a noise reduction problem, where NN tries to reduce noise of signals emitting from sensors in the presence of gas mixture. Hence, it was obvious that the number of outputs equals the number of inputs. In this application, we were designing a system that detects five gases. It is because there were five input signals to NN that leads to five processing units at input layer and five processing units at output layer. In three-layer NN configuration, the network configuration optimization reduces to the scope of regulating number of processing units (neurons) at hidden layer. Keeping other parameters fixed to certain values, the number of nodes at hidden layer was regulated from 1 to 8 to monitor influence of NN configuration on performance of BP algorithm. Parameter setup was as follows: number of iterations was set to thousand, η to 0.5, β to 0.1, $\chi \in [-1.5, 1.5]$ and training volume to fifty training examples.

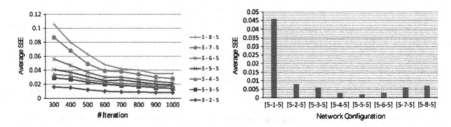

Fig. 7 (*left*) Convergence trajectory at NN Configuration; (*right*) SSE at various NN Configurations

Figure 7 demonstrates performance of BP algorithm based on network configuration, where horizontal axis (*X*-axis) indicates changes in number of nodes at hidden layer, while vertical axis (*Y*-axis) indicates average values of SSE obtained against various network configurations. From Fig. 7, it is evident that the algorithm performed best with a network configuration 5–4–5, where value 4 indicates number of nodes at hidden layer, whereas, for configuration higher than 5–5–5, performance of the algorithm was becoming poorer and poorer. Figure 7 illustrates BP algorithm convergence trajectory for its various network configurations. It may be observed that convergence speed was slower for higher network configurations than that of lower network configurations, and it is because of the fact that the higher network configurations have large number of free parameters in comparison to the lower network configurations. Therefore, the higher configurations need more number of iteration to converge than the lower network configuration.

We may increase number of hidden layers to form four- or five-layered NN. For the sake of simplicity, number of nodes was kept same at each hidden layer. It has been observed that computational complexity is directly proportional to number of hidden layers in network. It was obvious to bear additional computational cost if the performance of the algorithm improves for increasing number of hidden layers. It has also been observed that performances of algorithm and network configuration are highly sensitive to training set volume. Performance study based on training set volume is discussed in Sect. 3.1.6.

3.1.3   Initialization of Free Parameters (Initial Search Space—χ)

Proper choice of χ contributes to the performance of BP algorithm. Free parameters of BP algorithm were initialized with some initial guess. The remaining parameters were kept fixed at certain values, and synaptic weights were initialized between [−0.5, 0.5] and [−2.0, 2.0] in order to monitor influence of synaptic weights' initialization on performance of BP algorithm. Parameters setting was as follows: number of iteration was set to thousand, η to 0.5, β to 0.1, network configuration to 5–5–5 and training volume to fifty training examples.

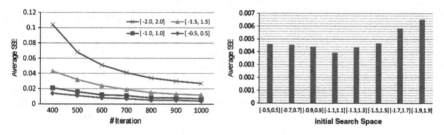

Fig. 8 (*left*) Convergence trajectory at χ; (*right*) SSE at various χ

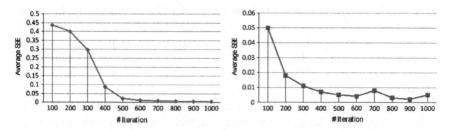

Fig. 9 (*left*) Convergence in batch mode; (*right*) Convergence in sequential mode

In Fig. 8, X-axis represents iteration, whereas, Y-axis represents average SSE values. The four continuous lines in Fig. 8 indicate convergence trajectory for various initialization ranges. Figure 8 demonstrates that large initial values lead to small local gradient that causes learning to be very slow. It was also observed that learning was good at somewhere between small and large initial values. In this case, ± 1.1 was a good choice of range χ.

3.1.4 Convergence Trajectory Analysis

In Fig. 9, X-axis indicates number of iteration, while Y-axis indicates average SSE achieved. Parameters setting was as follows: network configuration set at 5–5–5, number of iteration taken is 10,000, η taken is 0.5, β taken is 0.1 and training set volume is 50. Figure 9 indicates convergence trajectory of BP algorithm, where it may be observed that the performance of BP improves when iteration number was increased. For given training set and network configuration, average SSE gets reduced to 0.005 (SSE figures in this chapter are on normalized data) at iteration number 1000.

3.1.5 Learning Rate and Momentum Factor

Learning rate (η) is crucial for BP algorithm. It is evident from BP algorithm mentioned in Sect. 2.6.3 that weight changes in BP algorithm are proportional to derivative of error. Learning rate η controls weight changes. With larger η, larger

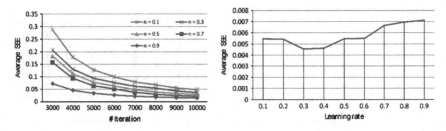

Fig. 10 (*left*) Convergence trajectories at η; (*right*) average SSE at various η

being weight changes in epochs. Training/Learning of NN is faster if η is larger and slower if η is lower. The size of η can influence networks ability to achieve a stable solution. A true gradient descent technique should take little steps to build solution. If η gets too large, then algorithm may disobey gradient descent procedure. Two distinct experiments have been done using η and their results are illustrated in Fig. 10.

In the first experiment, as shown in left of Fig. 10, network configuration was set to 5–5–5, number of iteration was set to 10,000, β was set to 0.1, training set volume was set to two-hundred and batch mode training was adopted. From this experiment, it was observed that for larger values of η, weight change ΔW_{ij} was larger and for smaller η, weight change ΔW_{ij} was smaller. At a fixed iteration 10,000 and at $\eta = 0.9$, network training speed was fast. We got SSE 0.02 at $\eta = 0.9$ and at $\eta = 0.1$, SSE was 0.05 which indicated that for small η, algorithm required more steps to converge, though convergence was guaranteed because small steps minimize the chance of escaping global minima. With larger values of η, algorithm may escape global minima that is clearly indicated in Fig. 10.

In the second experiment shown in right of Fig. 10, it may be observed that for lower η, i.e. 0.1 and 0.2, algorithm fell short to reach to a good result in limited number of iteration. However, at η 0.3 and 0.4, algorithm offered a good result, whereas, for learning rates higher than 0.4, algorithm tends to escape global optima.

For slight modification to BP weight update rule, additional momentum (β) term was introduced by Rumelhart [28]. Momentum factor was introduced to preserve weight change that should influence current search direction. The term momentum indicates that once weight starts moving in a particular direction, it should continue to move in that direction. Momentum can help in speeding-up learning and can help in escaping local minima. But too much speed may become unhealthy for a training process. Figure 11 indicates that for lower values of momentum rate, algorithm performs well. Therefore, β value should be increased as per learning speed requirements.

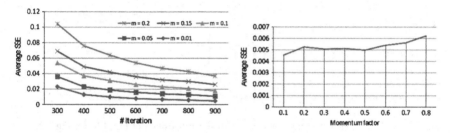

Fig. 11 (*left*) Convergence trajectories at β; (*right*) SSE at various β

3.1.6 Training Volume

We have already mentioned that performance of BP algorithm depends on its parameters and complexity of problem. We can regulate those parameters to improve the performance of the algorithm, but we cannot control the complexity of problem. Performance of an algorithm is highly influenced by the size of training set and the characteristic of data within training set. In Figs. 12 and 13, we present an observation based on volume of training set. In Fig. 12 (left), X-axis indicates volume of training set, whereas Y-axis indicates average SSE achieved. Figure 12 (left) indicates that average SSE has been plotted against different training set volumes feeding to networks of fixed size, trained at same iteration. From Fig. 12 (left), it may be observed that performance of the algorithm gets poorer on increasing volume of training set. To improve performance of the algorithm, we reconfigured NN to higher configuration and increased maximum training iteration values each time training set volume was increased. In Fig. 12 (right), first row values along X-axis indicate number of hidden layer nodes in a three-layered NN and second row values along X-axis indicate volume of training set, whereas Y-axis indicates average SSE achieved. In Fig. 13 (left), first row, second row and third row values along X-axis indicate maximum iteration, number of hidden layer nodes in three-layered NN and volume of training set, respectively, while Y-axis indicates average SSE achieved. From Fig. 12, it may be observed that increasing values of training set volume, average SSE was increased. Hence, the performance became

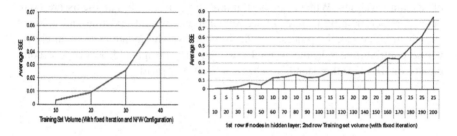

Fig. 12 (*left*) Fixed NN configuration and iteration (*right*) fixed iteration

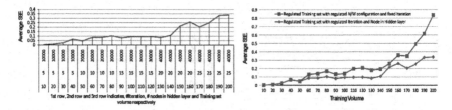

Fig. 13 (*left*) SSE versus training volume; (*right*) superimposition of Fig. 13 over Fig. 12

poorer, but as soon as network was reconfigured to a higher configuration, average SSE dips down. At that particular configuration, when training set volume was increased, SSE was also increased. Figure 13 (right) is a superimposition of Fig. 13 (left) over Fig. 12 (right). Figure 13 (right) indicates that iteration up-gradation and network reconfiguration together became necessary for improvement of algorithm when training set volume was increased.

3.2 Complexity Analysis

Time complexity of BP neural network algorithm depends on number of iteration it takes to converge, number of patterns in training data and time complexity needed to update synaptic weights. Hence, it is clear that time complexity of BP algorithm is problem dependent. Let n be the number of iterations required for convergence of BP algorithm and p be total number of patterns in training data. Synaptic weights of NN shown in Fig. 3 may be represented as a weight vector. Hence, to update synaptic weights, running time complexity required is $O(m)$, where m being the size of weight vector. To update synaptic weights, we need to compute gradient as per (5). The gradient computation for each pattern takes $O(n^2)$. NN weights may be updated either in sequential or batch mode as suggested in Sect. 3.1. In batch mode, weights are updated once in an epoch. One epoch training means updating NN weightes using entire trainning set at once. In sequential mode, weights are updated for each pattern presented to the network.

Let w be cost of gradient computation that is basically $O(m)$, equivalent to cost of updating weights. Whichever the training mode it may be, gradients are computed for each training pattern. In sequential mode, weight updating and gradient computation are parallel process. In batch mode, weight update once in an epoch, whose contribution is feeble in total cost and may be ignored. Hence, the complexity of BP algorithm stands to $O(p \times w \times n) = O(pwn)$.

3.3 Comparing BP with Hybrid NN Training Methods

We compared BP algorithm with three other intelligent techniques that were used for training of NN such as conjugate gradient (CG) method, genetic algorithm (GA) [30, 31] and particle swarm optimization (PSO) algorithm [32, 33]. This section provides a comprehensive performance study and comparison between these intelligent techniques that were applied for solving manhole gas detection problem.

3.3.1 Empirical Analysis

An empirical study is shown in Fig. 14, where X-axis indicates number of iterations while Y-axis indicates average SSE achieved against different iterations. Figure 14 (left) indicates convergence trajectory and Fig. 14 (right) indicates SSE obtained in various epochs in training process by each of the mentioned algorithms. It was observed that BP algorithm converged faster than the other algorithms. The convergence trajectory of CG method appeared smoother than that of BP algorithm, and its SSE value got reduced to value nearly equal to the value of SSE achieved by BP algorithm. Although the convergence trajectory of PSO approach was not as convincing as of BP algorithm and CG method, it was observed that PSO approach was efficient enough to ensure SSE nearer to the one achieved using classical NN training algorithms, BP and CG methods. Figure 14 indicates that GA was not as efficient as the other three approaches. GA quickly gets stuck into local optima as far as the present application was concerned.

3.3.2 Theoretical Analysis

The complexity analysis of BP algorithm is provided in Sect. 3.2. The cost met by line search in CG method was an additional computational cost in contrast with BP algorithm counterpart. The computational cost met by PSO and GA algorithm may be given as $O(pqwng)$, where q is size of population, n is number of iterations, p is

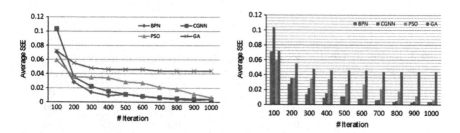

Fig. 14 (*left*) Convergence trajectory analysis; (*right*) SSE at various epochs

number of training examples, w is cost needed to update synaptic weights and g is cost of producing next generation. It may please be noted that the cost of g in PSO and GA depends on their own dynamics of producing next generation. In GA, it is based on selection, crossover and mutation operation [34], whereas PSO has simple non-derivative methods of producing next generation [35].

3.3.3 Statistical Analysis

We used Kolmogorov–Smirnov test (KS test), where KS test being nonparametric in nature, it does not make any assumption about data distribution. Two-sample KS test is useful for comparing two samples, as it is sensitive to differences in both location and shape of empirical cumulative distribution functions (*epcdf*) of two samples. Clearly speaking, KS test tries to determine if two datasets X and Y differ significantly [36]. KS test makes following hypothesis. Null hypothesis H_0 indicates that two underlying one-dimensional unknown probability distributions corresponding to X and Y are indistinguishable, i.e. datasets X and Y are statistically similar $(X \equiv Y)$. The alternative hypothesis H_t indicates that X and Y are distinguishable, i.e. datasets X and Y are statistically dissimilar. If it is alternative hypothesis, then their order (direction) becomes an important consideration. We need to determine whether the former (dataset X) is stochastically larger than or smaller than the latter one (dataset Y), i.e. $X > Y$ or $X < Y$. Such KS test is known as one-sided test that determines direction by computing distance D_{nm}^{+}, D_{nm}^{-}, and D_{nm}, where value n and m are cardinality of the set X and Y, respectively. The null hypothesis H_0 is rejected if $D_{nm}, > K\alpha$, where $K\alpha$ is known as critical value [37]. For n and m being 20, i.e. size of each sample, $K\alpha$ was found to be equal to 0.4301 for $\alpha = 0.05$ which indicates 95 % confidence in test. Readers may explore [37–39] to mitigate their more interest in KS test. KS test was conducted using twenty instances of SSEs produced by each mentioned algorithms. KS test was conducted in between BP and other intelligent algorithms. Hence, set X was prepared with SSEs of BP algorithm and three separate sets, and each representing set Y was prepared using SSE values of CG, PSO and GA algorithms. Outcome of KS test is provided in Table 3 where it is conclusive about significance of BP algorithm over the other algorithms.

Table 3 KS test: BP versus other intelligent algorithms

KS test type BP (X)	Intelligent algorithms (Y)		
	CGNN	PSO	GA
D_{nm}^{+}	0.10	0.15	1.00
D_{nm}^{-}	0.20	0.15	0.00
D_{nm}	0.20	0.15	1.00
Decision	$X \equiv Y$	$X \equiv Y$	$X > Y$

Table 4 System result presentation in ppm

Input gas	Responding unit			Safety limit (ppm)	Interpretation
	Sensor	NN	System (ppm)		
NH_3	0.260	0.016	80	25–40	Unsafe
CO	0.346	0.022	110	35–100	Unsafe
H_2S	0.240	0.023	115	50–100	Unsafe
CO_2	0.142	0.022	110	5000–8000	Safe
CH_4	0.843	0.993	4965	5000–10,000	Safe

4 Results and Discussion

Data sample was prepared as per the procedures mentioned in Sect. 2.5. The collected data sample was partitioned in two sets, training and test. The eighty percent of the original set was used as training set, and remaining twenty percent was used for testing purpose. After training, the system undergone a test using the test set. Output of the trained NN was de-normalized in order to present output in terms of concentration of gas components present in given test sample/gaseous mixture. We are providing a sample test result obtained for input sample 2, which is provided in Table 2. The predicted concentration value corresponding to the given input sample is shown in Table 4. In Table 4, interpretation column is based on comparison between de-normalized value of NN output and safety limits of the respective gases. Note that each of nodes at output layer is dedicated to a particular gas. Safety limits of manhole gases are as follows. Safety limit of NH_3 is in between 25 and 40 ppm (as per limit set by World Health Organization), CO is in between 35 and 100 ppm [40–42], H_2S is in between 50 and 100 ppm [43, 44]), CO_2 is in between 5000 and 8000 ppm [41, 45] and CH_4 is in between 5000 and 10000 ppm [46].

5 Conclusion

In this chapter, we have discussed design issues of an intelligent sensory system (ISS) comprising semiconductor-based GSA and NN regressor. BP algorithm was used for supervised training of NN. The proposed design of ISS offered a solution to manhole gas mixture detection problem that was viewed as a noise reduction/pattern recognition problem. We have discussed mechanisms involved in preparation and collection of data sample for the development of proposed ISS. Cross sensitivity was firmly addressed in this chapter. We have discussed the issues in training of NN using backpropagation (BP) algorithm. A comprehensive performance study of BP algorithm was provided in this chapter. Performance comparison in terms of empirical, theoretical and statistical sense between BP and various other hybrid intelligent approaches applied on said problem was provided in this chapter. A concise discussion on safety limits and systems' result presentation mechanism was presented in the remainder section of this chapter. Data sample mentioned in the scope of this research may not represent the entire spectrum of the mentioned manhole gas detection problem. Therefore, at present, it was a non-trivial task for NN regressor. Hence, it has offered a high-quality result. Therefore, an interesting study over larger dataset to examine how manhole gas detection problem can be framed as a classification problem using the available classifier tools.

Acknowledgments This work was supported by Department of Science & Technology (Govt. of India) for the financial supports vide Project No.: IDP/IND/02/2009 and the IPROCOM Marie Curie initial training network, funded through the People Programme (Marie Curie Actions) of the European Union's Seventh Framework Programme FP7/2007–2013/ under REA grant agreement No. 316555.

References

1. Gromicko N (2006) Sewer gases in the home. http://www.nachi.org/sewer-gases-home.html. Accessed on 15 July 2015
2. Lewis (1996) Dangerous properties of industrial materials, 9th edn. Van Nostrand Reinhold, New York, ISBN 0132047674
3. Whorton J (2001) The insidious foe–sewer gas. West J Med 175(6):427428, (PMC 1275984)
4. Hindu T (2014a) Deaths in the drains. http://www.thehindu.com/opinion/op-ed/deaths-in-the-drains/article5868090.ece?homepage=true. Accessed on 15 July 2015
5. Hindu T (2014b) Sewer deaths. http://www.thehindu.com/opinion/letters/sewer-deaths/article5873493.ece. Accessed on 15 July 2015
6. Hindu T (2014c) Supreme court orders states to abolish manual scavenging. http://www.thehindu.com/news/national/supreme-court-orders-states-to-abolish-manual-scavenging/article5840086.ece. Accessed on 15 July 2015
7. NIOSH (2011) Volunteer fire fighter dies during attempted rescue of utility worker from a confined space. http://www.cdc.gov/niosh/fire/reports/face201031.html. Accessed on 15 July 2015
8. Time of India (2014) Sewer deaths http://timesofindia.indiatimes.com/city/delhi/Panel-holds-DDA-guilty-for-sewer-death/articleshow/31916051.cms. Accessed on 15 July 2015
9. Li J (1993) A mixed gas sensor system based on thin film saw sensor array and neural network. IEEE, pp 179–181, 0-7803-0976-6/93
10. Srivastava AK, Srivastava SK, Shukla KK (2000a) In search of a good neuro-genetic computational paradigm. IEEE, pp 497–502, 0-78O3-5812-0/00
11. Srivastava AK, Srivastava SK, Shukla KK (2000b) On the design issue of intelligent electronic nose system. IEEE, pp 243–248, 0-7803-5812-0/00
12. Llobet E (2001) Multicomponent gas mixture analysis using a single tin oxide sensor and dynamic pattern recognition. IEEE Sens J 1(3):207–213, ISSN 1530437X/01
13. Liu J, Zhang Y, Zhang Y, Cheng M (2001) Cross sensitivity reduction of gas sensors using genetic algorithm neural network. Farquharson S (ed) Proceedings of SPIE Optical Methods for Industrial Processes, vol 4201
14. Tsirigotis G, Berry L, Gatzioni M (2003) Neural network based recognition, of co and nh3 reducing gases, using a metallic oxide gas sensor array. In: Scientific proceedings of RTU, telecommunications and electronics, Series 7, vol. 3, pp 6–10

15. Lee DS, Ban SW, Lee M, Lee DD (2005) Micro gas sensor array with neural network for recognizing combustible leakage gases. IEEE Sens J 5(3):530–536. doi:10.1109/JSEN.2005. 845186
16. Ambard M, Guo B, Martinez D, Bermak A (2008) A spiking neural network for gas discrimination using a tin oxide sensor array. In: International symposium on electronic design, test and applications, IEEE, pp 394–397, ISBN 0-7695-3110-5/08
17. Baha H, Dibi Z (2009) A novel neural network-based technique for smart gas sensors operating in a dynamic environment. Sensors 9:8944–8960, ISSN 1424–8220
18. Pan W, Li N, Liu P (2009) Application of electronic nose in gas mixture quantitative detection. In: Proceedings of IC-NIDC, IEEE, pp 976–980, ISBN 978-1-4244-4900-2/09
19. Wongchoosuka C, Wisitsoraatb A, Tuantranontb A, Kerdcharoena T (2010) Portable electronic nose based on carbon nanotube-sno2 gas sensors and its application for de- tection of methanol contamination in whiskeys. Sens Actuators B: Chem doi:10.1016/j.snb.2010.03.072
20. Zhang Q, Li H, Tang Z (2010) Knowledge-based genetic algorithms data fusion and its application in mine mixed-gas detection. IEEE, pp 1334–1338, 978-1-4244-5182-1/10
21. Won So D, Koo J (2010) The estimation of hazardous gas release rate using optical sensor and neural network. Pierucci S, Ferraris GB (eds) European Symposium on Computer Aided Process Engineering ESCAPE20, Elsevier B.V
22. Ojha VK, Duta P, Saha H, Ghosh S (2012c) Detection of proportion of different gas components present in manhole gas mixture using back-propagation neural network. In: International proceedings of computer science and information technology, vol 1, pp 11–15. ISBN 978-981-07-2068-1
23. Ojha VK, Duta P, Saha H, Ghosh S (2012) Linear regression based statistical approach for detecting proportion of component gases in man-hole gas mixture. In: International symposium on physics and technology of sensors, IEEE, doi:10.1109/ISPTS.2012.6260865
24. Ojha VK, Duta P, Saha H, Ghosh S (2012f) A novel neuro simulated annealing algorithm for detecting proportion of component gases in manhole gas mixture. International Conference on Advances in Computing and Communications, IEEE, pp 238–241, doi:10.1109/ICACC.2012.54
25. Ghosh S, Roy A, Singh S, Ojha VK, Dutta P, Saha H (2012a) Sensor array for manhole gas analysis. In: International Symposium on Physics and Technology of Sensors. IEEE, Pune, India, ISBN 978-1-4673-1040-6
26. Ghosh S, Roychaudhuri C, Saha H, Ojha VK, Dutta P (2012b) Portable sensor array system for intelligent recognizer of manhole gas. In: International conference on sensing technology (ICST), IEEE, India, pp 589–594, ISBN 978-1-4673-2246-1
27. Haykin S (2005) Neural networks a comprehensive foundation, 2nd edn. Pearson Prentice Hall, New Jersey, ISBN 81-7803-300-0
28. Rummelhart DE, Hinton GE, Williams RJ (1986) Learning representations by back-propagating errors. Nature 323(6088):533536. doi:10.1038/323533a0
29. Sivanadam SN, Deepa SN (2007) Principles of soft computing. Wiley, New Delhi, ISBN 81-265-1075-7
30. Ojha VK, Duta P, Saha H (2012) Performance analysis of neuro genetic algorithm applied on detecting proportion of components in manhole gas mixture. Int J Artif Intell Appl 3(4):83–98. doi:10.5121/IJAIA.2012.3406
31. Ojha VK, Duta P, Saha H, Ghosh S (2012b) Application of real valued neuro genetic algorithm in detection of components present in manhole gas mixture. Wyld DC (ed) Advances In Intelligent And Soft Computing, Springer, Vol. 166, pp 333–340, DOI: 10.1007/978-3-642-30157-5
32. Ojha VK, Duta P (2012) Performance analysis of neuro swarm optimization algorithm applied on detecting proportion of components in manhole gas mixture. Artif Intell Res 1:31–46. doi:10.5430/JNEP.V1N1PX
33. Ojha VK, Duta P, Saha H, Ghosh S (2012e) A neuro-swarm technique for the detection of proportion of components in manhole gas mixture. In: Proceedings of international conference on modeling, optimization and computing, vol 2. NI University, Kanyakumari, India, pp 1211–1218

34. Goldberg DE (2006) Genetic Algorithms in search, Optimization and machine learning, 1st edn. Pearson Education, Singapore, ISBN 81-7758-829-X
35. Kennedy J, Eberhart RC (2001) Swarm intelligence. Morgan Kaufmann Publishers, San Francisco, ISBN 1-55860-595-9
36. Dutta P, Dutta Majumder D (2012) Performance analysis of evolutionary algorithm. Lambert Academic Publishers, Saarbrücken, ISBN 978-3-659-18349-2
37. Gail MH, Green SB (1976) Critical values for the one-sided two-sample kolmogorov-Smirnov statistic. J Am Stat Assoc 71(355):757–760
38. Boes DC, Graybill FA, Mood AM (1974) Introduction to the theory of statistics, 3rd edn. McGraw-Hill, New York
39. Sheskin DJ (2003) Handbook of parametric and nonparametric statistical procedures, 3rd edn. CRC-Press, London, ISBN 1-58488-440-1
40. Ernst A, Zibrak JD (1998) Carbon monoxide poisoning. New Engl J Med 339(22) (PMID9828249)
41. Friedman D (2014) Toxicity of carbon dioxide gas exposure, CO_2 poisoning symptoms, carbon dioxide exposure limits, and links to toxic gas testing procedures. InspectAPedia http://www.inspectapedia.com/hazmat/CO2gashaz.htm#reviewers. Accessed on 15 July 2015
42. Goldstein M (2008) Carbon monoxide poisoning. J Emerg Nurs: JEN: Off Publ Emerg Dep Nurses Assoc (PMID 19022078)
43. USEPA (1980) Health and environmental effects problem for hydrogen sulfide. West J Med pp 118–8, eCAOCIN026A
44. Zenz C, Dickerson OB, EP Horvath (1994) Occupational medicine, 3rd edn. p 886
45. Shilpa G (2007) New insight into panic attacks: Carbon dioxide is the culprit. InspectAPedia
46. Fahey D (2002) Twenty questions and answers about the ozone layer

REFII Model as a Base for Data Mining Techniques Hybridization with Purpose of Time Series Pattern Recognition

Goran Klepac, Robert Kopal and Leo Mršić

Abstract The article will present the methodology for holistic time series analysis, based on time series transformation model REFII (REFII is an acronym for *Raise-Equal-Fall* and the model version is II or 2) Patel et al. (Mining motifs in massive time series databases, 2002) [1], Perng and Parker (SQL/LPP: a time series extension of SQL based on limited patience patterns, 1999) [2], Popivanov and Miller (Similarity search over time series data using wavelets, 2002) [3]. The main purpose of REFII model is to automate time series analysis through a unique transformation model of time series. The advantage of this approach to a time series analysis is the linkage of different methods for time series analysis linking traditional data mining tools in time series, and constructing new algorithms for analyzing time series. REFII model is not a closed system, which means that there is a finite set of methods. This is primarily a model used for transformation of values of time series, which prepares data used by different sets of methods based on the same model of transformation in the domain of problem space. REFII model gives a new approach in time series analysis based on a unique model of transformation, which is a base for all kind of time series analyses. In combination with elements of other methods, such as self-organizing maps or frequent-pattern trees, REFII models can make new hybrid methods for efficient time temporal data mining. Similar principle of hybridization could be used as a tool for time series temporal pattern recognition. The article describes real case study illustrating practical application of described methodology.

G. Klepac (✉)
Raiffeisenbank Austria, Zagreb, Croatia
e-mail: goran@goranklepac.com

R. Kopal · L. Mršić
IN2data Data Science Company, Zagreb, Croatia
e-mail: robert.kopal@in2data.hr

L. Mršić
e-mail: leo.mrsic@in2data.hr

© Springer India 2016
S. Bhattacharyya et al. (eds.), *Hybrid Soft Computing Approaches*,
Studies in Computational Intelligence 611,
DOI 10.1007/978-81-322-2544-7_8

237

Keywords Time series transformation · Pattern recognition · Data mining · REFII · Data mining and time series integration · Self-organizing maps · Decision tree · Frequent-pattern tree

1 Introduction

REFII, a unique model for time series transformation, enables the integration of different conceptual models of time series analysis, which improves upon the traditional way of using a series of unrelated methods during the analysis.

Also, the REFII model enables the analysis of time series by applying traditional data mining methods (decision trees, clustering, shopping cart analysis, etc.).

Combining different conceptual models of analysis within the REFII model can solve complex time series analysis tasks, such as market segmentation and detection of market regularities (behavioral models of market segments and entities).

The application of the REFII model improves the detection of market regularities in time series during the "ad hoc" market analysis by integrating a series of different methodological analysis procedures, unlike the traditional approach which uses a series of unrelated and incompatible methods that, sometimes, do not enable deeper analysis and solution modeling for nonstandard problems.

The *Improvement* concept means the integration of different methodological concepts in time series analysis, the application of traditional data mining methods on time series, the concatenation of time series analysis methods and the extension of the basic REFII model with basic algorithms in order to build a market segmentation model based on time series and market segments and subjects behavior estimates.

The *Deeper analysis* concept means the ability to implement different types of analyses on a reduced time series data set that is a result of processing preceding the application of a method or analytical procedure.

Integration concept is achieved through a single unified time series transformation model that is a basis for analytical procedures, analytical methods, and the development of new methods with respect to the field.

Earlier methods used for time series analysis were mainly focused on a specific problem and built for specific purpose (similar to term "black box," often used for systems without the ability of expansion). Based on time series as input parameters they output information such as seasonal fluctuations or recurring patterns and the like. What all of the previous methods have in common are their own time series transformation models that are usually adapted to specific issues and, in addition to being incompatible with each other, they cannot be extracted as the output values of the model.

REFII offers a completely new approach.

The REFII model's new approach is reflected through the transformation model being the basis of analysis as well as any following analytical methods.

Analytical methods are algorithmic procedures that are used to influence transformed values. Those algorithmic procedures can be basic algorithmic procedures that solve certain types of problems like time series-based market segmentation or they can be traditional data mining algorithms like decision trees or new time series analysis methods adapted to REFII syntax.

This focus on the transformation model as the starting point of the analysis allows for a more thorough analysis of temporal regularities by concatenating methods that influence a time series as well as modeling solutions for nonstandard problems that arise from the detection of market regularities and market segmentation.

The thoroughness notion in this context means the ability to implement different types of analyses on a reduced time series data set that is the result of processing before the application of a method or analytical procedure with the intention of solving complex tasks such as market segmentation based on time series.

As an example we will provide a hypothetical scenario of client segmentation using activity as criteria within a time period and the assessment of trend variations for each segment separately, within the bounds of the same system.

An additional advantage of this approach is reflected in the ability to define a desired degree of analysis reliability within a single model, which implies that the analyst determines if he wants to engage in the analysis of global trends or precise analysis, which, of course, depends on the nature of the problem being solved.

The REFII transformation model is designed with the aim of unambiguous description of the time series curve. The time series is transformed into a set of values that represent the area of the time slot below the curve, the angular coefficient of the time slot "curve" and its trend code.

The new algorithmic procedures for time series analysis can be performed over such transformation models, but that concept opens up the possibility of discovering new algorithms for time series analysis such as market segmentation on the basis of time series as well as the application of traditional data mining methods within the same basic transformation model.

What follows is the hybridization of methods that is derived from the REFII model as its base and it also supports time series analysis using data mining algorithms.

The scientific contribution is reflected in the presentation of a new concept in time series analysis in the field of research and market segmentation, which, unlike the traditional approach to these issues, focuses on the unambiguous time series transformation model.

The advantage of this approach is reflected in the synthesis of a variety of different approaches and methods of time series analysis which includes traditional data mining methods as well as the addition of new algorithmic analysis procedures.

This approach enables the implementation of different types of analyses on a reduced time series data set that is a result of the application of different methods of the analytical process.

This approach provides the answer on the way of conducting complex market analyses based on time series that are very difficult or nearly impossible to conduct using the traditional approach.

To illustrate our statement, solutions to the market segmentation and market regularity detection problems have been explored (sample data set of car insurance policy users that had a car accident) by using time series through the REFII model, which requires the use of a series of concatenated analytical procedures, both traditional and basic REFII procedures.

These analyses were very difficult or nearly impossible to implement because of the traditional approach to time series analysis by using unrelated methods, but their implementation is becoming a necessity in market regularities analysis by using time series.

2   Background

Time series have always been a very interesting area of application from the perspective of data mining. The observation of values of a characteristic in a time frame can be very useful for gaining knowledge about the observed phenomena. This knowledge can be used by classical statistics that offers methods like group time series index, average time series values, individual indexes, regression methods, trend estimation, etc. [4, 5].

These methods can be very useful in estimating phenomenon trends or certain indicators related to this matter, but if we look at the whole issue through the perspective of intelligent data analysis, we get a very limited toolset for performing a quality analysis of temporal characteristics.

With the development of the concept of data mining, the importance of time series as a field that can provide relevant analysis and information has been recognized. Time series analysis largely relies on the concept of neural networks in trend forecasting over time [6–8]. The application of the model of neural networks, primarily the application of the BPN algorithm, is most evident in predicting price trends of securities, financial ratios, company revenue, demand estimate, and so on [9–11].

Another branch of time series analysis uses the fuzzy logic methodology and tries to implement its resources to time series [12, 13]. Each of the described approaches has its own characteristics that can be applied in practice, but the basic characteristic of each of the approaches is the ambiguity of the description of time series or a certain segment of the time series. Recently, a trend of developing programming languages based on time series analysis has been emerging. Publications that arouse attention and a very meticulous approach to knowledge discovery from time series are [14–19]. The former is about using the data mining methodology approach to knowledge discovery through an algorithm that, in its essence, has elements of the function of distance, and can extract knowledge of corresponding elements from time series. The latter is about defining the episodes concept in time series whose points in time represent events. This concept will be presented in this work through the prism of the described model.

The time series issue is intertwined with the data preprocessing issue [20–23]. There are a number of methods that replace missing attribute values in time series. As some of the most common methods I will mention linear interpolation that has its roots in mathematical analysis, the mean value method, the similarity method that tries using distances to estimate the value of missing attributes, and the neural networks method.

The traditional approach to time series analysis, including the approach to time series analysis with the aim of market regularities detection, relies on the use of different independent methods to solve certain types of problems. In the detection of temporal market regularities we can use a variety of methods to detect seasonal fluctuations, detect cyclical fluctuations [8, 24, 25], methods for trend forecasting [22, 26], methods for episode detection within time series [17], and pattern discovery methods [27–29].

Authors that work on data preprocessing do not devote sufficient attention to time series preprocessing as an integral component of traditional data mining of transactional data [22, 28] whose goal is creating a unified transformation model which would enable the application of a whole variety of analytical methods to time series analysis through a single analysis system.

The mentioned authors focus on the traditional approach to time series analysis by using unrelated methods in detecting market regularities.

The importance of this problem is evident in these publications [17, 19, 28] which, after offering a solution to certain issues, as a possibility for future work, often mention issues that can be solved with existing data mining methods or by using other existing methods, but highlight the superiority of their model in a particular area of analysis.

That way, they still do not offer a solution to achieving "compatibility" between the methods from different areas of analysis and their own methods that offer an acceptable solution to a highly specialized area.

The issue grows even more when we want to analyze a time series by using data mining methods, such as decision trees and clustering, not to mention the analysis of seasonal fluctuations on the temporal clustering level.

The main problem is the inability to integrate a set of differing concepts of data analysis with a common analysis model and methodology as their starting point.

Thus, Manilla in [17] as part of his work on event and episode analysis within time series, where events are defined as, for example, sound of an alarm, theft, escape and the like, indicates the possibility of the analysis of a time series consisting of numerical values and warns that a way to describe events with a basis in numerical values should be devised.

Xsniaping in his publication [30] illustrates a method for pattern recognition within time series, and also mentions the development of a model for time slot trend probability estimation, based on his own model, as a focus for future work.

Han in [31] shows a model of pattern recognition based on the substitution of temporal values with alphanumerical codes and the calculation of occurrence frequency, but Pratt in [19] criticizes Han's vagueness of the model that is derived from the imprecise definition of time slots and offers his own model for pattern recognition based on minimum and maximum values. As possibilities for future work he proposes the application of Han's model to the methodology of cyclical fluctuations detection within his own model, which could be made even more precise by applying the transformation model as the starting point for analysis.

Different researchers [19, 20, 31–36] while solving issues from the time series analysis domain, do not offer a solution on how to methodologically integrate their analysis model with other successful models which would greatly contribute to the efficiency of the analysis in the field of marketing analyses.

All of the above solutions can be successfully applied in the field of research and analysis, where the relatedness of methods is the main issue of the field in question.

The problems that are often referred to as the subject of future research, as is in the case with Pratt and Xsniaping, can be solved by integrating the existing methodology with data mining methods or other existing methods, provided there is a single concept of the time series transformation model as the starting point for the analysis.

If we consider specific technical issues (event analysis in alarm systems, fluctuation analysis in financial data...), which were the impetus for the development of certain methods, then this problem is not so noticeable. However, if we look at these issues through the prism of analysis, segmentation and market regularity detection based on time series, and then this issue starts to show.

The problem of the relatedness of temporal methodologies for time series analysis and its implication for market analysis can be illustrated by an example. Say we want to determine if there is a cyclical regularity in the behavior of a certain group of clients, what would be the probability of an event (for example, the withdrawal of a large sum from an ATM) within a week after the occurrence of the cyclical regularity, if that regularity exists.

As is shown in this example, a series of difficulties arise from this kind of analysis if we approach it traditionally.

The problems arise when we want to define an event within a time series as well as when we want to describe a cyclical regularity as an episode, because that, naturally, depends on the method used to detect the cyclical fluctuations.

Another issue is how to use a Bayesian network to estimate the probability of an event after the occurrence of an episode. That is, the main problem occurs in the methodology of parameter initialization within the table of conditional probabilities.

If we choose the traditional approach to analysis that does not have a single model as a starting point, a number of technical issues appear and they are very hard or nearly impossible to solve. Even if we manage to solve these issues within the boundaries of traditional analysis with certain bridges between data and methodologies, those bridges will be of little to no use while solving a new and complex problem from the time series domain.

Here, the marketing analyst is facing a serious problem about the way in which he will integrate sets of methods, which is particularly pronounced when we want to apply traditional data mining methods to a time series, such as decision trees, clustering, market basket analysis, and the like.

The application of these methods can be of great importance when we, for example, want to cluster clients based on series and want to discover which group of clients across all clusters is more inclined to use a service of type X in the first half of the week.

It has been shown in practice that a quality time series analysis is done by using more than one method. Sometimes, after we recognize a pattern within a time series, we want to know if it is a recurring pattern, does it precede an event, how will an event influence that pattern, what is the probability of the occurrence of the pattern in a certain month or week, are clients with certain characteristics more or less involved in the event, etc.

These kinds of analyses can hardly be done by using the traditional approach.

Using the traditional approach to solve the mentioned issues would be hard or impossible, but if a number of different analysis models were integrated into a unified time series transformation model, those same issues could be solved by creating a precise (unambiguous) description of the time series which could then be used as a basis for further analyses.

Integrating different analysis models in time series analysis would lead not only to a synergistic effect, but would create an opportunity to design new algorithms that would be able to solve more complex issues in the area of marketing analysis, such as market segmentation based on time series.

The main reasons for the failure to integrate different concepts and methods in time series analysis arise from the main focus of authors being the discovery of a solution to a specific type of issue rather than focusing on the whole field of market analysis and the needs related to that field [37, 38].

The REFII model is designed to unify different concepts of time series analysis and traditional data mining methods in order to design new algorithmic procedures for market analysis, as well as finding solutions to the automatic time series pre-processing issue based on a new concept rather than a stronger hardware environment.

The reason for the development and implementation of the REFII model is derived from experience in market regularity analysis in the field of market segmentation and market segment behavior estimation.

The importance of the relatedness of methodological procedures of time series analysis was first noticed by Graham Williams in his publication [39] which provides an overview of data mining methods used for time series analysis. In that same publication he noted some challenges in the general theory of time series analysis that should be overcome in future research that should unify current and future research in the field of time series analysis by using data mining techniques.

He points to the state of relatedness and incompatibility of current methods as well as to the poor results of previous research in the field of integration of different analysis concepts in temporal data mining analysis.

As part of the solution he offers a theoretical concept called general hidden distribution-based analysis theory for temporal data mining, and points to a number of missing elements that should be the subject of future research. This solution would integrate his previous research [39–45] with the aim of creating a universal solution for time series analysis in the field of temporal data mining.

The main drawback of the aforementioned Williams' model stems from the fact that too little attention was given to the transformation model, which was put in the background when the solution was being defined.

Furthermore, too little attention was given to the flexibility of the analysis (flexible generation of variance categories by analysts), so this issue should be the subject of future research. The proposed solution also ignores the quantitative aspects of time series (area under the curve).

This is the reason for the announcement of future research focused on increasing flexibility and developing solutions for the integration of time series analysis and traditional data mining methods. Also, the issue of integrating fuzzy logic with time series within the proposed Williams' solution is mentioned as a subject for future research. Paradoxically, Williams intends to adapt traditional data mining methods in order to integrate them with time series analysis rather than develop a unified time series transformation model.

We define the unified transformation model as a starting point for the implementation of a number of analyses in the field of temporal data mining. The concept of unity here is synonymous with the versatility of time series transformation.

The main criticism of Williams' concept is related to:

- The stem of all issues is the sketchiness and the inflexibility of the time series transformation model that is worked out as the need arises [39–45]
- The neglecting of contingency in his approach to analysis
- There is no foreseen methodology for relating temporal and nontemporal attributes
- As a goal of future research the adaptation of existing data mining algorithms is stated, instead of building data-methodology bridges via the transformation model, which implies the existence of the issue with the application of fuzzy logic within the model

The REFII model has the solutions for the issues Williams stated would be the result of future research, and, in addition, offers a much more flexible approach to time series analysis.

Williams makes a common mistake while trying to unify methods and that is focusing on partial solutions to analytic tasks without a unified time series transformation model. Instead, he integrates a series of partial solutions via a sketchy transformation model, which should have a central role in the integration of differing concepts.

The main purpose of this paper is to point out the hidden potential for analysis in time series utilization and the variety of solutions that are the result of the application of a unified time series transformation model and its integral role in integrating differing methodological analysis procedures.

Also, this paper intends to point out the possibility of modeling business practice solutions through the use of the described model as well as pointing out a number of solutions that integrate it partially or completely.

The main characteristics of the model this publication is based on are compatibility and openness to other data mining methods. That characteristic enables further time series processing through algorithms such as standard statistical methods, neural networks, clustering, decision trees, memory-based resolution, and shopping cart method. This way, a time series can be analyzed by a variety of tools. The presented model, in addition to providing a number of possibilities for time series analysis, processing the values of either a segment or the complete series through available data mining methods and their concatenation, it provides additional options for knowledge extraction.

The goal of this research was exploring the possibilities for the extension of the REFII model, as well as its ability to synthesize existing algorithms for data mining and time series analysis into a single system for market regularity analysis and market segmentation based on time series.

Accordingly, the goal of this research is to offer solutions for various modalities of market segmentation and the detection of market regularities (behavior models of market segments and subjects) based on time series, through the use of REFII models.

3 Comparison with Other Methods

In traditional time series data mining analysis, there are lots of different methods, which solve particular kind of problems.

As a result, if we want to solve the problem of discovering patterns in time series, we can use different methods [19, 22, 31].

There are many different methods for solving seasonal oscillations, recognition time segments, similarity search, etc. A mutual characteristic of announced future work in those scientific papers is to discover methods for time series analysis, which are already discovered as a result of research done by other authors. Authors usually develop a qualitative model, which solves some kind of problem, but it is usually impossible to connect their model with the existing model for some other problem in the area of temporal data mining.

The reason why authors could not link other methods to their own methods can be found in the model of transformation.

Authors are, at first, focused on solving problems, and they do not care much about the model of transformation. Every developed model has its own model of transformation, which implies incompatibility among methods. Different model of transformation does not allow linking of different methods. Other big problem, which is based on transformation models, is linkage of time series with traditional data mining methods, automation in series data preparation, and construction of new analytical algorithms for time series analysis based on contingency.

For example, we cannot connect time series with decision trees using the traditional way of analysis. This could be interesting if we want to know the main characteristic and attributes of clients, which show the rising trend of using Automatic Teller Machines (ATM) during some period of time.

Another problem which emanates from this approach, and which is significant in time series analysis in the domain of marketing, is a problem of chaining methods for time series analysis.

In practice we have to connect two different models of transformation, and sometimes it is not so easy and elegant to be realized. Some methods use discrete Fourier transformation as a model of transformation.

Some methods use symbolic transformation, leg transformation, etc.

Those approaches imply that: "Series data has features that require more involvement by the miner in the preparation process than for non-series data. Where miner involvement is required, fully automated preparation tools cannot be used. The miner just has to be involved in the preparation and exercise judgment and experience." Solution to the mentioned problem Pyle [22] sees in more powerful, low-cost computer systems.

Suggested solution for all mentioned problems are REFII model, as a fully automated preparation tool which gives solution for problems such as:

- Discover seasonal oscillation
- Discover cyclic oscillation
- Discover rules from time series
- Discover episodes from time series
- Discover similarity of time segments
- Discover correlation between time segments
- Discover rules from in domain of finances from time series
- Connect time series and standard data mining methods

This kind of approach is different from traditional approach which is based on using a different model of transformation for each type of analysis. In case of using traditional approach of analysis, there are many problems in chaining different sets of methods, using traditional data mining methods in time series analysis and automation in series data preparation. The task of REFII model is linking and chaining different methods for time series analysis.

As a unique model of transformation REFII model provides an opportunity to use different concepts in time series analysis. As a result, we are able to make more complex analysis which demand chaining of methods for time series analysis, using traditional data mining methods on time series, and making new algorithms for solving contingency problem in the domain of time series analysis.

With REFII model as transformation model for time series, it is possible to create systems for automated series data preparation and systems for time series data mining based on different types of algorithms.

Instead of using many different incompatible methods for knowledge extraction from time series, with REFII model as a base it is possible to chain different conceptions, and using traditional data mining methods in time series. Main advantage of this approach is a possibility for constructing powerful time series query languages, which enables a unification of different methods, and construction of new algorithms for time series analysis.

4 The Definition of the REFII Model

In the following chapter term *functional gain* (changes in function output values) will be explained. We have a continuous function f on the I interval where a belongs to the I interval.

If t is such that

$$a + \Delta t$$

belongs to the I interval, then the value of

$$\Delta f = f(a + \Delta t) - f(a), a, a + \Delta t$$

is called the *function gain*, as can be seen in Fig. 1 [46–48].

The derivation of a continuous function at point t is defined as follows [46–48]:

$$f'(t) = \lim_{\Delta t \to 0} \frac{f(t + \Delta t) - f(t)}{\Delta t} = \frac{\Delta f}{\Delta t}$$

Indicators like function extremes and null points can be calculated by mathematically analyzing continuous functions.

In data mining and time series analysis in the field of detection of market regularities we come across time series represented with quantified values. Unlike the continuous function like the square function or the angled shot function, market phenomena are viewed in time intervals on the data level, mostly in uniform time intervals and are, like other functions that enter the data mining process of time series analysis, discrete functions. Similarly, business events and market activities

Fig. 1 Continuous function gain

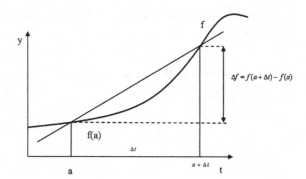

Fig. 2 Discrete gain function

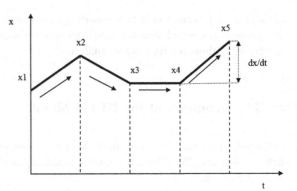

are also played out in discrete time intervals. Taking that into account, the REFII model and time series analysis data mining models are based on discrete functions as is shown on Fig. 2.

The REF component in the REFII model is defined as shown in Table 1.

While using the REFII model in light of the common logic, the REF codes signify the trends of growth.

Within the REFII discrete model, values are normalized to the <0, 1> interval. The following min–max normalization function is used:

$$x\_normalized = \frac{x\_denormalized - x\_min}{x\_max - x\_min}$$

where:

$x\_normalized$—normalized value within the <0, 1> interval
$x\_denormalized$—value to be normalized
x_max—maximum value of time series in question
x_min—minimum value of time series in question
Angular deflection coefficients are calculated by the formula:

$$\frac{dx}{dt} = |x\_normalized_t - x\_normalized_{t-1}|$$

Based on the angular deflection coefficients we supplement the REF model as is shown in Table 2.

Table 1 The definition of the REF codes based on the discrete gain function

dx/dt	REF code
$=0$	E
>0	R
<0	F

Table 2 Example of the REF model with a basic pattern

REF dx/dt	REF code	Angular deflection coefficient dx /dt	Angular deflection coefficient category (crisp)	Segment area (normalized)	Area category
=0	E	0	Equal	0–0.3	Small
				0.3–0.6	Medium
				0.6–1	Big
>0	R	0–0.2	Mild growth	0–0.3	Small
				0.3–0.6	Medium
				0.6–1	Big
		0.2–0.6	Medium growth	0–0.3	Small
				0.3–0.6	Medium
				0.6–1	Big
		0.6–1	Sharp growth	0–0.3	Small
				0.3–0.6	Medium
				0.6–1	Big
<0	F	0–0.2	Mild decline	0–0.3	Small
				0.3–0.6	Medium
				0.6–1	Big
		0.2–0.6	Medium decline	0–0.3	Small
				0.3–0.6	Medium
				0.6–1	Big
		0.6–1	Sharp decline	0–0.3	Small
				0.3–0.6	Medium
				0.6–1	Big

4.1 Area Below the Curve

When calculating the area below continuous functions, we use the integration function (Fig. 3).

The area of the segment is calculated according to the formula of numeric integration by the rectangle theory [46, 48].

$$p = \frac{(x_n * \Delta t) + (x_{n+1} * \Delta t)}{2}$$

The area expressed this way can be interpreted as the average quantitative value of a phenomenon between two intervals. For example, the area of the segment between two time intervals (for example, 2 months), where the amount of funds on an account is observed, can be interpreted as the average amount of funds on the account between two periods of observation (for example, months "May" and "June").

Fig. 3 Numerical integration

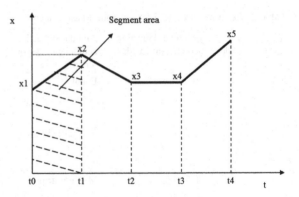

The area below the curve within the interval $<t_a, t_n>$ is calculated as the sum of segment areas according to the following formula:

$$P = \sum_{i=a}^{n} p_i$$

where the time distance, expressed as Δt, is equal for all time segments in the model and represents the measure of time which can be expressed as months, days, hours, minutes, etc.

The area on the level of time segments can also be normalized. When we introduce the area into the REFII model as the third element, we get the following table.

The highlighted values in the table show a sample of the basic pattern definition based on empirical values.

The table is also a unified schematic overview of basic patterns that constitute the REFII model and which relates the variance of the discrete function with the area, with the purpose of building the basic patterns.

The basic patterns can be analyzed separately for the purpose of analysis and can enter the process as separate values, or they can be viewed as a series of unrelated values.

The table also provides a form for segment classification.

4.2   The Definition of Basic Patterns

The REFII model combines the trends of discrete functions and the area on the time segment level and creates a basic pattern. The basic pattern is represented with three core values:

- Growth trend code (REF)
- Angular deflection coefficient that can be classified into categories with the aid of the classical crisp or fuzzy logic
- The time segment area can be classified into categories with the aid of the classical crisp or fuzzy logic

It is important to note that classification into categories within the REFII model depends on the nature of the issue being solved, which means that it is possible to use source values of the area and angular deflection coefficient in the analysis process, as can be seen in the empirical research portion of this paper.

Basic patterns can form complex structures of series of samples, and as such can be part of the analysis process. This procedure is visible in the empirical portion of the research that is focused on pattern recognition as well as similarity and motif detection.

The basic pattern defined through the REFII model is its fundamental part.

4.3 The Algorithmic Interpretation of the REFII Model

The final algorithm must integrate all of the three fields and create a basis for the application of analytical procedures.

The algorithm for time series transformation into the REFII model is done in several steps.

A time series can be defined as a series of values $S(s_1,...s_n)$ where S represents a time series and $(s_1,...s_n)$ represents the elements of series S.

Step 1: Time interpolation

Format of an independent time series $Vi(vi_1,...,vi_n)$. On the interval $<1...n>$ (days, weeks, months, quarters, years) with values of 0. It is necessary to implement the interpolation of values missing from $S(s_1,...,s_n)$. Based on the series Vi. The result of this process is the series S $(s_1,...,s_n)$ with interpolated values from the $Vi(vi_1,... vi_n)$ series.

Step 2: Time granulation

In this step we define the degree of summarization of the time series $S(s_1,...s_n)$ that is located within a basic unit of time (day, week, month...). Time series elements are summarized by using statistical functions like sum, mean, or mode on the level of granular slot. That way, the granulation degree of the time series can be increased (days to weeks, weeks to months) and the result is a time series $S(s_1,...s_n)$ with a greater degree of granulation.

We can return to this step during the analysis process to fulfill the analysis goals, and that includes the mandatory repetition of this process in the following steps:

Step 3: Normalization

The normalization procedure implies the transformation of a time series $S(s_1,...s_n)$ into a time series $T(t_1,...t_n)$, where each element of the array is subject to a min–max normalization procedure to the <0, 1> interval.

1. Time series T is made up of elements $(t_1,...t_n)$ where t_i is calculated as $t_i = \frac{s_i - \min(s)}{\max(s) - \min(s)}$, where $\min(s)$ and $\max(s)$ are the minimum and maximum values of time series T.
2. The time shift between basic patterns (a measure of time complexity) in a time slot on the X axis is defined as $d(t_i, t_{i+1}) = a$.

Step 4: Transformation to REF notation

According to the formula
$T_r = t_{i+1} - t_i$, $T_r > 0 \rightarrow R$; $T_r > 0 \rightarrow F$; $T_r = 0 \rightarrow E$; where the Y_i elements are members of the N_s series.

Step 5: Slope calculation based on the angle

Angular deflection coefficient:

- $T_r > 0 \rightarrow R$ Coefficient $= t_{i+1} - t_1$
- $T_r < 0 \rightarrow F$ Coefficient $= t_i - t_{i+1}$
- $T_r = 0 \rightarrow E$ Coefficient $= 0$

Step 6: Calculation of the area below the curve

Numerical integration by the rectangle theory:

$$p = \frac{(t_i * a) + (t_i * a)}{2}$$

Step 7: Creating time indices

Creating a hierarchical index tree depends on the nature of the analysis, where the element of the structured index can be located and an attribute such as the client's code.

Step 8: Category creation

Creating derived attribute values based on the area below the curve and the deflection angles.

It is possible to create categories by applying crisp and fuzzy logic.

Step 9: Connecting the REFII model's transformation tables with the relational tables that contain attributes with no time dimension.

These nine basic steps are the foundation of the algorithmic procedure underlying the REFII model whose ultimate goal is the formation of the transformation matrix. The transformation matrix is the foundation for performing future analytical procedures whose goal is time series analysis.

5   Pattern Recognition in Time Series

The inspiration for construction of algorithm for pattern recognition in time series comes from self-organizing maps.

The main idea is based on forming temporal grids for pattern filtration.

Initial length of observed pattern is one unit of time. After each cycle, safety coefficients are calculated and candidates with adequate amount of safety coefficient enter the next cycle. In the next cycle, the length of pattern is increased by one unit of time. The procedure is repeated as long as there are candidates in the next cycle or until length of pattern is equal to length of the time series.

This problem requires two-layer hybrid architecture of self-organizing maps and use of frequent-pattern tree authored by Han [28, 31, 36, 49–51]. Except for non-sequential patterns, Han uses frequent-pattern tree to search for sequential patterns [36, 51]. As it will be shown in this paper, frequent-pattern tree in seasonal detection model and in pattern recognition in time series are adapted to REFII model and objective of analysis (Fig. 4).

As can be seen from schematic diagram, the underlying strategy is to calculate values at each interval and use those values as elements of pattern comparison.

Unlike traditional approach to pattern recognition, very important element in discovering seasonal oscillation is the time detection of event and this element is part of analysis.

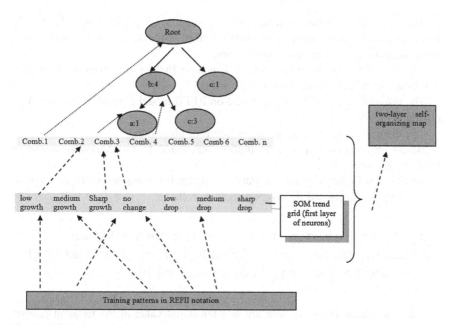

Fig. 4 REFII schematic diagrams

This element is declared as time index in REFII model and pattern occurrence is observed from perspective of the time indices and repeat cycles.

An important measure in analysis which affects the final result is reliability coefficient. It defines to which occurrence frequency is pattern taken as relevant t with respect to time index. For example, if reliability coefficient is 0.95 and "sharp growth" occurs in less than 95 % of cases with respect to the time index, the hypothesis of existence of seasonal oscillation for this time period will be rejected.

If hypothesis of existence of seasonal oscillation is accepted, generated pattern enters as candidate for the next processing.

Generating candidates and checking pattern relevance "2 or more" is performed because patterns "A" and "B," for which hypothesis of existence of seasonal oscillation is accepted, do not necessarily satisfy reliability coefficient criteria in pairs or for longer patterns.

For example, if patterns "A" and "B" of length 1, with time indices t and $t + 1$ have reliability coefficients 0.9 and 0.91, and the minimum required reliability coefficient is 0.9, reliability coefficient of a pair AB has to be calculated.

If reliability coefficient of a pair A and B of length 2 is at least 0.9, hypothesis of existence of seasonal oscillation "AB" on the interval $<t, t + 1>$ is accepted.

Furthermore, we can calculate probability of occurrence of pattern B if pattern A has occurred using the following formula for conditional probability:

$$P(B|A) = \frac{P(A \cap B)}{P(A)}$$

The algorithm generates candidates using time window principle [17, 35, 52] where potential candidates are generated—pattern of length 1 to $n - 1$, where n is the total number of elements in time series.

Classes written in the programming language Python are given in Appendix "Module for pattern recognition in time series."

The issue of pattern recognition based on REFII model is resolved as shown in diagram (Fig. 5).

The diagram represents simplified solution of algorithm.

Based on algorithm, program solution has made in programming language Python (see Appendix).

As a final solution program generates tree which is represented by layers of lists.

The tree is represented by layers of lists as follows:

```
Layer n: [head pattern, observed element of
pattern, occurrence frequency of the pattern,
[the whole pattern to layer n]]]
```

The constructor in the module *init* sets the initial value of the frequent-pattern tree, and auxiliary lists for processing.

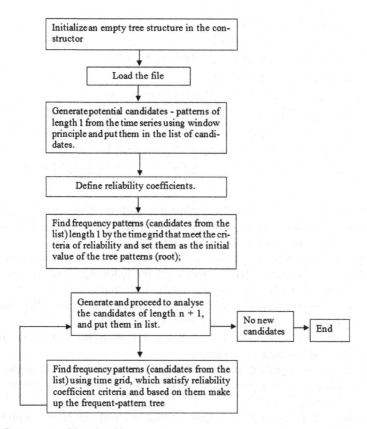

Fig. 5 Pattern recognition algorithm

Class *datoteka* reads the file and stores it in the initial list from which pattern will be generated.

The main class *prolaz* calls class *kroz_listu* that generates candidates based on the time window, and then calls the class *sito* that generates frequency patterns for the pattern of length 1.

Class kroz_listu carries out further testing using class *brisi_nefrekventne* and infrequent patterns are deleted calling the class *filtriraj_listu*.

Class *kroz_listu* further carries out pattern recognition through class *međuslojno_sito*.

Class *medjuslojno_sito* forms leaf of tree with the element of the tail if there is a "head" of the pattern is identical to the current pattern and there is no "tail" of the pattern as an element of the tree, and then initializes the counter of the pattern to 1. If there is no "head" of pattern identical to the current pattern and there is a "tail" of the pattern as an element of the tree, counter of the pattern is increased by 1.

A program that calls a module with classes and searches for patterns is given by the following instructions:

```
import REFII
stablo=Uzorci()
stablo.datoteka("niz.txt")
stablo.prolaz()
print "cvor", stablo.sloj0
print "prvi sloj ", stablo.sloj1
print "drugi sloj ", stablo.sloj2
print "treci sloj ", stablo.sloj3
print "cetvrti sloj ", stablo.sloj4
```

After loading modules using the statement import, creating objects and downloading files, program executes statement *stablo.prolaz* () that searches for patterns. Print command displays the layers of frequent-pattern tree.

The process of pattern recognition can be explained by the following example in which time series is given in REFII model with the following elements: (low drop, low growth, low drop, low growth, sharp drop, low drop).

The first step determines desired level of pattern reliability. After that, the occurrence frequency of pattern of length 1 is calculated.

Frequencies of occurrences are as follows: low drop $3/6 = 0.5$, low growth $2/6 = 0.33333$; sharp drop $1/6 = 0.1666$.

The next step is to generate patterns of length two. List of obtained candidates is (low drop–low growth, low growth–low drop, drop–low growth, low growth–sharp drop, sharp drop–low drop).

If required reliability coefficient is greater than or equal to 0.3, roots with values low drop and low growth are formed.

Once the list of candidates that in head of pattern have revealed frequent patterns of length 1 is isolated, frequency pattern of length 2 are calculated using self-organizing maps. As addition to shown example, self-organizing map can also include area as category among dimensions.

After calculating frequency of length 2, the following results are obtained: (Low drop–low rise $2/5 = 0.4$, low growth–low notch $1/5 = 0.2$, low growth–sharp drop in $1/5 = 0.2$).

Since reliability coefficient has to be greater than or equal to 0.3, pattern low drop–low growth $2/5 = 0.4$ is the only pattern that satisfies that condition and it enters in next cycles for generating candidates of length 3. The procedure runs as long as there are candidates in list whose reliability coefficient is greater than required reliability coefficient.

This algorithmic procedure is based on Han's postulates [28, 36], which states that if the head of the pattern does not satisfy reliability coefficient criteria, then entire pattern does not satisfy reliability coefficient criteria either. Saving time by reducing combinatorial explosion is achieved by using the time grid on the model of self-organizing maps, which in each subsequent step perform calculations of frequency patterns stored in frequent-pattern tree.

6 Results of Data Processing

After running described algorithms on empirical data sample of car incurred damage, for the reliability coefficient 0.2 the following results were obtained on a population time series of car accidents in the year 2013

```
Node [['0', '1', 158, ['1']], ['0', '4', 168,
['4']]]
```

```
First layer [['1', '4', 85, ['1', '4']], ['4',
'1', 88, ['4', '1']]]
```

Pattern tree based on results looks as it is shown in the diagram.
Reliability of the pattern is calculated using formula (Class *brisi_nefrekventne*):

```
the number of frequency pattern / the number
of candidates of patterns for testing,
```

where the number of patterns for testing is calculated as:

```
the number of candidates in list − (the number
of elements in pattern - 1).
```

The frequency, expressed as a value in the brackets in Fig. 6, indicates the occurrence of the whole pattern, where the number of levels represents the length of the sample. For example, the pattern "low growth" length of pattern 1 has a frequency of 158 occurrences and 43.40 % occurrence or reliability coefficient of 0.43, while the pattern "low growth"–"low drop" has a frequency of occurrences 85 and 23.41 % occurrence or reliability coefficient of 0.2341.
If reliability coefficient is decreased to 0.1, we obtain the following results:

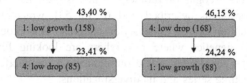

Fig. 6 Discovered patterns for reliability coefficient 0.2

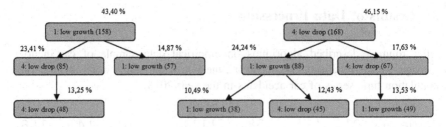

Fig. 7 Discovered patterns for reliability coefficient 0.1

Root [['0', '1', 158, ['1']], ['0', '4', 168, ['4']]]

First layer [['1', '1', 57, ['1', '1']], ['1', '4', 85, ['1', '4']], ['4', '1', 88, ['4', '1']], ['4', '4', 67, ['4', '4']]]

Second layer [['1', '1', 38, ['4', '1', '1']], ['4', '4', 48, ['1', '4', '4']], ['4', '1', 49, ['4', '4', '1']], ['1', '4', 45, ['4', '1', '4']]]

Results are shown in Fig. 7.

More branched tree can be obtained if reliability coefficient is decreased. The program is shown in Appendix "Module for pattern recognition in time series." Reliability and frequency of frequent-pattern tree is calculated as recommended by Han's work [28, 36] on forming a frequent-pattern tree. The population size is reduced in each new step in which the pattern length increases.

7 Mutation Recognition in Pattern

Thus resolved pattern recognition problem, which is presented using lists, is suitable for solving problems described in [26, 36, 53], which is referred to the searching of "mutation" in the time series patterns. This problem can be, for example, found in genetics where researchers are looking for gene mutation modalities or can be also found in telecommunications as interference estimation influence on signal time series and its diversifications.

Each of these papers in this area is reduced to its own incomplete modality of solutions that solves only that problem.

In terms of REFII model, we can answer to question of existence of mutations within patterns of the time series by analyzing frequent-pattern tree.

Basic procedure can be described as intervals comparison between beginning of sample ("head") and end of sample ("tail"). If those samples are identical, and middle part ("body") can be made of different samples, we are talking about mutations like shown in the next picture which represents empirical data which were used to show samples in the previous analysis (Fig. 8).

Algorithmic procedure is shown in Fig. 9.

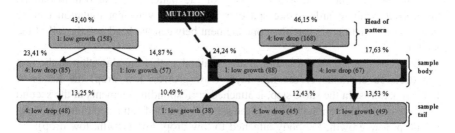

Fig. 8 Mutation recognition in pattern for reliability coefficient 0.1

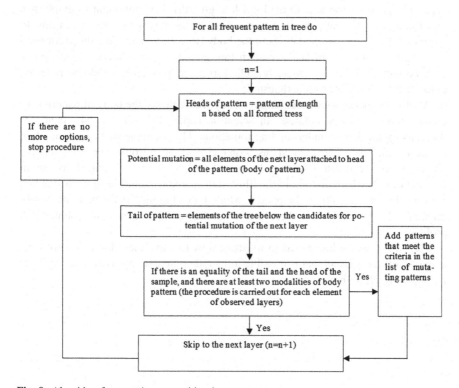

Fig. 9 Algorithm for mutation recognition in pattern

The terms head, tail, and body of pattern are related to hierarchy of the tree, as can be seen from the figure above.

Mutation recognition is reduced to comparing the interval equality of the head and tail of the pattern, where there may be n heads of pattern and n tails of pattern. If intervals comparison shows that they are identical then we can name all elements in between (if there is more than one modality) as mutations. In our example of car accidents, between modalities "low drop" as the head of the pattern and "low growth" as the tail of the pattern, there are inter-elements "low growth" and "low drop" called mutations of pattern. This solution sees mutation as a variation of one element between the head and the body of the pattern. The algorithm can be customized, with regards to the space problem, in a way that mutation can be viewed as a variation of more than one element between the head and the tail of the pattern.

Mutations that are shown in Fig. 8 are obtained so that the algorithm in the first step declared elements of the first layer (low growth, low drop) the heads of mutated patterns. In the same step, it attached corresponding elements in the second layer and declared them the bodies of the samples (the body attached to low growth: low drop, low growth; the body attached to low drop: low growth, low drop).

Furthermore, the tails are formed next to the body, next to body low drop was formed tail low drop, next to body slow growth was formed the tail low growth, next to body low drop was formed tail low growth). It is important to emphasize that the bodies and tails retain a connection to the heads of tree. According to algorithm, there are two modalities in the body (low growth and low drop) for head of pattern low drop, with tail of pattern low growth, so the pattern is declared as mutated pattern. In case of more branched tree, this procedure would be repeated until all the possibilities are exhausted.

As already mentioned, depending on analytical requests, the body of pattern may contain more than one element. In scientific papers [36, 49, 53, 54], there are different approaches to understanding mutations. This example starts with mutation estimation and "tails" are created based on elements taken from layer below sample "body" used to estimate mutation. In quoted papers are also explained options in forming of tails which, in light of the illustrated example, would include all attached elements below the body of the pattern. Mutation recognition in the patterns is very popular scientific field and there are a number of approaches for solving this problem.

The aim of this solution was to illustrate how to solve these kinds of problems using REFII model, and that is possible to include different approaches to this issue by adjusting algorithm.

8   The Integration of Data Mining Methods with Time Series

One of the most significant problems in the field of temporal data mining is to combine time series with traditional data mining. REFII model allows direct application of data mining algorithms on time series by its very conception, where the REFII model is bridge that links traditional data mining algorithms and time series.

The need to combine the traditional data mining methods with time series derives from the potential opportunities related to segmentation, estimation of influence of time and non-time component on trends of time series from which we can model behavior of market subjects and segments.

REFII model allows us to introduce time-independent attributes in time series analysis. The basic principle stems from the fact that trend of some phenomena (trend of timeslot) is affected by subjects that can be described by standard attributes (e.g., gender, age, region, income, level of education). It is necessary to carry out the procedure of expansion of timeslots with the nontemporal attributes so that these attributes could enter in the analysis process. Such data can be analyzed using statistical methods and traditional data mining algorithms. Expansion of timeslots is not a prerequisite for the application of data mining algorithms on time series. Analysis can be carried out using the elements of transformation matrix.

In integration of data mining methods and time series, current scientific activity is reduced to solving specific problems. Transformation model of time series is constructed with the aim to solve specific problem and uses one specific data mining method [55–58].

The current approach of using data mining methods in time series is reduced primarily to the use of clustering and conditional probabilities. The fundamental problem of these solutions is mutual incompatibility and solving particular problems, where the time series is transformed in order to apply specific data mining methods to achieve goals such as grouping (clustering). REFII model offers openness to data mining methods using current available algorithms hybridization procedure.

The following table shows some data mining methods and capabilities of analysis with respect to available elements of REFII model (Table 3).

In the table below, fields marked with "*" represent a link between data mining methods and elements of REFII model.

For example, standard statistical methods can be applied using angular deflection coefficients, area under the curve, and interval values of the model.

Clustering can be applied using angular deflection coefficients or area under the curve as elements of time series.

Difference between current available approach and approach using REFII is mostly related to data mining method selection depending on analysis goals. During that process researcher tends to make temporal data comparable (similar) to non-temporal data in terms of data mining usage. Using process called hybridization, current data mining methods are being adopted or can be used over temporal data in original form.

Table 3 Data mining methods and application to REFII model

	REFII				
	Primary transformation			Secondary transformation	
	Primary index	Area below curve (value/class)	Angular deflection coefficient (value/class)	Restructured index	Elements of secondary transformation (episodes, motives, series similarity)
Classic statistics	*	*	*	*	*
Associative algorithms	*	*	*	*	*
Clustering		*	*		*
Link analysis		*	*		*
Bayes network	*	*	*	*	*
Decision tree	*	*	*	*	*
SOM		*	*	*	*

For example, if we want to segment customers who responded to our offer and those who did not by age, gender, territory and similar attributes, there is no reason not to apply decision tree in order to segment the population.

If we want to examine what are the main factors that affect formation of certain detected episodes in the time series and we want to also include trends before/after examined episode, taking into account the nontemporal data such as age, sex, region, and similar characteristics of subjects who formed detected episodes, in classical approach we should form a separate model for solving this problem.

On the other hand, if we use REFII model, after extraction of episodes in time series, it is sufficient to apply the decision tree on elements of REFII model to determine main factors.

It is important to note that in the case of using decision tree for time series analysis and integration nontemporal elements in the model, it is necessary to expand the timeslots.

We can define time series sample expansion as joining original attribute values to with every unique case which affects trend by joining original trend values. Expansion can be illustrated as shown in the following figure (Fig. 10).

As in our example, REFII model tends to flexibility and problem solving according to the model shown in the figure, without creating specific solutions so that data mining methods could be applied.

The goal is to have an open system that is capable to absorb these specific algorithms applicable to REFII model in order to detect regularities in time series, but also strive to be opened to classical data mining algorithms.

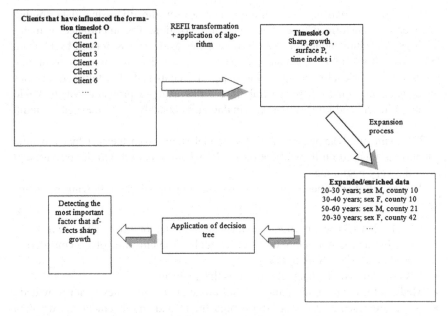

Fig. 10 Expansion of timeslots using nontemporal attributes

In a similar way as decision tree method is applied, Bayesian networks can be applied on the elements of primary and secondary transformations of time series, where, depending on requirement and purpose of the analysis, in the analytical process may enter classes of values and indexes created during the primary and secondary transformations.

Application of other methods on time series is shown in table.

It is hard to list all possibilities of integration and synergy effects between data mining models and REFII model. As already mentioned, they depend on ongoing analysis and problem space that needs to be solved. This approach allows us to achieve synergy effects based on linking bridges between two conceptions that, as an ultimate potential, provide combining time series and data mining methods without creating separate and mutually incompatible models.

9 Conclusion

The REFII model is designed to combine various concepts in time series analysis and traditional data mining methods, in order to construct a new algorithmic procedure in the field of market analysis and finding solutions to the problem of automatic preprocessing time series. These solutions can be realized using hybridization of existing data mining methods or directly using data mining methods on transformed time series in REFII notation. The importance of methodology procedures nonuniqueness in time series analysis was noticed by Graham Williams in his paper [39], which gives an overview of data mining methods in time series domain. He also noted in his paper the challenges for future researches (challenge questions) where general time series analysis problem, which would combine current and future researches in data mining analysis of time series, should be resolved.

The main disadvantage of Williams' model stems from the fact that too little attention is focused on transformation model which is placed into the background from the very beginning.

As a result of such neglecting of transformation model, the following problems arise:

- All problems arise because transformation model of time series analysis is inflexible and is not elaborated well enough. Problems are solved considering the purpose of analysis [39–45].
- Disregarding contingency approach to the analysis
- Methodology for linking temporal and nontemporal attributes is not provided
- As future research we can advise adjustment and improvement known data mining algorithms instead of using transformation as bridge toward comparison layer which also implicates general problem with fuzzy logic usage inside model

REFII model is a solution for such problems which are mentioned as results for future researches in Williams' work and gives much more flexible approach to time series analysis than the original idea represented by Williams.

The main goal of this research was to explore capability to upgrade REFII model and its capabilities of synthesizing existing data mining algorithms and time series analysis in a unique system for analyzing market behavior and market segmentation based on time series.

Through this work are given modalities upgrade of REFII model and its capability to synthesize existing methodological procedures in time series analysis in a unique system of data mining analysis of time series which combines data mining techniques, pattern recognition, motive recognition, derivation of mutation in algorithm, seasonal oscillations discovery, and event pattern recognition.

All these methods are performed on the same data set, using a unique transformation model of the time series.

Solutions are given for various modalities of market segmentation and market event pattern based on time series using REFII model on car insurance and database records. Unique transformation model of the time series (REFII) allows us to combine various conceptual models of time series analysis, which is an improvement compared to traditional way of using series of unrelated methods in analysis.

REFII model also allows applying traditional data mining methods in analysis of time series (decision tree, clustering, market basket analysis, etc.)

Combining various conceptual model of analysis within REFII model, we can solve much more complex analytical problems based on time series such as segmentation and detection of market behavior.

Chapter explains synergy from information usage which was extracted using iterative analytical methods and which are showing knowledge as follows:

Unlike traditional approach in which we use a series of unrelated and incompatible methods which sometimes do not allow us complex analysis and modeling solutions for nonstandard problem combining a number of different methodological approaches of analysis, REFII model provides an improvement in detection of market laws from time series during the "ad hoc" analysis of market problems.

Improvement conceptually means combining different methodological concepts of time series analysis, application of traditional data mining methods to the time series, combining methods for the time series analysis and upgrading the basic REFII model with basic algorithms in order to construct models for market segmentation on the basis of time series and estimation of behavior of market segments and subjects.

Deeper analysis conceptually means the capability to carry out different types of analysis on the reduced data set of time series, which comes as a result of processing before applying methods or analytical procedure.

Combining is achieved through a unique transformation model of time series which is the basis of analytical procedures, methods and development of new methods with respect to the problem space.

The advantage of this approach is the synthesis of different approaches and methods in the time series analysis which also include traditional data mining methods, as well as upgrade to new algorithmic analysis procedures.

This approach allows implementation of different types of analysis on reduced data set of time series which has already derived as a result of treatment, applied methods, or analytical method.

These options have so far been impossible or very difficult in traditional approaches to analysis of time series due to application of many unrelated method (general linking strategy did not exist), and appear as a necessity in the analysis of market principles through the time series.

Appendix: Module for Pattern Recognition in Time Series

```python
class Uzorci:
    def __init__(self):
        self.data = []
        self.stablo=[]
        self.nova=[]
        self.uzorak=[]
        self.sloj0=[]
        self.sloj1=[]
        self.sloj2=[]
        self.sloj3=[]
        self.sloj4=[]
        self.sloj5=[]
        self.sloj6=[]
        self.sloj7=[]
        self.sloj8=[]

    def add(self, x):
        self.data.append(x)
    def addtwice(self, x):
        self.add(x)
        self.add(x)
    def datoteka(self,dat):
        for line in open(dat):
            line=line.strip('\n')
            line=line.strip('"')
            self.add(line)

    def kroz_listu(self,duzina):

        for x in range(len(self.data)-(duzina-1)):

            self.uzorak=[]
            for y in range(x,x+duzina):
                self.uzorak.append(self.data[y])
                if duzina==1:
                    self.sito(self.data[y])

            if duzina==2:
                self.brisi_nefrekventne(self.sloj0,0.25,1)
                self.sloj0=self.filtriraj_listu(self.sloj0)
                self.medjuslojno_sito(self.sloj0,self.uzorak,self.sloj1)

            if duzina==3:
                self.brisi_nefrekventne(self.sloj1,0.1,2)
                self.sloj1=self.filtriraj_listu(self.sloj1)
                self.medjuslojno_sito(self.sloj1,self.uzorak,self.sloj2)

            if duzina==4:
                self.brisi_nefrekventne(self.sloj2,0.1,3)
                self.sloj2=self.filtriraj_listu(self.sloj2)
                self.medjuslojno_sito(self.sloj2,self.uzorak,self.sloj3)

            if duzina==5:
                self.brisi_nefrekventne(self.sloj3,0.1,4)
                self.sloj3=self.filtriraj_listu(self.sloj3)
                self.medjuslojno_sito(self.sloj3,self.uzorak,self.sloj4)

            if duzina==5:
                self.brisi_nefrekventne(self.sloj4,0.1,4)
                self.sloj4=self.filtriraj_listu(self.sloj4)
                self.medjuslojno_sito(self.sloj4,self.uzorak,self.sloj5)

    def prolaz(self):
        brojac=1
        while brojac<=6:
            self.kroz_listu(brojac)
            brojac=brojac+1

    def sito(self,uzorak):
        indikator=0
```

```
        for c in range(len(self.sloj0)):
            if self.sloj0[c][1]==uzorak:
                indikator=1
                self.sloj0[c][2]=self.sloj0[c][2]+1
        if indikator==0:
            pomoc=[]
            p1=[uzorak]
            p1.append
            pomoc=['0',uzorak,1,p1]
            self.sloj0.append(pomoc)

    def brisi_nefrekventne(self,l,ponavljanje,duz):
        for b in range(len(l)):
            if (float(l[b][2])/(len(self.data)-(duz-1)))<ponavljanje:
                l[b][2]=0

    def filtriraj_listu(self,l):
        lnova=[]
        for b in range(len(l)):
            if l[b][2]<>0:
                lnova.append(l[b])
        return lnova

    def medjuslojno_sito(self,lista_upper,uzorak,lista_izlaz):
        indikator=0
        for c1 in range(len(lista_izlaz)):
            if lista_izlaz[c1][1]==uzorak[-1] and lista_izlaz[c1][3]==uzorak:
                indikator=1
                lista_izlaz[c1][2]=lista_izlaz[c1][2]+1

        if indikator==0:
            for b1 in range(len(lista_upper)):
                if                         lista_upper[b1][1]==uzorak[-2]          and
lista_upper[b1][3]==uzorak[0:(len(uzorak)-1)]:
                    pomoc=[]
                    pomoc=[uzorak[-2],uzorak[-1],1,uzorak]
                    lista_izlaz.append(pomoc)
```

References

1. Patel P, Keogh E, Lin J, Lonardi S (2002) Mining motifs in massive time series databases. In: Proceedings of the 2002 IEEE international conference on data mining, Maebashi City, Japan, 9–12 Dec 2002
2. Perng CS, Parker DS (1999) SQL/LPP: a time series extension of SQL based on limited patience patterns. Technical report 980034 UCLA, Computer Science
3. Popivanov I, Miller RJ (2002) Similarity search over time series data using wavelets. In: Proceedings of the 18th international conference on data engineering, San Jose, CA, 26 Feb 26 Mar 1 to appear
4. Cha J, Cho BR (2014) Classical statistical inference extended to truncated populations for continuous process improvement: test statistics, P-values, and confidence intervals. In: Quality and Reliability Engineering International
5. Šošić I, Serdar V (1990) Uvod u statistiku, Školska knjiga-Zagreb
6. Apostolos-Paul R (1996) Neural networks in capital markets. Wiley, New York
7. Bisoi R, Dash PK (2014) A hybrid evolutionary dynamic neural network for stock market trend analysis and prediction using unscented Kalman filter. Appl Soft Comput 19:41–56
8. Taylor JG (1996) Neural networks and their applications. Wiley, London
9. Gaxiola F, Melin P, Valdez F, Castillo O (2014) Interval type-2 fuzzy weight adjustment for back propagation neural networks with application in time series prediction. Inf Sci 260:1–14
10. Jha GK, Sinha K (2014) Time-delay neural networks for time series prediction: an application to the monthly wholesale price of oilseeds in India. Neural Comput Appl 24(3–4):563–571

11. Kliček B, Zekić-Sušac M (2002) A nonlinear strategy of selecting NN architectures for stock return predictions, finance. In: Proceedings from the 50th anniversary financial conference svishtov, Bulgaria, ABAGAR, Veliko Tarnovo, pp 325–355, Apr 11–12 2002
12. Lin C-J (1997) SISO nonlinear system identification using a fuzzy-neural hybrid system. Int J Neural Syst 8(3):325–337
13. Dostál P, Zelinka I, Guanrong C, Rössler OE, Snasel V, Abraham A (2013) Nostradamus 2013: prediction, modeling and analysis of complex systems. In: Forecasting of time series with fuzzy logic, pp 155–161
14. Lin J, Keogh E, Patel P, Lonardi S (2001) Finding motifs in time series. In: 1st workshop on temporal data mining at 7 th ACM SIGKDD International conference of knowledge discovery and data mining, Edmont Alberta Canada, 27–30 July 27–30
15. Lin J, Keogh E, Patel, P, Lonardi S (2002) Clustering of time series subsequences in meaningless: implications for previous and future research. In: 2nd workshop on temporal data mining at 8 th ACM SIGKDD international conference of knowledge discovery and data mining, Edmont Alberta Canada, 23–26 July
16. Lin J, Keogh E, Lonardi S, Chiu B (2003) A symbolic representation of time series, with implications for streaming algorithms. In: Proceedings of the 8th ACM SIGMOD workshop on research issues in data mining and knowledge discovery, San Diego, CA, 13 June 2003
17. Heikki M, Hanu T, Verkamo I (1997) Discovery of frequent episodes in event sequences. University of Helsinki Finland, Report C-1997-15
18. Heikki M, Gunopulos D, Das G (2001) Finding similar time series. Technical Report D-2001-4, University of Helsinki Finland
19. Pratt K (2001) Locating patterns in discrete time series. University of South Florida, M.sc these, 2001
20. Bang, Y-K, Lee C-H (2008) Fuzzy time series prediction with data preprocessing and error compensation based on correlation analysis. In: Proceedings of the 2008 third international conference on convergence and hybrid information technology, pp 714–721, 11–13 Nov 2008
21. Fong S, Lan K, Wong R (2013) Classifying human voices by using hybrid SFX time-series preprocessing and ensemble feature selection. BioMed Res Int 2013:27
22. Pyle D (1999) Data preparation for data mining. Morgan Kaufmann Publishers, Inc., New York
23. Wu CL, Chau KW, Fan C (2010) Prediction of rainfall time series using modular artificial neural networks coupled with data-preprocessing techniques. J Hydrol 389(1):146–167
24. Sasaki H, Fujita S (2014) Pro-shareholder income distribution, debt accumulation, and cyclical fluctuations in a post-Keynesian model with labor supply constraints. Eur J Econ Econ Policies: Interv 11(1):10–30
25. Westphal C, Blaxton T (1998) Data mining solutions–methods and tools for solving real world problems. Wiley, New York
26. Lao W, Wang Y, Peng C, Ye C, Zhang Y (2014) Time series forecasting via weighted combination of trend and seasonality respectively with linearly declining increments and multiple sine functions. In: 2014 international joint conference on neural networks (IJCNN), IEEE, pp 832–837
27. Dougherty ER, Giardina CR (1988) Mathematical methods for artificial intelligence and autonomous systems. Prentice-Hall, Englewood Cliff
28. Han J, Pei J, Yin J (2000) Mining frequent patterns without candidate generation. In: Proceedings of ACM SIGMOID, pp 1–12
29. Wang C, Wang XS (2000) Supporting content-based searches on time series via approximation. In: Proceedings of the 12th international conference on scientific and statistical database management, Berlin, Germany, pp 69–81, 26–28 July 26–28 2000
30. Xsniaping G (1998) Pattern matching in financial time series data. http://www.datalab.uci.edu/people/xge/chart/
31. Han J, Kamber M (2011) Data mining-concepts and techniques. Morgan Kaufmann Publishers, San Francisco

32. Caraça-Valente JP, Lopez-Chavarrias I (2000) Discovering similar patterns in time series. In: Proceedings of the 6th ACM SIGKDD international conference on knowledge discovery and data mining, Boston, MA, pp 497–505, 20–23 Aug 2000

33. Chiu B, Keogh E, Lonardi S (2003) Probabilistic discovery of time series motifs. In: The 9th ACM SIGKDD international conference on knowledge discovery and data mining, Washington, DC, USA, pp 493–498, 24–27 Aug 2003

34. Craven MW (1997) Understanding time series networks: a case study in rule extraction. Int J Neural Syst 8(4):373–384

35. Das G, Gunopulos D, Mannila H (1997) Finding similar time series. In: Proceedings of principles of data mining and knowledge discovery, 1st European symposium, Trondheim, Norway, pp 88–100, 24–27 Jun 1997

36. Aggarwal CC, Han J (2014) Frequent pattern mining. Springer International Publishing, Switzerland

37. Klepac G, Kopal R, Mršić L (2014) Developing churn models using data mining techniques and social network analysis. IGI Global. doi:10.4018/978-1-4666-6288-9, ISBN13: 9781466662889, ISBN10: 1466662883, EISBN13: 9781466662896

38. Mršić L (2012) Decision support model in retail based on time series transformation methodology (REFII) and Bayes network. Doctoral thesis, Faculty of Humanities and Social Sciences, Zagreb, Croatia

39. Williams, JG, Weiqiang L, Mehmet AO (2002) An overview of temporal data mining. In: Simeon JS, Graham JW, Markus H (eds) Proceedings of the 1st Australian data mining workshop (ADM02), University of Technology, Sydney, Canberra, Australia, pp 83–90, ISBN 0-9750075-0-5

40. Williams JG, Weiqiang L, Mehmet O (2000) Temporal data mining using multi-level local polynomial models. In: Proceedings of the 2nd international conference on intelligent data engineering and automated learning (IDEAL00). Lecture Notes in Computer Science, vol 1983. Springer Hong Kong

41. Williams, JG, Weiqian L, Mehmet O (2001) Temporal data mining using hidden markov-local polynomial models. In: David C, Graham W, Qing L (eds) Proceedings of the 5th Pacific Asia conference on knowledge discovery and data mining (PAKDD01). Lecture Notes in Artificial Intelligence, vol 2035. Springer, Hong Kong, China

42. Williams, JG, Rohan B, Graham W, Hongxing H (2001) Feature selection for temporal health records, advances in knowledge discovery and data mining. David C, Graham W, Qing L (eds) Proceedings of the 5th Pacific Asia conference on knowledge discovery and data mining (PAKDD01). Lecture Notes in Artificial Intelligence, vol 2035. Springer, Hong Kong, China

43. Williams, JG, Weiqiang L, Mehmut O (2002) Mining temporal patterns from health care data. In: Proceedings of the 4th international conference on data warehousing and knowledge discovery (DaWaK02). Lecture Notes in Computer Science, vol 2454. Springer, Pages 221–231, ISBN 3-540-44123-9

44. Williams, JG (2003) Mining the data stream, invited plenary. In: International conference on hybrid intelligent systems melbourne, Australia, Dec 2003

45. Williams JG, Chris K, Rohan B, Lifang G, Simon H, Hongxing H, Chris R, Deanne V (2003) Temporal event mining of linked medical claims data. In: Proceedings of the PAKDD03 workshop on data mining for actionable knowledge DMAK-2003 Seoul, Korea, Apr 2003

46. Fanchi, JR (2000) Math refresher for scientists and engineers, 2nd edn. Wiley-IEEE Press, Hoboken

47. Javor P (1988) Uvod u matematičku analizu, Školska knjiga- Zagreb

48. Mardešić S (1977) Matematička analiza I, Školska Knjiga

49. Han J, Wang W, Yu SP, Yang J (2002) Mining long patterns in a noisy environment. In; ACM SIGMOID, Madison USA, June 2002

50. Han J, Xifeng Y, Ashfar J (2003) CloSpan: mining closed sequential patterns in large datasets. NSF IIS-02-09199, University of Illinois

51. Han J, Pei B, Mortazavi-Asl J, Wang H, Pinto Q, Chen U Dayal, Hsu M-C (2004) Mining sequential patterns by pattern-growth: the prefix span approach. IEEE Trans Knowl Data Eng 16(10):2004

52. Wu J, Wan L, Xu Z (2012) Algorithms to discover complete frequent episodes in sequences. New frontiers in applied data mining. Springer, Berlin, pp 267–278

53. Brazma A, Jonassen I, Vilo J, Ukkonen E (1998) Pattern discovery in biosequences. Lecture Notes in Artificial Intelligence, vol 1433. Springer, New York, pp 256–270

54. Wang W, Yang J, Yu P (2001) Mining long sequential patterns in a noisy environment. IBM Research Report 2001

55. Keogh E, Smyth P (1997) A probabilistic approach to fast pattern matching in time series databases. In: Proceedings of the 3rd international conference on knowledge discovery and data mining, Newport Beach, CA, pp 20–24, 14–17 Aug 1997

56. Keogh E, Pazzani M (1998) An enhanced representation of time series which allows fast and accurate classification, clustering and relevance feedback. In: Proceedings of the 4th international conference on knowledge discovery and data mining, New York, NY, pp 239–241, 27–31 Aug 1998

57. Keogh E, Chu S, Hart D, Pazzani M (2001) An online algorithm for segmenting time series. In: IEEE

58. Keogh EG (2011) Data, mining time series data. International encyclopedia of statistical science. Springer, Berlin, pp 339–342

A Soft Computing Approach for Targeted Product Promotion on Social Networks

Monika Bajaj, Shikha Mehta and Hema Banati

Abstract Soft computing techniques such as nature-inspired algorithms have always been a great source of inspiration for researchers in developing intelligent systems. One of the prominent nature inspired algorithms is firefly algorithm. The algorithm simulates the attraction approach of real fireflies where each firefly has specific agenda and coordinates with other fireflies in the group (swarm) to achieve the same. The presented work employs this attraction mechanism of firefly algorithm for product promotion at global scale through social networking sites. The algorithm is strategically employed to capture the user interest toward product features and explore the social network of prospective consumers to identify the best initial seeds for efficient product promotion. The strategy is divided into three phases. In the first phase, the market analysis phase, the often changing market demands for a product feature and user preferences for the same are captured. Based on these preferences users are grouped into homogeneous segments in the second phase, i.e., market segmentation. Thereafter, for targeted product promotion the most potential segment(s) with respect to the product to be promoted is selected in the third phase, i.e., targeted product promotion. Experimental studies are conducted to evaluate the performance of each phase individually and subsequently overall strategy is evaluated on epinion dataset. The results reveal the supremacy of firefly algorithm-based approach over other algorithms and substantiate the potential of presented plan to target wide and right range of audience by employing small fraction of advertising budget.

M. Bajaj (✉)
Department of Computer Science, University of Delhi, New Delhi, India
e-mail: mbajaj48@gmail.com

S. Mehta
Department of Computer Science and Engineering,
Jaypee Institute of Information Technology, Noida, India
e-mail: mehtshikha@gmail.com

H. Banati
Department of Computer Science, Dyal Singh College,
University of Delhi, New Delhi, India
e-mail: banatihema@hotmail.com

© Springer India 2016
S. Bhattacharyya et al. (eds.), *Hybrid Soft Computing Approaches*,
Studies in Computational Intelligence 611,
DOI 10.1007/978-81-322-2544-7_9

271

Keywords Soft computing · Evolutionary algorithms · Social networking · Opinion mining · Online product promotion · Targeted e-marketing

1   Introduction

Research is currently being directed toward simulation of human intelligence into mathematical models to solve large, complex, and dynamic real-world problems. The size and the complexity of these problems are beyond the capability of conventional (hard) computing such as mathematical optimization algorithms, direct search method, dynamic programming, etc. (Deb 2001). In addition to this, these techniques are neither flexible enough to adapt the algorithm accordingly nor capable to model the problem closer to reality (Janga 2006). The need to replace traditional time-consuming and complex techniques of hard computing with more intelligent processing techniques lead to the emergence of the term soft computing. Soft computing (Zadeh 1994) is defined as a fusion of intelligent paradigms that deal with the real practical situations in the similar way as humans deal with them. The main aim of soft computing is to exploit the tolerance for imprecision and uncertainty to achieve tractability, robustness, and low solution cost (Zadeh 1994). Its principal constituents are fuzzy systems, probabilistic reasoning, neural networks, and evolutionary computation. Soft computing is not just a mixture of these ingredients, but a discipline in which each constituent contributes a distinct methodology for addressing problems in its domain, in a complementary rather than competitive way. Fuzzy logic [1] refers to imprecision, approximate reasoning, and representation of aspects that are only qualitatively known. Probabilistic reasoning (Nillson 1986) is associated with uncertainty and belief propagation. Neural networks [2] focus on the understanding of neural network and learning system, self-organizing structures, and implementation of models from available data.

Evolutionary computation provides approaches to computing based on analogues of natural creatures. It is a term used to describe algorithms which were inspired by "survival of the fittest" or "natural selection" principles (Holland 1975). The evolutionary computation begins with problem-specific population. Each individual in the population represents a randomly generated solution to the problem and has fitness value that is determined by the value of the function to be optimized. These individuals evolve over generations and reproduce better solutions. During each generation new population is produced by applying operators inspired by nature and biological processes. This iterative process continues till the convergence criteria are satisfied. Evolution in the evolutionary computation is carried out with the operators borrowed from natural genetics such as selection, recombination, mutation, neighborhood, locality, migration, and attraction. Based on these operating processes the evolutionary computation is broadly classified into evolutionary algorithm (EA) and swarm intelligence (SI) algorithms. In evolutionary algorithms each individual in the population is called chromosome. These chromosomes evolve over generations and produce new population by applying crossover and mutation operators on current population. The idea behind crossover is that two or more individuals with high fitness values will create one or more new offspring (children) that inherits the features of their parents. Mutation is another reproduction operator used for finding new points in the search space. It is performed by randomly selecting one chromosome from the population and then arbitrarily changing some of its information. It is generally used with a crossover operator to diversify the population. It helps in avoiding premature convergence to a suboptimal solution by introducing new genetic material to the evolutionary process. In contrast to this swarm intelligence is a term used to describe the algorithms and distributed problem-solvers which is inspired by the cooperative group intelligence of swarm or collective behavior of insect colonies and other animal societies [3]. Similar to evolutionary algorithm swarm intelligence models are population-based methods. The system is initialized with a population of individuals (i.e., potential solutions). These individuals then evolve over many generations by way of mimicking the social behavior of insects or animals, in an effort to find the optimum solution. Unlike EA, SI models do not use evolutionary operators such as crossover and mutation. Each individual adapts itself according to its relationship with other individuals in the population and the environment. The main paradigms of nature and biological inspired algorithms include genetic algorithm (GA), differential algorithm (DE), ant colony optimization (ACO), particle swarm optimization (PSO), artificial bee colony (ABC), firefly algorithm (FA), etc. Due to their immense capability to model complex and dynamic real-world problems in diverse domains these algorithms have been used in varied applications of science and engineering [4, 5, 6], management [7], marketing [8], and finance [9]. Researchers also studied the potential of evolutionary computation for various web-based applications such as web page designing [10], information retrieval [11, 12], and web mining to extract the user interest [13], for personalization [14] and intelligent e-business [15] etc.

This chapter presents online product promotion plan that employs an evolutionary approach to optimize the resources and spread the influence at global scale. One of the most promising online communication medium currently used for product awareness at low cost is the social media. Social media has emerged as a common means of connecting and communication with the outside world. It allows the web users to share their knowledge and experiences about the products or services with millions of users across the globe. These interactions in the form of online discussions, expressing opinions, chats, blogs, etc., are considered as an electronic word of mouth (eWOM) and have the potential to influence large number of users. This aspect of information shared over social media is tapped by organizations for market campaigning. The next section explores the significance of social media to enhance the marketing returns. The third section discusses firefly a nature inspired algorithm used for optimization. Section 4 presents the firefly-based online marketing strategy. The chapter concludes by highlighting the experimental results and the contribution of the work.

2 Social Media

Social media as a concept was first introduced in 2004 when LinkedIn (https://in.linkedin.com/) created its social networking application. The application was primarily an online technology tool that allowed people to communicate easily and share information over Internet. Social media [16] is defined as a group of Internet-based applications built on the technological foundations of Web 2.0 that allows the creation and exchange of user-generated content. Kietzmann and Hermkens [17] defined social media with respect to seven (07) prime functional building blocks. These parameters are identity, conversation, sharing, presence, relationship, reputation, and groups. Each parameter represents a specific facet of social media and its implications for firms. With growth of Internet and related network technologies the popularity of social media has increased manyfold. It has emerged as an ideal platform for interactions, experience sharing, and recommendation among electronic peers. Sharing of information and relationship over social network provides rich source of information about user interest, preferences, and friends. It allows the user to indicate whom they trust and distrust, creating links in the network. It has been observed that, in online systems, people chatting with each other (using instant messaging) are more likely to have a common interest (their web searches) [18, Beth et al. 1994; Kautz et al. 1997]. This proliferating significance of social media data motivated various researchers to utilize social media as an e-marketing tool to boost e-commerce.

Social media marketing refers to the process of gaining attention of web users through social media sites (Mangold and Faulds 2009). The social media-based marketing strategies create and distribute content to attract attention and encourage readers to share it with others. Social media has become a hybrid element of promotion mix because in a traditional sense it enables companies to talk to their customers, while in a nontraditional sense it enables customers to talk directly to one another (Mangold and Faulds 2009). Thus strategic use of social media improves company's profitability and its position in the industry by facilitating brand popularity (Vries et al. 2012), sales promotions (Bonilla-Warford 2010), customer relationship (Sashi 2012), find new customers (Weston 2008), and other marketing practices (Samuel 2012). It also provides an inexpensive platform for communicating a marketing message (Skul 2008). Referral Web (Kautz et al. 1997) was one of the first systems that suggested the combination of direct social relations and collaborative filtering (CF) to enhance searching for documents and people. Analysis of social bookmarking site "Delicious" revealed high similarity between the tag vector of a URL and its keyword vector extracted from the corresponding web page (Guo and Zhao 2008). Sinha and Swearingen (2001) established that recommendations from friends are more preferable to collaborative filtering recommender systems in terms of quality and usefulness. It was also emphasized that tag-based profiles produce better personalized recommendation of tracks within the popular music portal "Last.Fm" than conventional ones. Vatturi et al. (2008) presented a personalized bookmark recommendation system using a CF approach that leveraged tags. It was based on the assumption that users would be interested in pages annotated with tags similar to ones they have already. Studies also established that incorporation of explicit social network information in CF systems improves the quality of recommendation in various domains such as music (Konstas et al. 2009), clubs (Groh and Ehmig 2007), and news stories (Leman 2007). However, all these studies extracted user preferences for a product by observing the explicit social network information. Users often indicate their interests and preferences via opinions and reviews, expressed for a product or service. These reviews express the strengths and weaknesses of each product with respect to functionality of its features or attributes thus cause direct or indirect influence spread over network [19]. This social influence is tapped by companies to market new products by maximizing the influence spread through the effect of "word of mouth" in the social network [20, 21, 22]. The "word of mouth" based marketing strategies are much more cost-effective as compared to conventional methods because the promotional efforts are carried out by the customers themselves [20]. These models are based on the assumption that consumers propagate only positive information about the products [23]. One of the prominent strategies that adopt similar kind of methodology for promoting new product is "viral marketing."

Viral marketing is defined as "marketing technique that seek to exploit preexisting social networks to produce exponential increases in brand awareness, through processes similar to the spread of an epidemic" [24]. It targets limited number of initial users called seeds and distribute free or discounted product to them. These seeds share their experiences with their friends who then pass on the message in their circle. Thus the message takes the form of a virus that spread through contacts with others and over a period of time it covers a substantial portion of network. Thus these seeds activate a chain reaction of influence that is driven by eWOM on social network (Fig. 1).

The impact of this influence spread depends upon the sum of probability of seeds to influence their acquaintances and in turn their influence on others and so on recursively until the entire network is covered.

For example according to Fig. 2a if user "I" is selected as a seed the message will pass to only user C. In contrast to this if user "D" is selected as a seed the strategy will be able to cover the whole network as shown in Fig. 2b. Apart from the connectivity of seeds other important parameters which can impact the influence

Fig. 1 eWOM on social network

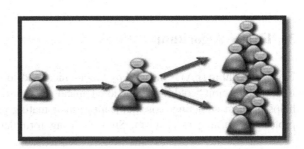

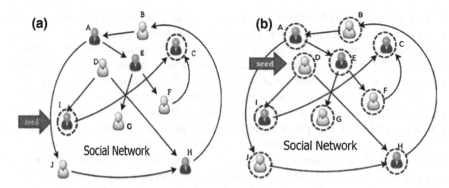

Fig. 2 a Social network of seed "I." **b** Social network of seed "D"

spread are the "*liking*" and the "*willingness*" to purchase the product. Seeds possessing high connectivity but a severe dislike for a product can have a negative impact on the market and hence marketing to them should be avoided [21]. Dissatisfied customers disseminate their bad experiences and feelings which are more powerful as compared to the positive one [25].

Therefore a need is felt for a strategic approach which not only catapults the users to the product but also builds a positive rapport for the product in an optimal manner. The approach should be in synchronization with the needs of an effective marketing activity; optimum usage of resources such as cost, human resources, etc. Hence the issue then filters down to identifying the set of users who can spread the maximum influence with an optimum usage of resources. The selection of targeted users, so as to maximize the net profit is a well-defined optimization problem and has been proved an NP-hard problem [22]. Thus targeting a specific group of customers with their appropriate interests is also an optimization problem. In addition, nature-inspired algorithms (Zhang et al. 2010) have become an indispensable tool for solving such optimizations. Notable nature-inspired algorithms which have been applied for such scenarios are feature selection [26–28], clustering [3, 7, 29, 30, Murthy and Chowdury 1996], etc. This chapter explores a prominent nature-inspired algorithm, "Firefly algorithm," for solving the optimization problem presented above. The following section presents the basic behavior of firefly algorithm.

3 Firefly Algorithm

Firefly algorithm (FA) is inspired by biochemical and social aspects of real fireflies. Real fireflies emit "cold light" due to chemical reaction called bioluminescence that helps them in attracting (communicating) their mating partners and also serves as protective warning mechanism. Shin Xi Yang formulated this flashing behavior with the objective function of the problem to be optimized [31]. The basic formulation of FA algorithm is based on the following three assumptions:

- All fireflies are unisex so that fireflies will attract each other regardless of their sex.
- Attractiveness is proportional to their brightness, thus for any two flashing fireflies, the less brighter one will move toward the brighter one. The attractiveness is proportional to the brightness and both decrease as distance increases between two flies. In case it is unable to detect brighter one it will move randomly.
- The brightness of a firefly is determined by the landscape of the objective function.

Let us consider an optimization problem where the task is to minimize objective function $f(x)$ for, i.e., find x^* such as Eq. (1)

$$f(x^*) = \min_{x \in S} f(x) \tag{1}$$

The swarm of N fireflies are initially dislocated in search space S (randomly or employing some deterministic strategy). Each firefly x_i, $i = 1, 2, \ldots, N$ represent a solution for problem at given iteration that determines fitness $f(x_i)$. Each firefly finds its mating partner based on the attractiveness of other members in the space and move toward that partner in order to improve its fitness. The attractiveness β is proportional to the light intensity seen by adjacent fireflies and is defined as Eq. (2).

$$\beta = \beta_0 e^{-\gamma r_{ij}} \tag{2}$$

where β_0 is the attractiveness at $r = 0$; $r_{ij} = d(x_i, x_j)$ is a distance between two fireflies i and j calculated as Eq. (3).

$$r_{ij} = \left\| x_i - x_j \right\| = \sqrt{\sum_{k=1}^{d} (x_{i,k} - x_{j,k})^2} \tag{3}$$

A fixed length absorption coefficient γ characterizes the variation of attractiveness value of firefly and determining the speed of convergence. In theory, $\gamma \in [0, \infty]$, but in practice γ is determined by the characteristics of the problem to be optimized. The movement of ith firefly toward the more attractive (brighter) jth firefly is determined by Eq. (4)

$$x_i = x_i + \beta_0 e^{-\gamma r_{ij}} (x_j - x_i) + \alpha \left(rand - \frac{1}{2} \right) \tag{4}$$

Fig. 3 Firefly algorithm

Objective function $f(x)$, $x = (x_1, \ldots, x_4)^T$
Generate initial population of fireflies x_i (i=1,2,...,n)
Light intensity I_i of x_i firefly is determined by $f(x_i)$
Define light absorption coefficient γ
While (t>MaxGeneration)
   for i=1:n all n fireflies
     for j=1:i all n fireflies
       if (I_j>I_i),
          Move firefly i towards j in d-dimension; using Equation 4
      Evaluate new solution and update light intensity; using Equation 2
     end for j
   end for i
end while

It considers two factors first, the attractiveness of jth firefly. Second is a randomization with $\alpha \in [0, 1]$. Figure 3 depicted the algorithmic step for firefly algorithm.

The optimization capability of firefly algorithm has already been established in various applications Hönig (2010), applied firefly algorithm for task graph scheduling. Jati and Suyanto (2011) addressed the application of firefly algorithm (FA) to traveling salesman problem (TSP). Khadwilard et al. (2011) employed Firefly Algorithm (FA) for solving job shop scheduling problem and Senthilnath et al. (2011) measured the performance of FA with respect to supervised clustering problem.

The presented work employs the attraction mechanism of firefly algorithm to dynamically capture the user interest toward product features. The algorithm is strategically employed to explore the social network of prospective consumers to identify the most influential customers termed as "initial seeds" for effective product promotion. The next section explains the firefly algorithm-based product promotion plan on social networks.

4   Frame Work for Targeted Product Promotion on Social Networks

The presented product promotion plan [32] captures the preferences and interest of the users to identify the group of users having substantial interest in a product. The social connectedness of these users helps to identify the most influential users in the group. Targeting these users helps the organization in building a positive image of the product in the group. The complete strategy for targeted product promotion on social networks (Fig. 4) is divided into following three distinct phases:

- Market analysis
- Market segmentation
- Targeted product promotion

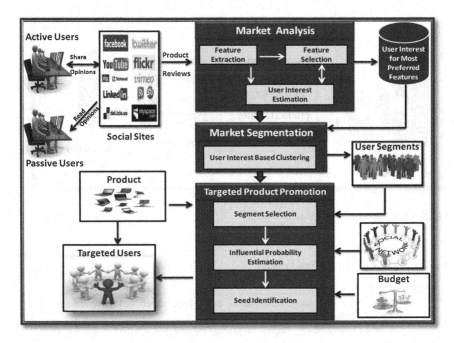

Fig. 4 Framework for targeted product promotion on social networks

4.1 Market Analysis

The market analysis phase of the presented strategy captures the changing market demands for a product feature and user preferences for these. The user preferences are subject to frequent change due to technological advancements in due course of time. For example, a decade ago the size of a mobile phone was a primary decisive factor during its purchase. However, with evolution in mobile technology the focus has shifted toward its camera pixels or audio quality (Fig. 5).

Due to the advent of online e-commerce, users tend to project their product preference online through product opinion/reviews. The approach works on mining these online product opinions/reviews for extracting the product features that are of interest to prospective customers. The following sections discuss the techniques used to extract the relevant feature set.

4.1.1 Syntactic Analysis-Based Product Feature Extraction

Online reviews for a product are expressed in varying ways. Table 1 shows the pattern of reviews extracted from epinion.com site along with rating given by the users for a product "digital camera." These reviews express the strengths and weaknesses of each product with respect to functionality of its features or attributes.

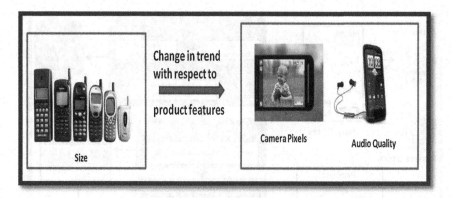

Fig. 5 Changing market demands

Table 1 Review pattern for digital camera

User_id	Reviews
U$_1$	**Pros**: Price, size, image stabilization, looks, large screen, fast operation, great photos, dedicated view button **Cons**: Slightly blurry corners, no manual control over aperture or shutter speed
U$_2$	**Pros**: Simple to use, Great photos, Auto mode does a great job with low price, compact size **Cons**: Slight shutter lag, Trouble taking clear photos in low light
U$_3$	**Pros**: intuitive menu, nice LCD, good picture quality, extremely portable, good ISO 400 images, fashionable **Cons**: blurry edges, minimal flash, difficult to press buttons
U$_4$	**Pros**: Excellent picture quality and flexibility **Cons**: While light, it will not easily fit in pockets

The features are described either explicitly or implicitly in the review sentences. For example, in a review of digital camera: *"Excellent picture quality, and flexibility." the* user explicitly express opinion about the "picture quality" and the "flexibility" of camera. In contrast to an expression such as *"While light, it will not easily fit in pockets."* describes the feature "weight" and "size" but these words are not mentioned in the sentence. These implicit features are hard to identify and need semantic understanding of language. It is often observed that reviewers usually provide explicit reviews rather than implicit ones. The presented feature extraction approach therefore focuses on features that appear explicitly in the review sentences. The process is initiated by collecting a set of reviews on specific product as the input and preparing them for further processing.

- Preprocessing

Each collected review is preprocessed, that includes removal of ellipsis (…) and arrangement of one review per sentence. Product features are usually noun or noun phrases in review sentences, hence each review is parsed through part-of-speech (POS) tagger to extract these terms from a sentence. The linguistic parser "NLProcessor" [33] is used to parse each sentence. It yields the part-of-speech tag for each word (whether the word is a noun, verb, adjective, etc) and identifies simple noun and verb groups (syntactic chunking) by generating XML output. For instance, when a sentence *"Excellent picture quality, and the flexibility"* is passed to POS it specifies the role/kind of each word in sentence as shown in Fig. 6 where <W C = 'NN'> indicates a noun and <NG> indicates a noun group/noun phrase that contains one or more noun terms. The XML output generated by NLProcessor is further processed.

- Syntactic Rules

The features explicitly mentioned in review sentences are either nouns or noun phrases and mostly follow an adjective or determinant such as "good picture quality" or "the flexibility." Moreover, these features are usually expressed either as single word such as resolution, size, price, design, etc., or as a combination of two or more words, e.g., "picture quality," "screen resolution," etc. Each noun group (NG) is therefore analyzed to count the number of elements in each NG. Based on this count specialized syntactic rules are formulated to extract the various NN/NNS terms [34].

One-word features are extracted through (rule 1 and 2) depicted in Fig. 7. Similarly rule 3 and 4 are formulated to extract two-word features depicted in Fig. 8. Figure 9 depicts the possible phrases that can be handled with these rules. The terms extracted by applying these rules are the candidate features and are stored in two separate databases, viz., Database1 (DB1) to store one-word features and database2 (DB2) to store two-word features.

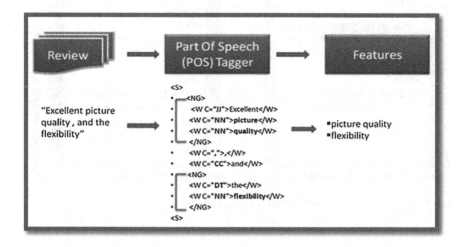

Fig. 6 Functioning of POS tagger

Fig. 7 Syntactic rules for one-word feature extraction

Rule 1:
if count ==1
    if first element tag == NN/NNS
        DB1 = DB1 + first element
Rule 2:
if count ==2
    if first element tag == JJ/JJS/DT
        if second element tag == NN/NNS
            DB1 = DB1 + second element

Fig. 8 Syntactic rules for two-word feature extraction

Rule 3:

if count ==2
    if first element tag == NN/NNS
        if second element tag == NN/NNS
            DB2 = DB2 + first element second element
Rule4:
if count ==3
    if first element tag == JJ/JJS/DT
        if second element tag == NN/NNS
            if third element tag == NN/NNS
            DB2 = DB2 + second element third element

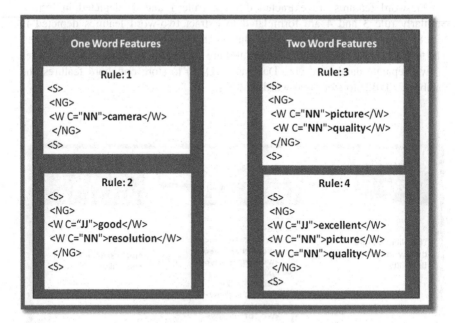

One Word Features

Rule: 1
```
<S>
<NG>
<W C="NN">camera</W>
</NG>
<S>
```

Rule: 2
```
<S>
<NG>
<W C="JJ">good</W>
<W C="NN">resolution</W>
</NG>
<S>
```

Two Word Features

Rule: 3
```
<S>
<NG>
<W C="NN">picture</W>
<W C="NN">quality</W>
</NG>
<S>
```

Rule: 4
```
<S>
<NG>
<W C="JJ">excellent</W>
<W C="NN">picture</W>
<W C="NN">quality</W>
</NG>
<S>
```

Fig. 9 Example of syntactic analysis based feature extraction

Extracted Features
aperture, battery, body, button, edge, flash, image stabilization, issue, light, look, menu, movie mode, operation, photo, picture quality, playback, price, review, screen, screen display, shot, shutter lag, shutter speed, size, speed, view button, view finder.

Fig. 10 Pattern of extracted product features

Every candidate feature in DB2 is in the form of XY where X and Y are two consecutive nouns. This eliminates the overhead of determining the co-occurrence between words and distance between them. A further pruning of each candidate feature is done to eliminate stop words, correction of spellings and lemmatization. However, all relevant features extracted from syntactic analysis-based feature extraction approach may not be important, hence considering all of them for further processing will be cumbersome (Fig. 10). A dimensionality reduction approach is, therefore, applied to select only the most significant features. The prominent data mining method, feature selection, is used to reduce the dimensionality of data.

4.1.2 Feature Selection

Feature selection is a technique to select optimal subset of features that represent the original features in problem domain with high accuracy [35]. The feature selection process can proceed in either forward (forward selection method) or backward direction (backward selection method) [36]. The forward selection method [36] begins with blank set and progresses by adding one feature at each step, reducing the error in each step, till any further addition does not significantly decrease the error. The backward selection method [36], as the name suggests, begins with set of all features and reduces error by eliminating one feature at each step till any further removal change the error significantly. One of the most important mathematical tools to find a subset (termed as reduct) of the original features that characterizes the basic properties of original features in problem domain is Rough Set Theory (RST) [37–40]. Rough set is approximation of a vague concept by a pair of precise concepts, called lower and upper approximations, which are informally a classification of the domain of interest into disjoint categories [38]. Thus objects belonging to the same category characterized by the same attributes are not distinguishable. It discovers data dependencies and reduces the number of attributes contained in dataset by purely structural methods [41, 42].

Rough set-based attribute reduction (RSAR) [41] approach provides a theoretical background to feature selection problem. It employs a rough set to remove redundant conditional attributes from discrete-valued datasets, while retaining their information content. The reduction of attributes called "*Reduct*" can be computed either by considering the degree of dependency or with discernibility matrix. Table 2 presents an example dataset for computing reduct on the basis of degree of dependency.

Table 2 Example Dataset

$x \in U$	A	B	C	D	E
0	1	0	2	2	0
1	0	1	1	1	2
2	2	0	0	1	1
3	1	1	0	2	2
4	1	0	2	0	1
5	2	2	0	1	1
6	2	1	1	1	2
7	0	1	1	0	1

Let $I = (U, A)$ be an information system where:

U—be a non empty finite set of *objects* called the universe,
A—be a non empty finite set of *attributes*,
V—be a set of attribute values such that $a: U \rightarrow V_a$ for every $a \subset A$.

Any $P \subseteq A$ determines an equivalence relation over U that is defined as an indiscernibility relation and denoted as $IND(P)$ and can be computed using Eq. (5).

$$INP(P) = \{(x, y) \subset U^2 | \forall a \subset Pa(x) = a(y)\} \tag{5}$$

The partition of U, generated by $IND(P)$, i.e., U/P and is calculated by Eq. (6).

$$U/P = \otimes \{a \subset P : U/IND(\{a\})\} \tag{6}$$

where
$A \otimes B = \{X \bigcap Y : \forall X \subset A, \forall Y \subset B, X \bigcap Y \neq \emptyset\}$

If $(x.y) \subset IND(P)$, then x and y are indiscernible by attributes from P. The equivalence classes of the P-indiscernibility relation are denoted as $[x]_p$. Referencing to Table 2, if $P = \{b, c\}$, then objects 1, 6, and 7 are indiscernible as are objects 0 and 4. $IND(P)$ creates the following partition of U

$U/IND(P) = U/IND(\{b\}) \otimes U/IND(\{c\})$
$\{\{0, 2, 4\}, \{1, 3, 6, 7\}, \{5\}\} \otimes \{\{2, 3, 5\}, \{1, 6, 7\}, \{0, 4\}\}$
$\{\{2\}, \{0, 4\}, \{3\}, \{1, 6, 7\}, \{5\}\}$

Let $X \subseteq U$. X can be approximated using only the information contained within p by constructing the P-*lower* approximation and P-*upper* approximation of X using Eqs. (7) and (8) respectively.

$$\underline{P}X = \{x | [x]_p \subseteq X\}. \tag{7}$$

$$\overline{P}X = X | [x]_p \cap X \neq \emptyset \tag{8}$$

Let P and Q be equivalence relations over U, then the positive regions can be defined by Eq. (9)

$$\text{POS}_p(Q) = \bigcup_{x \in U/Q} \underline{P}X \tag{9}$$

The positive region contains all objects of U that can be classified to class of U/Q using the information in attributes P. For example, let $P = \{b, c\}$ and $Q = \{e\}$, then $\text{POS}(Q) = \bigcup\{\emptyset, \{2, 5\}, \{3\}\} = \{2, 3, 5\}$.

An important issue in data analysis is discovering dependencies between attributes. Intuitively, a set of attributes Q depends totally on a set of attributes P, denoted $P \Rightarrow Q$, if all attribute values from Q are uniquely determined by values of attributes from P. If there exists a functional dependency between values of Q and P, then Q depends totally on P. Using the definition of positive region, the rough set degree of dependency of a set of attributes Q on a set of attributes P is defined in the following way:

For $P, Q \subset A$, it is said that Q depends on P in a degree k ($0 \leq k \leq 1$), denoted $P_k \Rightarrow Q$, and computed by Eq. (10).

$$k = \gamma_p(Q) = \frac{|\text{POS}_p(Q)|}{|U|} \tag{10}$$

If $k = 1$, Q totally depends on P, if $0 < k < 1$ Q depends partially (in a degree k) on P, and if $k = 0$ then Q does not depend on P. In the example, the degree of dependency of attribute $\{e\}$ from the attributes $\{b, c\}$.

$$k = \gamma_{b,c}(e) = \frac{|\text{POS}_{b,c}(e)|}{|U|}$$

$$k = \frac{|2, 3, 5|}{|0, 1, 2, 3, 4, 5, 6, 7|} = \frac{3}{8}$$

The reduction of attributes is achieved by comparing equivalence relations generated by sets of attributes. Attributes are removed so that the reduced set (termed as reduct) provides the same predictive capability of the decision feature as the original. A *reduct* is defined as subset R of the conditional attribute set C such that $\gamma_R(D) = \gamma_C(D)$. A given dataset may have many attribute reduct sets, so the set R of all reduct is defined by Eq. (11)

$$R = \{X : X \subseteq C, \gamma_X(D) = \gamma_C(D)\} \tag{11}$$

The intersection of all the sets in R is called the *core*, the elements of which are those attributes that cannot be eliminated without introducing more contradictions to the dataset. In RSAR, a reduct with minimum cardinality is searched for; in other words an attempt is made to locate a single element of the minimal reduct set $R_{\min} \subseteq R$ (Eq. 12)

$$R_{\min} = \{X : X \subset R, \forall Y \subset R, |X| \leq |Y|\} \tag{12}$$

The most basic solution to obtain minimal reduct is to generate all possible reducts and choose any one with minimal cardinality. Exploring all the possible reducts and selecting the best one is an expensive task and a practically impossible for large dataset and this needs and exhaustive search. Therefore heuristic and random search strategies-based methods such as quick reduct [41] and entropy-based reduct method [35] are capable of avoiding such complexity. These methods find reduct in less time but do not guarantee to find a minimal and optimal reduct. They generate a close-to-minimal reduct, thus reducing dataset dimensionality significantly. Therefore efforts have been directed to develop a stochastic approach by incorporating the nature inspired algorithm such as genetic algorithm [43], ant colony optimization [42], particle swarm optimization [28], and bee colony optimization [27] into RSAR approach to improve the performance. These methods did improve the performance but consumed more time [26, 27]. All these existing methods achieved success in either of two critical aspects of feature selection problem, viz., the degree of optimality (in terms of subset size and corresponding dependency degree) and time required to achieve this optimality. [44] improved the efficiency of basic RSAR by incorporating the flashing behavior of firefly algorithm. Figure 11 presents the FA_RSAR algorithm that takes C, the set of all conditional features

```
void FA_RSAR(C,D)
{
    Generate initial population of fireflies x_i (i=1,2,...n) corresponding to each conditional feature
    Light intensity I_i at x_i is determined by I_xi=γ_xi(D)
    F=C
    while (γ_x(D)!=γ_c(D))
        F'=F
        F = [ ]
        for i=1:F' fireflies
          for j=1:F' fireflies
                find the best matting partner j for i that satisfies the following conditions
                i.  Intensity of j is greater than intensity of i, i.e. (I_j>I_i)
                ii. Distance between i and j should be minimum in terms of distance between
                    γ_(xi,xj)(D) and γ_c(D)
                iii. Movement of i towards j increases the intensity of j i.e. γ_(xi,xj)(D)>I_xj and
          end for j
          Move firefly i towards j i.e x_ij.
          I_ij= γ_(xi,xj)(D)
          F = F ∪ x_ij
        end for i
        Evaluate each x_ij in F for dependency i.e. γ x_ij(D) = = γ_c(D) and minimality
    end while
```

Fig. 11 FA_RSAR algorithm for feature selection

and D, the set of decision features as input and to obtained $R = \{X : X \subseteq C, \gamma_X(D) = \gamma_C(D)\}$.

The performance of the FA_RSAR approach was evaluated with respect to other heuristic, random search techniques such as quick reduct and entropy-based reduct method. The efficacy was also evaluated with respect to other nature-inspired algorithms such as GenRSAR, AntRSAR, PSO-RSAR, and BeeRSAR based on genetic algorithm, ant colony optimization, particle swarm optimization, and bee colony optimization respectively. The results in terms of size of the reduct obtained were compared with the results established in [27]. The experimental results substantiated the consistency and better performance of FA_RSAR as compared to other methods. These results formed the motivation for adoption of FA_RSAR method to reduce the dimensionality of features extracted by syntactic analysis-based feature extraction approach. The subset of features generated by FA_RSAR approach formed the set of features which are most in demand. The next step as per the strategy is to assess user interest toward this set of most preferred features (Fig. 12).

4.1.3 User Interest Estimation

Users can be classified into two categories in accordance with the level of activity on social medium as passive users and active users. Passive users only read and consider these reviews while purchasing a product. User interested in a product or product feature can be estimated by the time spent on the review of that particular product. Various studies in literature established the positive correlation between user interest and article reading time [45, Konstan et al. 1997; Ding et al. 2002;

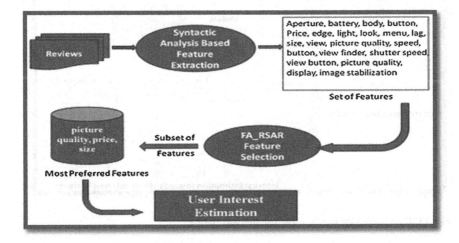

Fig. 12 Techniques used for market analysis

Kellar et al. 2004; Hofgesang and Amsterdam 2006]. Based on these studies metrics are formulated [46] to ascertain the interest of a passive user in a product feature.

In contrast to this active users participate and continuously share their experiences and opinions for a product. These opinions are pointers to their likes and dislikes and hence can be mined to build a feature-based profile for every individual. Relevant features for each user is extracted through syntactic analysis-based feature extraction approach discussed in Sect. 4.1.1. Association rule mining is then applied on these relevant features to generate two-word frequent features set and one-word frequent features set respectively [34]. Feature-based user's profile is thus generated by storing top N features (Fig. 13). It is a strong predictor of current preferences of user at a given instant of time. However, preferences are dynamic in nature. They change with changing interest of the user/arrival of new technology/change in product type. The presented strategy takes into consideration this dynamic change in user opinion. It allows for periodic updation of features stored in the user profile and subsequent dynamic change in the selected feature set. For instance, over a period of time the inclination toward "flash" feature of digital camera might increase. This would lead to increase in occurrence of this feature in user profile and hence would cause an updation in the "selected features" list. This estimation of users' interest toward most preferred feature set is utilized to group them into homogeneous segments as discussed in the second phase, i.e., market segmentation.

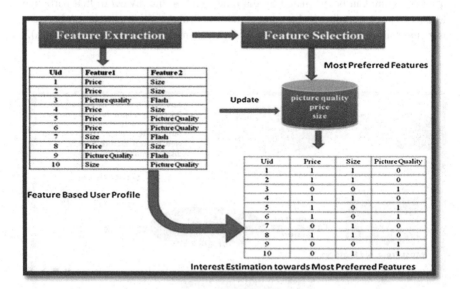

Fig. 13 User interest estimation using feature-based profile

4.2 Market Segmentation

Market segmentation can help the decision-makers to determine suitable competitive strategy [47]. The market segmentation phase segments large and dynamic e-markets into separate submarkets where customers in the same group are similar with respect to a given set of characteristics. It enables marketers to understand and characterize the diversity of demand that individuals bring to market place. The segments can then be effectively targeted by a dedicated customized marketing strategy to forge a close relationship with customers. However, this efficacy of market segmentation depends on the effectiveness of methods employed to achieve its objectives.

Literature studies indicate [48, 49, 50, 51, Torre et al. 2006] that the goals of market segmentation may be achieved by clustering. Clustering is a technique employed for partitioning a set of objects into groups such that each group is homogeneous with respect to certain attributes based on the specific criterion. This purpose of clustering makes it a popular tool for market segmentation.

4.2.1 Clustering

Clustering refers to partitioning the unlabeled data objects into certain number of clusters. The primary objective of clustering is to achieve homogeneity within cluster, i.e., objects belonging to the same cluster should be as similar as possible. Secondly, to maintain heterogeneity amongst the clusters, i.e., objects belonging to different clusters should be as different as possible. Thus the algorithmic task has been stated as an optimization problem [52].

Clustering is a process of partitioning a given set of n points into K groups or clusters based on some similarity (distance) metric. Let $O = \{\overrightarrow{o_1}, \overrightarrow{o_2}, \ldots, \overrightarrow{o_N}\}$ be a set of N objects to be clustered into K groups and each object is represented as $\overrightarrow{o_i} = \{o_{i1}, o_{i2}, \ldots, o_{id}\}$ where o_{id} represents a value of object i at dimension d. The goal of clustering algorithm is to determine a set $C = \{C_i, \ldots, C_k\}$ with K partitions such that $C_i \neq \emptyset$ and $C_i \cap C_j \equiv \emptyset$, for $i = 1, \ldots, k, j = 1, \ldots, k$ and $i \neq j$ and $\cup_{k=1}^{k} C_k = O$.

Each cluster is represented by a cluster center calculated on the basis of distance metric, then $Z = \{\overrightarrow{z_1}, \overrightarrow{z_2}, \ldots, \overrightarrow{z_k}\}$ is a set of cluster centers to which data objects are assigned. Equation (13) calculates $\overrightarrow{z_i}$ using Euclidean distance where n is number of objects in cluster.

$$\overrightarrow{z_i} = \frac{1}{n_i} \sum_{o_i \in c_i} \overrightarrow{O_i} \tag{13}$$

The objective functions of clustering problems are usually nonlinear and non-convex so some clustering algorithms may fall into local optimum. Moreover, they possess exponential complexity in terms of number of clusters and become an NP-hard problem when the number of clusters exceeds [53]. Clustering is usually performed when no information is available concerning the membership of data items to predefine classes. For this reason clustering is traditionally seen as a part of unsupervised learning. The simplest and the most popular clustering algorithm is *k-means* algorithm. It is very efficient, due to its linear time complexity but the deterministic local search used in algorithm, may converge to the nearest local optima. Various metaheuristics and swarm intelligence-based evolutionary algorithms such as tabu search (Al-Sultan 1995), simulated annealing (Bandyopadhyay et al. 1998; Selim and Al-Sultan 1991), genetic algorithms (GA) [29, 30, Bandyopadhyay et al. 1995, 1999; Murthy and Chowdury 1996), ant colony optimization (ACO) (Shelokar et al. 2004), particle swarm optimization (PSO) (Karthi et al. 2008), and differential evolution (DE) (Paterlini and Krink 2006) are applied to clustering problem. However, these algorithms successfully applied heuristic and evolutionary approaches to avoid convergence to local optima with limited success rate. Therefore the prospects of firefly algorithm for clustering problem were studied. Figure 14 presents the FClust algorithm for market segmentation.

The performance of FClust is evaluated with respect to Particle Swarm Optimization (PSO) and Differential Evolution (DE) based clustering approaches. The experimental study proves FClust has a higher probability to achieve the optimality as compare to PSO and DE. The analysis has been carried out to study the stagnation behavior and convergence speed of algorithms. The results indicate that

```
void FClust()
{
        Randomly create  an initial population of n fireflies that represent the      solutions within k
        dimensional search space xik, i=1,2,.......,N and k=1,2,....,K. where K  is  the number of
        segments.
        Evaluate the fitness of each individual based on homogeneity within the segment and
        heterogeneity between the segments using equation 13.
        t=0
        while(t<=max_generations)
        {
            for each individual i
                {
                find most attractive  partner j
                move i towards j in order to improve its fitness
                evaluate new position of i i.e new f(i)
                if new f(i)> f(i)
                  replace i with new i
                else
                  move i randomly
                }
            t=t+1}}
```

Fig. 14 FCLust algorithm for market segmentation

the convergence of the FClust is fast as compared to PSO and DE and the performance of FClust is remarkable. Thus FClust approach is used for market segmentation. Segments generated in this phase are used to target the best users for efficient product promotion in the next phase.

4.3 Targeted Product Promotion

Promotion of right products to the right customers depends on the identification of suitable customers. After dividing the market into homogeneous segments, the next phase of the presented strategy is to identify potential segment(s) for the product to be promoted. The better the suitability of the target segment(s) chosen, the higher the probability of success of the product.

This phase selects the most potential segment(s) offline to persuasively communicate marketing message. The selection of segment and the number of segments to be considered for product promotion depends on the type of the product to be promoted as well as the marketing budget. The key concept of the presented model is that it does not promote products to all or some random number of users of the selected segments. It takes the advantage of social influence spread through social sites. For instance, if user 'v' performed an action 'a' at time 't' and if the same action is performed at time $t + \Delta$ by user 'u' a friend of 'v' then it can be considered that action is propagated from user 'v' to user 'u' (Fig. 15). If this happens frequently for many different actions, then it can be concluded that user 'v' is indeed

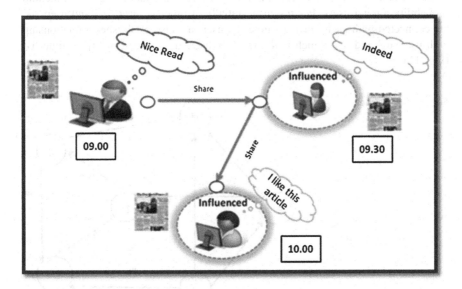

Fig. 15 Influence spread in social network

exerting some influence on user 'u.' The basic assumption is that when users see their social contacts performing an action they may decide to perform the action themselves. The presented strategy leverages this aspect of human social behavior and encourages the faster adoption of product by targeting most influential users called "seeds" in the network.

The best seeds are identified by exploring the social connectedness among users of selected segment(s). A Social network is considered as a social graph $G(V, E)$ (Fig. 16) where each individual in network (namely a, b, c, d, e and f) is represented as a node (V) and each edge (E) represents the interactions or link between individuals. The weight w_{uv} associated with each edge indicates the probability of node u to influence node v. However in practice, social network does not provide any explicit information about the influential probabilities among users. Goyal et al. [54] developed static and time dependent models to capture influence from the log of past propagations.

These models computed the prediction about the performance of an action by a user. They also predicted the time by which an influenced user would perform an action after their neighbors have performed the action. Although these methods efficiently generate the influence matrix by exploiting explicit log of action but issue of optimization of the number of scans required over action log was still not addressed. In contrast to utilizing the explicit log of actions, the presented work estimates the influential probability by implicitly capturing the propagation of social actions among users.

4.3.1 Influential Probability Estimation

Social behavior over social networks forms the core for estimating the influential probability among users. For instance, initially at time 't_0' user 'u_1' and user 'u_2' are connected with each other. At time 't_1' user 'u_1' also establishes a relationship with user 'u_3' and 'u_4' which is followed by user 'u_2' a friend of 'u_1' at time 't_2' (Fig. 17).

Fig. 16 Social graph

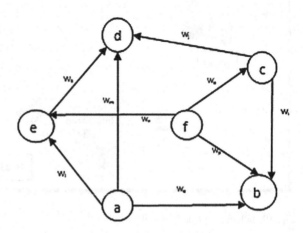

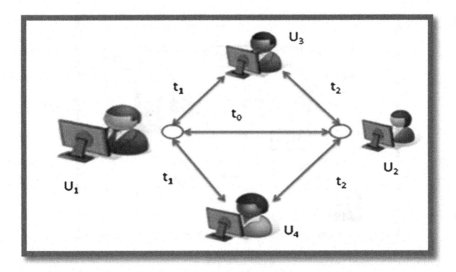

Fig. 17 Propagation of social actions on social networks

This propagation of social behavior over network can be measured from the explicit labeling of trust/friends or distrust/enemy over other users. The presented work utilized this information and formulated a social behavior-based similarity metric (SBsim) to implicitly predict the probability of propagating an action. The metric measured the similarity between two users on the basis of "how similar they are in their social relationship."

- **Social Behavior-Based Similarity SBsim ($w_{i,j}$):** It computes the similarity weight between user i and j, based on their social relationships which is defined as the size of the intersection divided by the size of the union of friend sets as given in Eq. (14) [55].

$$\text{SBsim}\left(w_{ij}\right) = \frac{F(U_i) \bigcap F(U_j)}{F(U_i) \bigcup F(U_j)} \tag{14}$$

However, the propagation of action from one user to other depends on their interest similarity. Users are generally influenced by other users if they have similar mindset [56, 57]. They like and forward only those messages that are of their interest as shown in Fig. 18.

Therefore the user interest-based similarity (UIsim) computed from their rating behavior is incorporated into social behavior based similarity (SBsim) and hybrid similarity metric (HBsim) is formulated to estimate influential probabilities among users (Fig. 19).

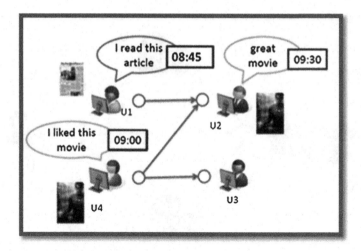

Fig. 18 Interest-based propagation on social networks

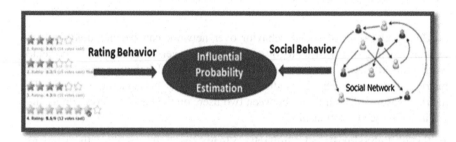

Fig. 19 Influential probability estimation

- **User Interest Based Similarity UIsim($w_{i,j}$):** This is similarity metric commonly used for recommending a product in conventional recommender systems [58]. It computes the similarity weight $w_{i,j}$ between user i and user j using adjusted cosine function given in Eq. (15) as follows:

$$\text{UIsim}(w_{i,j}) = \frac{\sum_{n=1}^{n\_\text{items}}(U_{in} - \overline{o(n)})(U_{jn} - \overline{o(n)})}{\sqrt{\sum_{n=1}^{no\_\text{items}}(U_{in} - \overline{o(n)})^2}\sqrt{\sum_{n=1}^{no\_\text{items}}(U_{in} - \overline{o(n)})^2}} \tag{15}$$

In Eq. (6), U_i and U_j are the vectors containing ratings given by user i and j for item n. $\overline{o(n)}$ refers to the average of the ratings given by all users on the nth item.

- **Hybrid Similarity HBsim($w_{i,j}$):** This metric integrates both the interest-based similarity metric and social behavior-based similarity metric to estimate the overall influential probabilities among users as follows (Eq. 16) [55]:

$$\text{HBsim}\left(w_{ij}\right) = \alpha^* \text{UIsim}\left(w_{ij}\right) + \beta^* \text{SBsim}\left(w_{ij}\right) \tag{16}$$

where α and β are the weights given to two similarity measures respectively according to the type of application.

The efficacy of presented hybrid metric has been evaluated over conventional collaborative filtering (CF) recommender systems. Collaborative filtering recommender systems formulate a model of the community preferences on the basis of past rating behavior. Subsequently, the preferences of all users belonging to the community whose profile matches with the active user are considered for recommending the items. The results established the efficacy of proposed metrics and their potential to generate more accurate recommendations [55]. Therefore the weight computed with the HBsim metric is used to compute the influential probabilities. These influential probabilities are then used to obtain optimal seeds (influential users) that can maximize the influence spread.

4.3.2 Optimal Seed Identification for Targeted Product Promotion

Identification of target seeds having strong influence in the network can be accomplished by exploring the social graph of the organization. However, the decision regarding the number of seeds, i.e., k to be employed depends upon the marketing budget of an organization. The presented work aims to identify k initial users that maximize the profit within a given seeding budget. Maximizing influence spread with the limited seeding budget (k) in large network is denoted as k-Max influence problem and has been proven to be NP-hard [21, 22].

- **k-Max Influence Problem**

  Domingos and Richardson [21] introduced k-Max influence problem posing the influence maximization as an optimization problem of selecting the best k seeds.

  Given a social graph $G(V, E)$ and an integer k, find k seeds such that the incurred influence is maximized [22].

  The influence process starts with the set S V of nodes called seeds and activates them. These seeds in turn activate some of their neighbors. These newly active nodes then influence some of their neighbors, and so on. Nodes that are influenced by a product are called active and others are called inactive. Hence the influence starts from the set S and cascades in the graph through the outgoing edges of the active nodes. The aim of the influence maximization problem is to choose the initial seed set S so that final influence (i.e., the number of active nodes at the end of the cascade) is maximized.

- **Greedy Approach for k-Max Influence Problem**

  Kempe [22] applied greedy strategy to tackle k-Max influence problem. It takes the graph G and number k (i.e., marketing budget) as input. The algorithm generates a seed set S of cardinality k, with the intention that the expected number of vertices influenced by the seed set S is maximum. The algorithm adds one vertex into the set S with each iteration i such that this vertex together with current set S maximizes the influence spread. Thus the vertex selected at iteration i is the one that maximizes the incremental influence spread during that iteration. The working of greedy algorithm is illustrated with the help of social graph shown in Fig. 20.

The graph has six nodes (users) namely a, b, c, d, e, and f respectively. Each edge (u, v) represents the social relationship between u and v and the weight w_{uv} associated with each edge indicates the probability of node u to influence node v. The main objective is to identify the set $S \subseteq V$ of cardinality k where $k = 2$ with the intention that the expected number of vertices influenced by the seed set S is maximum. Figure 21 describes the algorithmic steps of greedy approach.

The algorithm first calculates the expected influencing value $E_{a,b,c,d,e,f} = \{2.45, 1, 2.0, 1, 1.5, 2.25\}$ of each node using diffusion model and probability information. For instance, the influencing value for node, i.e., E_a is computed by considering all cascades through node, i.e., sum of influence on the nodes $E_a = 1 + 0.9 + (0.1 + (1 - 0.1) * 0.3 * 0.5) + 0.3 \approx 2.45$; where the first term is due to a's influence on itself while the third enumerates expectation on two possible paths from a to d since node has highest influencing value so it is selected as the first seed. The second seed is selected from rest of the nodes, i.e., and the corresponding influences are $\{2.25, 2.0, 1.74, 1.38\}$. These values are conditional to already being selected as a seed. Thus the second selection is and final seed set is with the expected influence of 4.1. The

Fig. 20 Weighted social graph

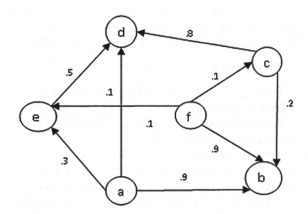

Fig. 21 Greedy algorithm for k-max influence problem

```
Void Greedy ( )
{
    initialize S=∅, t=0
    while (t<=k)
        for each vertex v ∈ V\S
            S_v=0
            calculate Influence Spread S_v= {SU{v}}
        S = SU {arg max_{v∈V\S} {S_v}}
        t=t+1
}
```

greedy algorithm although significantly outperforms the classic degree and centrality-based heuristics in influence spread but suffers from high computation cost on estimating the influence function. Therefore "Cost-Effective Lazy Forward" (CELF) scheme for seed selection was proposed [59]. However, this improved algorithm also takes hours to complete a graph with few tens of thousands of vertices, so it is not suitable for large-scale networks. Therefore efforts were made to define a local viral marketing problem (LVMP) which is opposed to global viral marketing problem (GVMP) [60]. Sarkar et al. [61] studied how innovations diffuse across a variety of different topologies whereas [62] studied the effect of word-of-mouth on product adoption. All these techniques did succeed in finding the best seeds but required global knowledge of the network, i.e., knowledge about every node in the network and its connection with every other node which is unrealistic in many real-world cases and time-consuming.

The presented work tackles these issues with an evolutionary approach. Banati and Bajaj [55] studied the viability of evolutionary algorithms, i.e., Differential Evolution (DE) and Firefly (FA) for k-Max influence problem. The results revealed that both evolutionary approaches DE and FA perform better as compared to greedy approach with respect to maximum influence incurred as well as gain achieved by increasing the value of k. Amongst the evolutionary approaches FA maintains consistency in its results and has higher probability to score over DE and greedy-based algorithms. On the basis of these observations the presented strategy employed firefly algorithm-based evolutionary approach to identify the most influential seeds from the social network. All the users are divided into homogenous groups by observing their interest and subsequently the best segments are selected as target according to the characteristics of product to be promoted. The social networks of these selected segments are then explored and the most influential seeds are identified for efficient product promotion (Banati and Bajaj 2013c). The strength of this product promotion plan is substantiated through a case study discussed in the next section.

5 Case Study for Targeted Production Promotion Strategy on Social Networks

To assess the efficacy of presented product promotion plan, a case study is conducted over epinion dataset obtained from epinion.com website. Dataset consisted of 2378 users who rated 16,861 different products belonging to 6 categories. For experimental study, only those users are considered who have rated at least 10 products. All the products rated by these users are divided into test and training sets. The division of dataset is made on the basis of products such that products of the test dataset did not participate in the training model. Hence, the system promotes new products to the users. Results are averaged over twenty five independent runs with the following settings (Table 3).

Table 3 Details of training and test set

Training (%)	Test (%)
95	5
90	10
85	15
80	20

According to the presented product promotional plan users are divided into three homogenous segments on the basis of their product rating information using firefly algorithm-based FClust clustering approach. For efficient segment selection each segment is associated with a weight for each category that is computed using Eq. (17).

$$W(sc) = \frac{\text{Total number of users in a segment rated products of category } c}{\text{Total number of users in a segment } s}$$

(17)

This weight forms the basis for the segment selection. The higher the value of segment with respect to a given category, the better the chances of its selection for promoting product belonging to that category. Subsequently, the social connectedness among the users of selected segments is converted into social weighted graph $G(V, E)$ where weight to each edge is computed using hybrid similarity metric. Finally, the best influential seeds are identified using firefly evolutionary approach to promote products in test set. This configuration of targeted approach is compared with the conventional viral approach. The conventional promotion model considers all the users and identifies the targeted seeds from the social network consists of these users. Other configurations for both the approaches are same except the targeted approach that first identifies the best group of users and then selects the best seeds from the same group whereas the conventional approach selects the seeds from whole market. The efficacy of these approaches is measured using the following evaluating metrics:

5.1 Evaluation Metric

The following evaluation metrics are used to evaluate the performance of the approach:

- **Precision**: It measures a percentage of influenced users in terms that they actually rated the product promoted to them from the total number of users targeted through seeds, i.e.,

$$\text{Precision} = \frac{\text{Number of targeted users actually rated the product}}{\text{Number of users targeted through seeds}}$$

- **Recall**: It measures a percentage of influenced users in terms that they actually rated the product promoted to them from the total number of users in the market actually rated the product, i.e,.

$$Recall = \frac{Number\ of\ targeted\ users\ actually\ rated\ the\ product}{Total\ Number\ of\ users\ actually\ rated\ the\ product}$$

- **F1 Measure**: It combines both the recall and precision, i.e.,

$$F1 = 2 * \frac{Precision * Recall}{Precision + Recall}$$

5.2 Performance Analysis

The performances of these approaches are measured with respect to the value of marketing budget (k), i.e., 2 and 3 in experiments. Figure 22 shows the precision obtained with marketing budget $k = 2$. Higher precision value of the targeted approach indicated its ability to target right customers as compared to the conventional approach. It is observed that a higher recall value of targeted approach (Fig. 23) also proved its ability to cover large right audience with $k = 2$. F1 measure (Fig. 24) depicts the harmonic balance between both precision and recall values. Results reveal that F1 measure values are better for the targeted approach as compared to the conventional approach.

Fig. 22 Precision value with $k = 2$

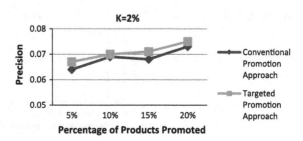

Fig. 23 Recall value with $k = 2$

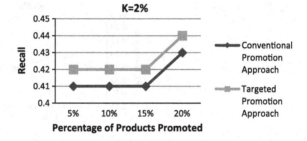

Fig. 24 F1 measure with
$k = 2$

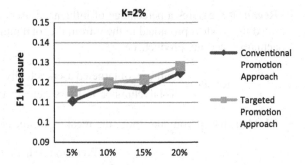

Another study is performed to evaluate the effect of increase in marketing budget from 2 to 3. The results with respect to precision, recall, and F1 measure are shown in Figs. 25, 26 and 27 respectively. It can be observed from these results that by raising the marketing budget, the targeted strategy is able to convince more customers. The recall value depicts the efficiency of presented strategy.

Fig. 25 Precision value with
$k = 3$

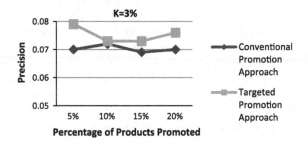

Fig. 26 Recall value with
$k = 3$

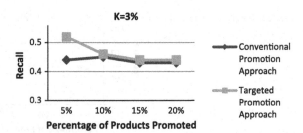

Fig. 27 F1 measure with
$k = 3$

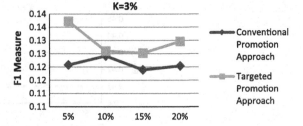

All these results establish that the presented framework to promote the products over social networks has more potential to enhance the number of influenced users and thereby increase the sales of products. Targeted product promotion using Firefly algorithm thus leads to optimal utilization of resources such as cost, effort, etc., to target large number of potential customers. Therefore, this approach may serve as a boon to the product industry in augmenting their sales as well as profit graph.

6 Conclusion

The chapter presented an effective e-marketing strategy that incorporated the flashing behavior of fireflies and took the advantage of quick spread behavior of eWOM for marketing campaigning. The functionality of proposed strategy divided into three phases, viz., market analysis, market segmentation, and targeted product promotion. Each phase was optimized by "firefly" a nature inspired algorithm and subsequently experiments were conducted to evaluate the same. An initial with market analysis of current trend in terms of relevant product features is done by mining user opinions and extract user interest toward specific product features. Subsequently market segmentation was done by clustering similar users and finally the best initial seeds were identified for product promotion. The strategy exploited the social connectedness among users in the selected segment to identify the best initial seeds to spread the influence at global scale. The experimental result substantiated the ability of the approach to cover large right audience by employing a small fraction of advertising budget. Thus the presented product promotion plan can be a boon in the marketing domain where organizations can use optimal company resources and efforts with limited budget to reach out to a vast potential segment.

References

1. Zadeh LA (1965) Fuzzy sets. J Inf Control 8(3):338–353
2. Hopfield JJ, Tank DW (1985) Neural computation of decisions in optimization problems. J Biol Cybern 52(3):141–152
3. Bandyopadhyay S, Murthy CA, Pal SK (1999) Theoretic performance of genetic pattern classifier. J Franklin Inst 336:387–422
4. Dasgupta D, Michalewicz Z (1997) Evolutionary algorithms in engineering applications. Springer Science & Business Media, pp 1–557
5. Mehta S (2013) Bio-inspired approach to solve chemical equations. In: Proceedings of international conference on contemporary computing, pp 461–466
6. Zhang J, Lo WL, Chung H (2006) Pseudocoevolutionary genetic algorithms for power electronic circuits optimization. IEEE Trans Syst Man Cybern Part C 36(4):590–598
7. Bandyopadhyay S, Murthy CA, Pal SK (1995) Pattern classification with genetic algorithms. Pattern Recogn Lett 16:801–808

8. Hurley S, Moutinho L, Stephens NM (1995) Solving marketing optimization problems using genetic algorithms. Eur J Mark 29(4):39–56

9. Potvina JY, Sorianoa P, Maxime V (2004) Generating trading rules on the stock markets with genetic programming. Comput Oper Res 31:1033–1047

10. Fuchs MM (2000) An evolutionary approach to support web page design. In: Proceedings of congress on evolutionary computation, pp 1312–1319

11. Cutler M, Deng H, Maniccam SS, Meng W (1999) A new study on using HTML structures to improve retrieval. In: Proceedings of 11th IEEE international conference on tools with artificial intelligence, pp 406–409

12. Salton G, Buckley C (1998) Term-weighting approaches in automatic text retrieval. J Inf Proc Manage 24:513–523

13. Atsumi M (1997) Extraction of user's interests from web pages based on genetic algorithm. English version of the original Japanese paper published in IPSJ SIG notes (information processing society of Japan, the special interest groups notes), vol 97, no. 51, pp 13–18

14. Morgan JJ, Kilgour AC (1996) Personalising on-line information retrieval support with a genetic algorithm. In: Proceedings of PolyModel 16: applications of artificial intelligence, pp 142–149

15. Akkermans H (2001) Intelligent E-business: from technology to value. In: Guest Editor's Introduction. IEEE intelligent systems, July/August 2001, pp 8–10

16. Kaplan AM, Haenlein M (2010) Users of the world, unite! The challenges and opportunities of social media. Bus Horiz 53(1):59–68

17. Kietzmann JH, Hermkens K et al (2011) Social media? Get serious! Understanding the functional building blocks of social media. Bus Horiz 54(3):241–251

18. Yu Z, Zhaoqing W, Chaolun X (2010) Identifying key users for targeted marketing by mining online social network. In: Proceedings of 24th international conference on advanced information networking and applications workshops (WAINA), pp 644–649

19. Pritam G, Huan L (2012) Mining social media: a brief introduction. In: Tutorials in operation research, INFORMS, pp 1–17

20. Domingos P, Richardson M (2001) Mining the network value of customers. In: Proceedings of knowledge discovery and data mining, pp 57–66

21. Domingos P, Richardson M (2002) Mining knowledge-sharing sites for viral marketing. In: Proceedings of the eight ACM SIGKDD international conference on knowledge discovery and data mining, pp 61–70

22. Kempe D, Kleinberg J, Tardos E (2003) Maximizing the spread of influence through a social network. In: Proceedings of knowledge discovery and data mining, pp 137–146

23. Fowler JH, Christakis NA (2008) Dynamic spread of happiness in a large social network: longitudinal analysis over 20 years in the framingham heart study. Br Med J 337(768):1–23

24. Datta P, Chowdhury D, Chakraborty B (2005) Viral marketing: new form of word of mouth through internet. Bus Rev Cambridge 3(2):69–75

25. Breazele M (2009) Word of mouse—an assessment of electronic word-of-mouth research. Int J Market Res 51(3):297–318

26. Jensen R, Shen Q (2003) Finding rough set reducts with ant colony optimization. In: Proceedings of 6th annual UK workshop on computational intelligence, pp 15–22

27. Suguna N, Thanuskodi K (2010) A novel rough set reduct algorithm for medical domain based on bee colony optimization. J Comput 2(6). ISSN:2151-9617

28. Yue B et al (2007) A new rough set reduct algorithm based on particle swarm optimization. In: Proceedings of IWINAC, Part I, LNCS, vol. 4527, pp 397–406

29. Bandyopadhyay S, Maulik U (2002) An evolutionary technique based on k-means algorithm for optimal clustering. J Inform Sci 146:221–237

30. Bandyopadhyay S, Maulik U (2002) Genetic clustering for automatic evolution of clusters and application to image classification. Pattern Recogn 35:1197–1208

31. Yang XS (2008) Nature inspired metaheuristic algorithms. Luniver Press, Beckington

32. Banati H, Bajaj M (2012) Promoting products online using firefly algorithm. In: Proceedings of 12th international conference on intelligent system design and applications (ISDA), November 27–29, 2012, India, pp 580–585
33. NLProcessor–TextAnalysisToolkit (2000). http://www.infogistics.com/textanalysis.html
34. Banati H, Bajaj M (2011b) Dynamic user profile generation by mining user opinion. In: Proceedings of 3rd international conference on computer engineering and technology (ICCET-11), June 17–19, 2011, Malaysia, pp 540–545
35. Dash M, Liu H (1997) Feature selection for classification. J Intell Data Anal 1(3):131–156
36. Devijver PA, Kittler J (1982) Pattern recognition: a statistical approach. Prentice Hall, UK
37. Pawlak Z (1982) Rough sets. Int J Comput Inform Sci 11:341–356
38. Pawlak Z (1991) Rough sets: theoretical aspects of reasoning about data. Kluwer Academic Publishers
39. Pawlak Z (1993) Rough sets: present state and the future. Found Comput Decis Sci 18:157–166
40. Pawlak Z (2002) Rough Sets and Intelligent Data Analysis. J Inform Sci 147:1–12
41. Chouchoulas A, Shen Q (2001) Rough set-aided keyword reduction for text categorization. Appl Artif Intell 15(9):843–873
42. Jensen R, Shen Q (2001) A rough set-aided system for sorting WWW bookmarks. In: Zhong N et al (eds) Web intelligence: research and development. pp 95–105
43. Liu H, Abraham A, Li Y (2009) Nature inspired population-based heuristics for rough set reduction, rough set theory, SCI, vol. 174. Springer, pp 261–278
44. Banati H, Bajaj M (2011a) Firefly based feature selection approach. Int J Comput Sci Issues (IJCSI) 8(4/2):473–480
45. Morita M Shinoda Y (1994) Information filtering based on user behavior analysis and best match text retrieval. In: Proceedings of the 17th annual International ACM SIGIR conference, pp 272–281
46. Banati H, Bajaj M (2009) Feature based implicit user modeling. In: Proceedings of 4th Indian international conference on artificial intelligence (IICAI-09), December 16–18, 2009, India, pp 1077–1084
47. Aaker DA (2001) Strategic market management. Wiley, New York
48. Andrade RD (1978) U-statistic hierarchical clustering. Psychometrika 58–67
49. D'Urso P, Giovanni LD (2008) Temporal self-organizing maps for telecommunications market segmentation. Neurocomputing 71:2880–2892
50. Huang J, Tzeng G, Ong C (2007) Marketing segmentation using support vector clustering. J Expert Syst Appl 32(2):313–317
51. Kiang MY, Hu MY, Fisher DM (2006) An extended self-organizing map network for market segmentation—a telecommunication example. J Decis Support Syst 42:36–47
52. Brucker P (1978) On the complexity of clustering problems. In: Beckmenn M, Kunzi HP (eds) Optimization and operations research. Lecture notes in economics and mathematical systems, vol 157. Springer, Berlin, pp 45–54
53. Welch JW (1983) Algorithmic complexity: three NP-hard problems in computational statistics. J Stat Comput Simul 15:17–25
54. Goyal A, Bonchi F, Lakshmanan LVS (2010) Learning influence probabilities in social networks. In: Proceedings of the third ACM international conference on web search and data mining, pp 241–250
55. Banati H, Mehta M, Bajaj M (2014) Social behaviour based metrics to enhance collaborative filtering. Int J Comput Inf Syst Ind Manage Appl 6:217–226
56. Aggarwal CC, Wolf JL, Wu K-L, Yu PS (1999) Horting hatches an egg: a new graph-theoretic approach to collaborative filtering. In: Proceedings of the fifth ACM SIGKDD international conference on knowledge discovery and data mining, pp 201–212
57. Goldberg Nichols D, Oki BM, Terry D (1992) Using collaborative filtering to weave an information tapestry. Commun ACM 35(12):61–70

58. Shardanand U, Maes P (2005) Social information filtering: algorithms for automating "word of mouth". In: Proceedings of ACM conference on human factors in computing systems. ACM Press, pp 210–217

59. Leskovec J, Krause A, Guestrin C, Faloutsos C, VanBriesen J, Glance N (2007) Cost-effective outbreak detection in networks. In: Proceedings of the 13th ACM SIGKDD international conference on knowledge discovery and data mining, pp 420–429

60. Stonedahl F, Rand W, Wilensky U (2010) Evolving viral marketing strategies. In: Proceedings of genetic and evolutionary computation conference (GECCO), pp 1195–1202

61. Sarker R, Mohammadian M, Yao X (2002) Evolutionary optimization. In: International series in operations research and management science, vol 48. Kluwer Academic, Boston, pp 87–113

62. Chevallier J, Maylin D (2006) The Effect of Word-of-Mouth on Sales: Online book Reviews. J Mark Res 43(3):345–354

63. Balasubramanian S, Mahajan V (2001) The economic leverage of the virtual community. Int J Electron Commer 5(3):103–138

64. Banati H, Bajaj M (2013) Performance analysis of firefly algorithm for data clustering. Int J Swarm Intell 1(1):19–35

65. Banati H, Bajaj M (2013) An evolutionary approach for k-max-influence problem. Int J Netw Innovative Comput 1(1):163–172

66. http://www.epinion.com

67. Nissen V, Biethahn J (1995) An introduction to evolutionary algorithms. In: Biethahn J, Nissen V (eds) Evolutionary algorithms in management applications. Springer, pp 3–97

68. Terano T, Ishino Y (1998) Interactive genetic algorithm based feature selection and its application to marketing data analysis. In: Liu H, Motoda H (eds) Feature extraction, construction and selection. pp 393–406

69. De la Torre F, Kanade T (2006) Discriminative cluster analysis. In: Proceedings of the 23rd international conference on machine learning. ACM, pp 241–248

Hybrid Rough-PSO Approach in Remote Sensing Imagery Analysis

Anasua Sarkar and Rajib Das

Abstract Pixel classification among overlapping land cover regions in remote sensing imagery is a very challenging task. Detection of uncertainty and vagueness are always the key features for classifying mixed pixels. This paper proposes an approach for pixel classification using a hybrid approach of rough set theory and particle swarm optimization methods. Rough set theory deals with incompleteness and vagueness among data, which property may be utilized for detecting arbitrarily shaped and sized clusters in satellite images. To enable fast automatic clustering of multispectral remote sensing imagery, in this article, we propose a rough set-based heuristical decision rule generation algorithm. For rough-set-theoretic decision rule generation, each cluster is classified using heuristically searched optimal reducts to overcome overlapping cluster problem. This proposed unsupervised algorithm is able to identify clusters utilizing particle swarm optimization based on rough set generated membership values. This approach addresses the overlapping regions in remote sensing images by uncertainties using rough set generated membership values. Particle swarm optimization is a population-based stochastic optimization technique, inspired from the social behavior of bird flock. Therefore, to predict pixel classification of remote sensing imagery, we propose a particle swarm optimization-based membership correction approach over rough set-based initial decision rule generation. We demonstrate our algorithm for segmenting a LANDSAT image of the catchment area of Ajoy River. The newly developed algorithm is compared with fuzzy C-means and K-means algorithms. The new algorithm generated clustered regions are verified with the available ground truth knowledge. The validity analysis is performed to demonstrate the superior performance of our new algorithms with K-means and fuzzy C-means algorithms.

A. Sarkar (✉)
SMIEEE, Government College of Engineering and Leather Technology,
Kolkata 98, West Bengal, India
e-mail: ashru2006@hotmail.com

R. Das
School of Water Resources Engineering, Jadavpur University,
Kolkata 700032, West Bengal, India
e-mail: rajibdas79@gmail.com

© Springer India 2016
S. Bhattacharyya et al. (eds.), *Hybrid Soft Computing Approaches*,
Studies in Computational Intelligence 611,
DOI 10.1007/978-81-322-2544-7_10

Keywords Remote sensing · Pixel classification · Rough set · Rough membership value · Particle swarm optimization

1 Introduction

In the analysis of remote sensing images, the pixel classification of a particular land cover region is often posed as clustering in the intensity space of multispectral satellite images [1]. There are small regions (like bridges and roads) within only a few pixels, while the vegetation regions are significantly large. Thus the detection of overlapping clusters considering neighborhoods becomes an important challenge in remote sensing pixel classification.

Clustering is the unsupervised classification method with maximum intraclass similarity and minimum interclass similarity. Several earlier works proposed clustering for pixel classification in remote sensing imagery like—self-organizing map (SOM) [2], K-means clustering [3, 4], simulated annealing [5], graph theoretic approach [6], fuzzy C-means clustering [7], and scattered object clustering [8]. Several other methods like clustering based on symmetry [9], supervised multi-objective learning approach [10], also may be applicable efficiently for detection of efficient land cover regions in remote sensing images.

The rough set theory deals with uncertainty, vagueness, and incompleteness in class definition. It computes lower and upper approximations of rough sets for overlapping partitions. Therefore, recently rough set theory is being used with clustering [1, 7, 11, 12]. Lingras [6, 7, 12] used rough set theory to develop interval representation of clusters. This model is useful when the clusters do not necessarily have crisp boundaries.

PSO is a population-based algorithm that uses a population of individuals to probe the best position in the search space. In PSO, the individual is called a particle, which moves stochastically in the direction of its own best previous position and the whole swarm's best previous position. Therefore, to predict pixel classification of remote sensing imagery, we propose a Rastrigin function-based PSO to rough set-based initial decision rule generation.

The present study focuses on the integration of rough-set-theoretic automatic optimal classification PSO-based membership correction for pixel classification in remote sensing imagery. Clusters are associated with indiscernibility classes containing different land cover regions that occur in remote sensing images. The most widely used clustering algorithms for pixel classification analysis are hierarchical clustering [13], K-means clustering [3, 4], and SOM [2].

We demonstrate the performance of the new distance metric in pixel classification of a chosen LANDSAT remote sensing image of the catchment area of Ajoy River. The quantitative evaluation over three existing validity indices indicates the satisfactory performance of our new hybrid rough set-based and particle swarm optimization (PSO) corrected algorithm (RPSO) to detect imprecise clusters. We compare our obtained solutions with those of K-means and FCM algorithms to verify with the ground truth knowledge. The statistical tests also demonstrate the significance of our new RPSO algorithm over K-means and FCM algorithms.

2 Rough Set Theory and PSO Method

2.1 Rough Set Theory

The theory of rough sets begins with the notion of an indiscernibility relation, which induces a pair $\langle U, R \rangle$, where U is a nonempty set (the universe of discourse) and R is an equivalence relation on U. R decomposes the set U into disjoint equivalence classes, denoted as $[x]_B$ (for some object x described by a set of attributes B). Let $U/R = \{X_1, X_2, \ldots, X_m\}$, where X_i is an equivalence class of $R, i = 1, 2, \ldots, m$.

The main goal of rough set analysis is to synthesize approximations of concepts from acquired data. Consequently, given an arbitrary set $X \subseteq U$ can be approximated by using the information contained in B by defining a pair of B—lower and B—upper approximations of X [14, 15], denoted by $\underline{B}X$ and $\overline{B}X$ respectively.

$$\underline{B}(X) = \{x \in U | [x]_B \subseteq X\} = \cup\{[x]_B \in U/R | [x]_B \subseteq X\} \tag{1}$$

$$\overline{B}(X) = \{x \in U | [x]_B \cap X = \phi\} = \cup\{[x]_B \in U/R | [x]_B \cap X = \phi\} \tag{2}$$

The lower approximation $\underline{B}X$ is the union of all the elementary sets, which are subsets of X and the upper approximation $\overline{B}X$ is the union of all the elementary sets, which have a nonempty intersection with X. The interval $| \underline{B}X, \overline{B}X |$ simply is called the rough set of X.

Pawlak discussed in [14, 15], two numerical characterizations for imprecision of X: accuracy and roughness. Accuracy of approximation for X, denoted by $\alpha_B(X)$, is the ratio of objects on its lower approximation to that on its upper approximation.

$$\alpha_B(X) = |\underline{B}X|/|\overline{B}X| \tag{3}$$

The roughness of X, denoted by $\rho_B(X)$, is defined as $\rho_B(X) = 1 - \alpha_B(X)$. Note that the lower the roughness of a subset, the better its approximation. Each decision d determines a classification of objects in X [14, 15] as $\text{CLASS}_R(d) = \left\{X_R^1, \ldots, X_R^{r(d)}\right\}$ of the universe U, where $X_R^k = \{x \in U | d(x) = v_d^k\}$ for $1 \leq k \leq r(d)$, where v_d^k denotes kth value for decision d and $r(d)$ denotes the rank of decision d. A classification can be characterized numerically by the quality of classification as the fitness function measure defined below:

$$\gamma_B(X) = |\underline{B}X \cup \underline{B}\neg X| / \ |U|, \tag{4}$$

where $\underline{B}\neg X$ is the lower approximation of set of objects that do not belong to X.

2.2   PSO Algorithm

PSO is a population-based algorithm that uses a population of individuals to probe the best position in the search space. In PSO, the individual is called a particle, which moves stochastically in the direction of its own best previous position and the whole swarm's best previous position. Suppose that the size of the swarm is N and the search space is M dimensional, then the position of the ith particle is presented as $X_i = \{x_{i1}, x_{i2}, \ldots, x_{iM}\}$. The velocity of this particle is presented as $V_i = \{v_{i1}, v_{i2}, \ldots, v_{iM}\}$. The best previous position of this particle is denoted as $P_i = \{p_{i1}, p_{i2}, \ldots, p_{iM}\}$.

Consequently, the best previous position discovered by the whole swarm is denoted as $P_S = \{p_{S1}, p_{S2}, \ldots, p_{SM}\}$. Let the maximum number of iteration be T and t be the present iteration. The unit time is denoted by $\Delta \tau$ Then the position of a particle and its velocity are changed following the constraints shown below [1, 5, 6]:

$$v_{im}^{t+1} = \omega^t * v_{im}^t + \frac{\text{rand}() * c_1 * \left(p_{im} - x_{im}^c\right)}{\Delta \tau} + \frac{\text{rand}() * c_2 * \left(p_{Sm} - x_{im}^c\right)}{\Delta \tau} \tag{5}$$

$$x_{im}^{t+1} = x_{im}^t + v_{im}^t * \Delta \tau \tag{6}$$

$$\omega^t = \omega_{\max} - \frac{t * (\omega_{\max} - \omega_{\min})}{T} \tag{7}$$

, where $1 \leq t \leq T$, $1 \leq m \leq M$, and rand() generates the random number with uniform distribution $U(0, 1)$. c_1 and c_2 are acceleration coefficients. ω is the inertia weight, with $\omega_{\max}$ and $\omega_{\min}$ as it's the maximum and minimum values respectively.

For the initial matrix, $X = \begin{bmatrix} x_{11} & x_{12} & x_{1M} \\ x_{21} & x_{22} & x_{2M} \\ x_{n1} & x_{n2} & x_{NM} \end{bmatrix}$, the equation to generate particle value is:

$$\begin{aligned} x_{\text{initial}} &= x_{im} = x_{\min} + (x_{\max} - x_{\min}) * \text{rand}(), \\ \forall m &= \{1, \ldots, M\}, n = \{1, \ldots, N\}. \end{aligned} \tag{8}$$

Then the boundary constraints for x_{im}^{t+1} and v_{im}^{t+1} are as follows:

$$x_{im}^{t+1} = \begin{cases} x_{im}^{t+1}, x_{min} \leq x_{im}^{t+1} \leq x_{max} \\ x_{initial}, x_{im}^{t+1} > x_{max} \\ x_{initial}, x_{im}^{t+1} < x_{min} \end{cases} \quad (9)$$

$$v_{im}^{t+1} = \begin{cases} v_{im}^{t+1}, -v_{max} \leq v_{im}^{t+1} \leq v_{max} \\ v_{max}, v_{im}^{t+1} > v_{max} \\ -v_{max}, v_{im}^{t+1} < -v_{max} \end{cases}, \quad (10)$$

where $\{v_{max}, v_{min}\}$ and $\{x_{max}, x_{min}\}$ are respectively maximum and minimum values for v and x, respectively.

2.3 Related Works

There is a canopy of existing works using fuzzy and rough set theory and PSO-based approaches on the problem of multispectral image segmentation on remote sensing images.

Dixon et al. [16] experiments SVM method to the Landsat Thematic Mapper (TM) image classification, comparing the results with the maximum likelihood classifier (MLC), the neural network classifier, and the decision tree classifier. They show SVM classifier obtain higher classification accuracy than those of the other classifiers [16]. Mountraki and Im make their experiments using several methods for analysis of airborne- and satellite-derived imagery which are proposed and assessed [17].

Foody works on the SVM algorithm to classify the airborne image, which shows that SVM method often provide a higher accuracy than those existing stat-of-the-art classification methods [18] for multiclass image segmentation problem. Chiang and Hao implement an SVM-based fuzzy inference system which exhibits reliable performance using fuzzy rule based modeling [19]. Further, Melgani and Bruzzone experiment on the problem of the classification of hyper spectral remote sensing images by support vector machines (SVMs) in their relevant work [20]. Zhang et al. [21] experiment with fuzzy topology integrated support vector machine approach on remote sensing image classification problem. They find the significant boundary and the interior parts of the classification using the fuzzy topology space. They demonstrate that their FTSVM method achieves higher classification accuracy than standard SVM and other classification methods [21].

In SVM classification, the accuracy depends on whether the training data is sufficient to provide a representative description of each class or not. In general, the higher number of training pixels, the higher the classification accuracy. However, due to low image resolution, complexity of ground substances, diversity of disturbance, etc., many mixed pixels exist in a remotely sensed image [21].

The SVM-FAHP method [22] divides 3PL provider selection into two stages. SVM is used in the first stage to classify all the enterprises for further election. Then fuzzy AHP is utilized to estimate the superior enterprises from those were selected in the first stage. In comparison with the classical methods, this model based on SVM-FAHP [22] improves the selection efficiency by reducing the computational cost during decision-making process and also decreases the cost of information collection simultaneously. The FAHP model uses the uncertainty problem very efficiently. The example study shows that the SVM-FAHP model is feasible and effective.

SVMs are particularly useful in the remote sensing field for their ability to generate outputs even with limited training samples. This is a frequent necessary restriction for remote sensing problems. However, SVM faces the parameter assignment problem which can significantly affect obtained results. The classification algorithm SVM is a supervised approach with several advantageous features. This algorithm supports self-malleability with speedy learning and limited needs on training samples. Therefore SVM has been proved to be very methodology in pattern recognition field on processing of data acquired through remote sensing devices. Several earlier researches on both real-world data and simulated datasets have shown that SVMs exhibit superiority over most of the state-of-the-art algorithms.

Liu et al. [23] experiment to solve the shortcomings of traditional linear SCDA assessment methods. They propose an improved SVM method combined with the multistage dynamic fuzzy decision. The algorithm takes as input the multistage fuzzy decision rules outputs as the sampling data. Then this algorithm uses the SVM algorithm to begin evaluation using those training points. This technique not only utilizes the advantages of multilayer SVM classifier, but also overcomes the problem of finding out the high grade training sample data.

This model utilizes the principle of structural risk minimization and therefore upgrades the accuracy and generalization ability of SVM much more. It utilizes the method of less learning over the problem of more learning, with an overall optimum solution. A very good feature of SVMs is that only the support vectors are of importance during training. This method also is superior to the neural network approach, as that method only provides partial optimal solution.

Li et al. [24] divide the categories in remote sensing classification based on two views—basic thought and novel categorization algorithms. According to their survey, the approaches for remote sensing classification have been changed from per-pixel multispectral-based approaches to multiscale object-based approaches. The new category of the categorization algorithms comprises of the SVM, fuzzy clustering (FCM) algorithm, evolutionary algorithm (EA), as well as implementation with artificial neural networks (ANNs). This redirection leads to the development of several new hybrid remote sensing image classification methods in the past years. The research works are combinations of the multi scale object-based approaches with existing categorization algorithms like, SVM, fuzzy clustering algorithm, EA, ANNs.

In the past decades, several experiments with remote sensing devices generate several multisource datasets. Therefore the scientific challenge comes to be how to use this multisource imagery, like data in formats of multispectral, hyper spectral, radar, LIDAR, optical infrared sensors. The requirement becomes to efficiently utilize all these data in remote sensing applications to improve the classification accuracy.

Qiu et al. [25] in their work, project the image categorization problem as an image texture learning problem. They view a remote sensing image as a collection of regions, each obtained from the output classes after image segmentation. These approaches provide efficient segment classes through a chosen metric distance function. Therefore, the segmentation problem becomes convertible to the regular categorization algorithm.

Sparse SVM [25] method has been developed to radically decrease the regions that are needed to classify in the remote sensing images. The chosen regions by a sparse SVM algorithm are utilized in the next phases as the target concepts in the traditional diverse density approaches. Therefore, the SVM classification method becomes to be very reliable in remote sensing image analysis problems. Several works show that the SVM approaches combined with fuzzy improvements can produce superior results than the nearest neighbor (NN) approaches in the category of the supervised classifications.

Surveying several approaches over recent works in hyperdimensional feature space reveals the potentialities of SVM classifiers. Three significant useful properties of different SVM approaches in remote sensing image classification problems are shown below.

SVMs are more efficient than other conventional nonparametric classifiers like RBF neural networks and the K-NN classifier. The SVM approaches provide more categorization accurateness in terms of validity indices, requires computational times, and utilizes a trend of constancy to parameter setting. Several research works reveal that SVMs are more effective than the traditional pattern recognition approaches, which incorporates a combination of existing feature extraction/ selection methods and a conventional classifier.

To search existing works on remote sensing applications, the study on several existing clustering algorithms is needed. The most common applied algorithms in the field of remote sensing image classifications are: hierarchical clustering algorithm, K-means algorithm, expectation maximization clustering algorithm, and SOMs algorithm. Different factors need to be considered while comparing these algorithms like—number of clusters, size of dataset, type of dataset, and type of software used to generate input dataset.

The hybrid FSVM (Fuzzy-SVM) method has been used to enhance the SVM in reducing the effect of outliers and noises (fuzziness) in data points. This method is very efficient in remote sensing applications, in which data points do not have any modeled characteristics [19]. Combing the advantages of traditional SVM framework and the fuzzy basis function inference system, Chiang and Hao [19] propose an SVM-based fuzzy inference system on remote sensing imagery. The method exhibit a reliable performance in for classification and prediction of remote sensing images. Consequently, Tsujinishi and Abe [26] solve unclassifiable regions for multiclass problems in remote sensing images. They utilize fuzzy LSSVMs to resolve this problem.

Fuzzy topology is an enhanced form of ordinary topology by introducing the concept of membership value in a fuzzy set using fuzzy logic theory. The combined method of fuzzy based SVM, named FSVM uses fuzzy topology. The FSVM algorithm imposes a fuzzy membership to each input point which may belong to multiple classes on the decision surface. By using different types of fuzzy membership definitions, they apply FSVM to solve different kinds of problems.

This enhances the effectiveness of SVM even in fuzzy domain [27]. In spite of the efficiency of the SVM approach, this method still has some classification limits for its theory. For each class, the SVM usually treats all training points of this class uniformly following the theory of SVM. In many real-world problems, the consequences of choosing the training points affect the classification outputs. It is frequently that some specific training points are more important than others in the classification problem of remote sensing images. Therefore to choose more important training points becomes very important.

Foody and Mathur [28] showed that only a quarter of the original training samples acquired from SPOT HRV satellite imagery was sufficient to produce an equally high accuracy for a two-crop classifier. Mantero et al. [29] estimated probability density of thematic classes using an SVM. The SVM based approach used are cursive procedure to generate prior probability estimates for known and unknown classes by adapting the Bayesian minimum-error decision rule. The approach was tested using synthetic data and two optical sensor data (i.e., Daedalus ATM and Landsat TM) and confirmed method effectiveness, especially when the availability of ground reference data was limited.

Bruzzone et al. [30] implement the transductive inference learning theory. They incorporate this method into an SVM for remote sensing classification. Their SVM-based method defines the separating hyperplanes according to their algorithm that integrates the unlabeled samples together with the training samples. Their experiments demonstrate that the proposed method is effective, for a set of ill-posed remote sensing classification problems with limited training samples.

Foody and Mathur [31] propose a method with on mixed pixel training samples over conventional pure pixel samples, for an SVM classifier. The analysis of a three-wave band multispectral SPOT HRV image showed the benefits of mixed pixel sampling on a crop-type classification task. Foody et al. [31] evaluate four different dataset reduction approaches for a one-class problem (cotton class vs. other classes) using SVMs. They work on LISS-III data and found that significant data reduction is feasible ($\sim$90 %) with minimal information loss. Sahoo et al. [32] incorporate localized, highly sensitive transformations to capture subtle changes in hyperspectral signatures. They compare the outputs of so called S-transform method with those of the classifiers without S-transform method. The results come satisfactory. The implemented algorithm is on an SVM which exhibits additional robustness for small data samples in a geological classification.

Blanzieri and Melgani [33] investigate a local k-nearest neighbor (k-NN) adaptation method to formulate localized variants of SVM approaches. Their results exhibit encouraging improvements, specifically with the integration of nonlinear kernel functions. Tuia and Camps-Valls [34] experiment the issue of kernel pre-determination by developing a regularization method to identify the kernel structure from the analysis of unlabeled samples. Camps-Valls et al. [35] experiment with an improved version of their method to assess kernel independence in various image types using the Hilbert–Schmidt independence criterion.

Marconcini et al. [36] develop an algorithm with the incorporation of spatial information through composite kernels. Their approach finds satisfactory improvements, however, with an additional computation cost. Camps-Valls et al. [37] experiment in another work develops a method using composite kernels for multi-temporal classification of remote sensing data from multiple sources. This method has been tested using both synthetic and real optical Landsat TM data. They demonstrate that the cross-information composite kernel was the best in general, but a simple summation kernel also exhibit similar improved performance. They work with composite kernels in their earlier work [38].

Chi et al. [39] develop a method, named primal SVM. This algorithm is capable of classify landcovers using areas with notably small amount of training examples. Their method experiment to replace the regularization-based earlier approaches using SVMs. The primal SVM vector development makes it possible to optimize directly on the primal representation, and therefore limits the number of samples in their approach. They evaluate their work on Hyperion imagery of the Okavango Delta (in Botswana) for vegetation classification. Primal SVM exhibits competitive accuracy results in comparison with the state-of-art alternative algorithms trained on larger datasets.

Gómez-Chova et al. [40] incorporate an addition of a regularization term on the geometry of both labeled and unlabeled samples on SVM. The variation is based on graph Laplacian, leading to a Laplacian SVM variant. Their semi-supervised classification approach offers new direction when compared with traditional SVMs with more efficient results. It shows its superiority in cases especially with small training datasets and for complex problems. Castillo et al. [41] develop a modified version of SVM algorithm, namely bootstrapped SVM. The training method adapted in this bootstrapped SVM is to training pool. An incorrectly classified training sample in a training step is removed from the training pool. It is reassigned to a correct label and reintroduced into proper class of the training set in the next training cycles. Their result shows the ability to capture data variability even in a highly biased binary dataset. With only 0.05 % of the total number of training pixels it can show to achieve about the same accuracy level as the standard SVM.

An interesting SVM adaptation was proposed by Wang and Jia [42], where the space between support vectors is considered to provide a soft classification in addition to the traditional hard classification. Demir and Erturk [43] offer an improved algorithm over hyperspectral SVM classifiers by incorporating border training samples in a two-step classification process.

Similarly, Song et al. [44] experiment with an SVM adaptation for Landsat-based vegetation monitoring. Their SVMs parameters are set using an integration of one and two class SVM sequential classification steps. Further, Mathur and Foody [45] experiment with the methods for efficient reduction of field data. They conclude that for cropland segmentation mapping classification, the good results can be obtained with one third of the original dataset assuming to be training points in SVM methods. In their experiment, at the 24 m ground pixel resolution acquired by the LISS-III sensor, their reduced dataset yield a small 1.34 % accuracy with a loss at 90.66 %.

Integration of a genetic algorithm (GA) and SVM for remote sensing classification has been experimented with a limited availability of training samples by Ghoggali et al. [46] in their remote sensing works. The experimental results exhibit an ability to improve classification accuracy with a small training sample size. However, the computational load becomes heavy primarily due to the slow GA convergence. Ghoggali and Melgani [46] combined genetic training into SVM classification in order to incorporate land cover transition rules in multi temporal classification. The results show an improved performance.

Bruzzone and Persello [30] develop a new context-sensitive semi-supervised SVM classification model, which they successfully use on chosen dataset when some of training data are not reliable. Their model explores the contextual information of the neighboring pixels of each training sample and improves the unreliable training data. They experiment their algorithm using Ikonos and Landsat TM data and compare their obtained results with those based on some of the most popular classification algorithms like the standard SVM, a progressive semi-supervised SVM, maximum likelihood (ML) and k-nearest neighbor (k-NN) algorithm. Their implemented SVM algorithm is superior to the other classification models in terms of robustness and effectiveness, particularly when nonfully reliable training samples are used.

Huang and Zhang [47] experiment with a multi-SVM method using traditional vector stacking techniques on high resolution urban mapping. Su [48] demonstrate training data reduction using a hierarchical clustering analysis on a Multi angle Imaging Spectro Radiometer (MISR) satellite data (250 m–1.1 km, 17 products) over a vegetation classification problem. It has been shown that a two thirds reduction of the dataset size is possible without significant accuracy degradation in combining approach of SVM and MLC methods.

Gomez-Chova et al. [40] implement a method to increase classification reliability and accuracy by combining labeled and unlabeled pixels using clustering and the mean map kernel methods. They experiment their approach to classify clouds using En visat's Medium Resolution Imaging Spectrometer (MERIS) data. Their experiment reveals that their method is specifically particularly successful when sample selection bias (i.e., training and test data follow different distributions) exists. Selecting an optimum SVM method for remote sensing classification is a very challenging task now a day. Foody and Mathur [18] implement a single multiclass SVM classification method while typical multiclass SVMs are based mainly on the use of multiple binary analyses. They evaluate the results of their approach with other classification methods, like discriminant analysis (DA), decision trees (DT), and neural networks (NN). They also exhibit the SVM-based method which is superior to the other methods with different sizes of training samples.

Bazi and Melgani [49] experiment on a most appropriate feature subspace and model selection based on a genetic optimization model. They use three feature selection methods including steepest ascent, recursive feature elimination technique, and the radius margin bound minimization method. They make constrain with two criteria—the simple support vector count and the radius margin bound. They use those two criteria to identify an optimum SVM-based classification method for hyperspectral remote sensing image classification problem. The genetically optimized SVM using the support vector count as a criterion demonstrate the best performance for both simulated and real-world AVIRIS hyperspectral data.

Mathur and Foody [45] experiment with the performance of SVMs in nonbinary classification tasks. Their results show their implemented one shot SVM classifier is superior to the binary-based multiple classifiers in terms of obtained accuracy but also in initial parameterization.

SVMs have also been used for feature selection. Pal [50] implement methods for feature selection based on SVMs. Showing the advantage of exhaustive search approaches for real-world problems, the scientists puts importance on the use of a non-exhaustive search procedure in selecting features with high discriminating power from large search spaces. SVM-based methods combined with GA are comparatively better than the random forest feature selection method, in land cover classification problems with hyperspectral data. Earlier works also exhibit their small benefits. Zhang and Ma [51] work on the issue of feature selection in SVM approaches. They implement a modified recursive SVM approach to classify hyperspectral AVIRIS data. The reduced dimensionality demonstrates slightly better results, however, their method has higher computational demands compared with others.

On the same subject Archibald and Fann [52] propose a hybrid integration of feature selection within the SVM classification approach. They obtain efficient accuracy while significantly reducing the computational load. Some studies show the improvements on the performance of SVM-based classification through algorithms and/or data fusion. Zhang et al. [53] define a pixel shape index approach describing the contextual information of nearby pixels. They evaluate its efficiency over land cover classification using QuickBird data based on SVMs. In their work, the pixel shape indices are combined with transformed spectral bands using methods like principal component analysis (PCA) or independent component analysis (ICA). They show that integration of spectral and shape features as well as the transformed spectral components in an SVM produce improved classification accuracy.

3 Hybrid Rough-PSO Approach for Pixel Classification

3.1 Rough Set-Based Automatic Pixel Classification Approach

To explicitly identify the initial automatic diversity in pixels of remote sensing image/data, we introduce the rough set based classification rules on patterns with PSO-based corrections in our RPSO algorithm. The patterns are grouped into automatically determined rough limit RL number of clusters using rough set-based classification rules in the initial phase.

3.1.1 Range Initialization

Given a decision system $\beta = (U, B \cup \{d\})$ in general B is called the set of conditional attributes and $d \notin B$ is called the decision attribute. To initialize, all patterns have been assigned a single class in the decision attribute. The regions of the spectral bands of the remote sensing image are transformed into the decision variables.

3.1.2 Compute Boundary Approximations

To build the feasible set boundaries, the following B—lower and B—upper approximations for each decision variable set X_i, denoted by $\underline{B}X_i$ and $\overline{B}X_i$ respectively, are computed: $\overline{B}X_i = \{x | [x]_B \subseteq X_i\}$ and $\overline{B}X_i = \{x | [x]_B \cap X_i \neq \phi\}$. Here $[x]_B$ denotes some value x described by a set of attributes B. For each set X_i, we compute the accuracy measurement in Eq. 3.

3.1.3 Compute Reducts and Generalized Decisions

From the indiscernibility relation R, we define $\text{ind}_\beta(A, x, d)$ to be the set of patterns, which are either indiscernible from the pattern x considering the attribute set A or are equivalent with x for the decision attribute d. From this definition, we derive the reduct set $\Re_A(x, d), \forall x \in A$ to be the minimal set of attributes B with similar discriminative power as A.

$$\Re_A(x, d) = \{B | B \subseteq A, (\text{ind}_\beta(B, x, d) == \text{ind}_\beta(A, x, d)\}.$$

These reducts synthesize the fundamental descriptor $\delta_A(x) = \{i | (\exists y \in U, (y \in \text{ind}_\beta(A, x, d) \text{ and } d(y) == i)\}$ to define the generalized decision rules. The minimal decision rules set in β constitutes

$$\Delta_\beta = \cup_{x \in U} \{\delta_B(x) \to \delta'_B(x) | B \in \Re_\beta(x, d)\}$$

3.1.4 Automatic Rough Limit and Decision Classes Generation

We utilize the strategy for generating *best* rules of the Rough Set Library (RSL) [54] to use for the estimation of the strength of the rules to automatically compute initial clusters. For estimating classificatory accuracy, we calculate the rough sets-based *quality of classification* coefficient as defined in Eq. 4. The maximum value for the *Strength of Rules* on the dynamically generated clusters is taken to be the automatic generation of *RoughLimit* (*RL*) to determine the initial number of clusters.

Each decision d determines a classification of objects in X [14, 15] as $\text{CLASS}_R(d) = \left\{X_R^1, \ldots, X_R^{r(d)}\right\}$ of the universe U, where $X_R^k = \{x \in U | d(x) = v_d^k\}$ for $1 \leq k \leq r(d)$, where v_d^k denotes kth value for decision d and $r(d)$ denotes the rank of decision d. the optimal decision rules also determines the partitions of the universe U. The set $X_B(u)$ denotes the decision class $\{x \in U | d(x) = d(u)\} \forall u \in U$, determined the decision d. This automatic class detection also enhances the quantitative efficiency of the result clusters with improved validity as we found in our experiments.

4 PSO Based Membership Correction Method

The new RPSO algorithm consists of two phases—initial automatic rough set based heuristical rule generation on the chosen remote sensing image to generate rough membership matrix U and finally using that membership, the membership-based PSO method to generate optimal pixel allocations for overlapping regions, as shown in Fig. 1.

Initial random assignment put N pixels in K clusters for initializing rough set-based rule generation algorithm, as described in the previous subsection. We initialize the membership degree matrix U from the initial allocations in rough set automatic decision rule generation algorithm.

After the first phase of RPSO algorithm, we obtain the membership matrix to generate the initial pixel positions in $M(= K)$ number of rough classes to denote overlapping regions for our hybrid membership-based PSO approach. The value $x_{\text{initial}} = U$ has been set and V_{initial} for our PSO approach is computed within the constraints $\{-v_{\max}, v_{\max}\} = \{0, 1\}$ using Rastrigrin function. The Rastrigrin function is shown below:

Fig. 1 The flowchart of
RPSO algorithm for remote
sensing classification

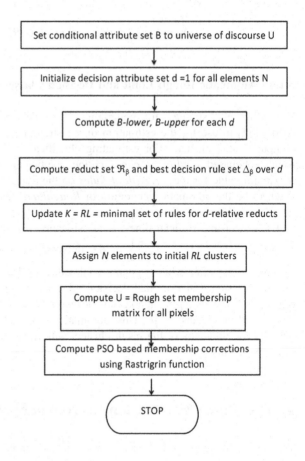

$$f(x_i) = \sum x_i^2 - 10 * \cos(2 * \pi * x_i) + 10 \,, \qquad (11)$$

$$x_{\min} = [0, 0, \ldots, \ 0], \qquad (12)$$

$$f x_{\min} = 0 \qquad (13)$$

Similarly, $\{x_{\max}, x_{\min}\} = \{0, 1\}$.

Using these new constraints, we generate new membership degree matrix $U_{\mathrm{SE}} = x_m^{t+1}$ and reassign the pixels to the clusters with maximum membership values. The cluster centroids are updated again and the iteration continues, until the convergence occurs with a difference between the old best particle in current population $f_{G\mathrm{Best}}$ and global minimum GM to be smaller than the terminating threshold $\in$. The validity indices are computed lastly over final PSO-optimized solutions.

5 Application to Pixel Classification

5.1 Experimental Framework

The new RPSO algorithm is implemented using MATLAB 7.0 and RSL library on MacBook dual-core processor. To compare well-known K-Means and FCM methods are also executed. Dunn [55], Davies–Bouldin (DB) [56] and Silhouette [57] validity indices evaluate the effectiveness of RPSO over K-Means and FCM quantitatively. The efficiency of RPSO is also verified visually from the clustered images considering ground truth information of land cover areas. We have used one remote sensing data set for evaluation purpose. A numeric image data contains some image pixels where land cover types denote class labels and band intensity values serve as feature values. Absence of pixel locations loses the spatial information. We have used data set, named 'IRS with 198 samples.' This data set has 198 samples and 3 bands: green, red, and near infrared (NIR) [2] with overlapping 6 classes.

5.2 Validity Indices

The fitness of a solution indicates the degree of goodness of the solution of the proposed algorithm [58]. In this article, three validity indices values, namely Davies-Bouldin (DB), Dunn and Silhouette indices, are used to determine the performance of the new hybrid algorithm. The validity indices are now described below.

'Davies—Bouldin index (DB)'—The Davies–Bouldin index (DB) [56] is a metric for evaluating clustering algorithms. This is an internal evaluation validity index. The best clustering solutions essentially minimizes the Davies–Bouldin Index.

'Dunn index'—The Dunn index (DI) [55] is another metric for evaluating clustering algorithms. This is an internal validity index. For one allocation of clusters, a higher Dunn index indicates better clustering. Let C_i be a cluster of vectors. If there are M clusters, then the Dunn Index for the set is defined as:

$$\mathrm{DI}_M = \min_{1 \leq i \leq M} \left\{ \min_{1 \leq j \leq m, j \neq i} \left\{ \frac{\delta(c_i, c_j)}{\max\limits_{1 < K < M} \Delta_K} \right\} \right\} \tag{14}$$

'Silhouette index s(C)'—Let a denotes the average distance of a point from other points of same cluster and b denotes the minimum of the average distances of that point from the points in other clusters. Then the Silhouette width (s) is defined as follows:

$$s = (b-a)/\max\{a, b\} \tag{15}$$

Silhouette index s(C) [57] is the average Silhouette width of all points, which reflects the compactness and separation of clusters. The value of s(C) varies from −1 to 1. For appropriate clustering s(C) should be high [57].

5.3 *Performance Analysis*

The new RPSO algorithm is implemented using MATLAB 7.0 on HP 2 quad processor with 2.40 GHz. To compare well-known K-Means and FCM methods are also executed. Dunn [55], Davies-Bouldin (DB) [56], and Silhouette [57] validity indices evaluate the effectiveness of RPSO over K-Means and FCM quantitatively. The efficiency of RPSO is also verified visually from the clustered images considering ground truth information of land cover areas.

The chosen LANDSAT image of the catchment region of Ajoy River, which has been extracted for further research works, is available in 3 bands viz. green, red, and blue bands with original image as shown in Fig. 2. Figure 2 shows the original LANDAST image of Ajoy River catchment with histogram equalization with 7 classes: turbid water (TW), pond water (PW), concrete (Concr.), vegetarian (Veg), habitation (Hab), open space (OS), and roads (including bridges) (B/R).

The river Ajoy cuts through the image in the middle of the catchment area. From the upper left corner of the catchment area the river is flowing through the middle part of the selected area. The river is shown a thin line in the middle of the catchment area of blue and river colors.

The segmented catchment area of Ajoy River images obtained by K-Means and FCM algorithms, respectively, are shown in Figs. 3 and 4 for ($K = 7$). In Fig. 3, K-Means algorithm fails to classify the catchment area from the background. FCM clustering solutions in Fig. 4 also fails to detect the catchment area properly from the background in the middle part. Some waterbodies part and the background are mixed in both K-Means and FCM clustering solutions in Figs. 3 and 4 respectively. However, our new RPSO algorithm is able to separate all catchment areas from the background as shown in (Fig. 5). These indicate that RPSO algorithm detects the overlapping arbitrary-shaped regions significantly with better efficiency than K-Means and FCM algorithms.

Fig. 2 Original image of the catchment area of Ajoy River

Fig. 3 Pixel classification of
Ajoy River catchment area
(partial) obtained by K-Means
algorithm (with $K = 7$)

Fig. 4 Pixel classification of
Ajoy River catchment area
(partial) obtained FCM
algorithm (with $K = 7$)

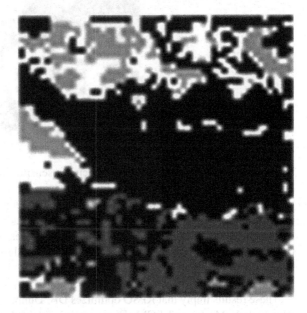

5.4 Quantitative Analysis

The clustering results have been evaluated objectively by measuring validity
measures Davies-Bouldin (DB), Dunn, and Silhouette index, as defined in [55, 56,
57] respectively, for K-Means, FCM, and RPSO algorithms on the remote sensing

Table 1 Validity indices values of the classified remote sensing image provided by K-means, FCM, and RPSO algorithms

Index	IRS 198			Ajoy catchment image		
	K-Means	FCM	RPSO	K-Means	FCM	RPSO
Davies-Bouldin index	0.582	2.993	0.708	0.402	0.667	4.243
Silhouette index	0.357	−0.341	0.579	0.400	0.170	0.235
Dunn index	0.472	0.144	2.664	1.964	2.109	73.163

Fig. 5 Pixel classification of Ajoy River catchment area (partial) obtained by RPSO algorithm

image data and Ajoy River catchment remote sensing image in Table 1. It can be noticed that, RPSO produces best final value on the catchment image for maximized Dunn index as 73.163, while K-Means obtains a Dunn value of 1.964 and FCM obtains 2.109. The maximizing Silhouette index values on Ajoy River catchment image for K-Means, FCM and RPSO are respectively, 0.400, 0.170, and 0.235. Similarly, the DB index produced by RPSO algorithm (minimizing DB) is 0.708 for the chosen remote sensing data, which is better value than FCM algorithm.

These results imply that RPSO optimizes DB, Dunn, and Silhouette indices more than both K-Means and FCM. Hence, it is evident that RPSO is comparable in goodness of solutions to K-Means and FCM algorithms and even RPSO sometimes outperform to obtain superior fuzzy clustering results.

5.5 Statistical Analysis

A nonparametric statistical significance test called Wilcoxon's rank sum for independent samples has been conducted at 5 % significance level [59]. Two groups have been created with the performance scores, Silhouette index values produced by 10 consecutive runs of FCM and RPSO algorithms on the chosen remote sensing Image. From the medians of each group on the dataset in Table 2, it is observed that RPSO provides better median values than FCM algorithm.

Table 3 shows the P-values and H-values produced by Wilcoxon's rank sum test for comparison of two groups, RPSO-K-Means and RPSO-FCM. All the P-values reported in the table are less than 0.005 (5 % significance level). For the chosen remote sensing Image on the catchment area of Ajoy, comparative P-value of rank sum test between RPSO and K-Means is very small 2.43E-005, indicating the performance metrics produced by RPSO to be statistically significant and not occurred by chance. Similar results are obtained for other group with FCM algorithm also. Hence, all results establish the significant superiority of RPSO over K-Means and FCM algorithms.

6 Future Research Works

As a scope of future research, the time efficiency of RPSO algorithm may be improved further by mapping it to the master-slave paradigm. Moreover, incorporation of spatial information in the feature vector as this is found to be effective in pixel classification, in lieu of intensity values at different spectral bands in RPSO method, constitutes an important direction for farther research.

Table 2 Median values of performance parameter Silhouette index over 10 consecutive runs on different algorithms

Data	Algorithms	
	FCM	RPSO
Catchment area of Ajoy River image	0.171	0.235
IRS 198	−0.786	−0.587

Table 3 P-values produced by rank sum while comparing RPSO with K-means and RPSO with fcm respectively

Algorithm	Ajoy catchment image		IRS 198	
	H	P-value	H	P-value
K-Means	1	2.43E-005	1	1.59E-005
FCM	1	2.00E-003	1	2.00E-003

7 Conclusions

In the realm of the remotely sensed imagery, the mixed pixel problems are common. This problem denotes the presence of multiple and partial class memberships for them. Therefore, the conventional crisp methodology fails to map land covers properly to different regions similar to the ground truth information. The soft computing theory may overcome this problem. Therefore rough set theory may be applied to map overlapping regions in the image.

The contribution of this article lies in better detection of overlapping land cover regions in the remote sensing image than other crisp partitioning methodology by utilizing new rough set-based automatic pixel in our RPSO clustering algorithm. The primary contributions are—to utilize one new rough set-based automatic initial classification in remote sensing imagery with PSO-based membership correction. The membership correction phase helps to correct the wrong allocation of a single pixel to multiple clusters. It verifies the overall allocations with respect to the rough set generated membership values to obtain improved land cover regions.

The efficiency of the new RPSO algorithm is demonstrated over one chosen remote sensing image of the catchment region of Ajoy River. Superiority of new RPSO clustering algorithm over the widely used K-Means and FCM algorithms is established quantitatively over three validity indices. The verification with ground truth information also shows significant efficiency of new RPSO algorithm over other two existing methods. Statistical tests also establish the statistical significance of RPSO over K-Means and FCM algorithms.

As a scope of future research, the time efficiency of RPSO algorithm may be improved further by mapping it to the distributed environment [10, 60, 61]. Moreover, in corporation of spatial information in the feature vector as this is found to be effective in pixel classification [62], in lieu of intensity values at different bands in RPSO method, constitutes an important direction for further research.

References

1. Gonzalez RC, Woods RE (1992) Digital image processing. Addison-Wesley, Massachusetts
2. Spang R (2003) Diagnostic signatures from microarrays, a bioinformatics concept for personalized medicine. BioSilico 1(2):64–68
3. Tavazoie S, Hughes J, Campbell M, Cho R, Church G (2001) Systematic determination of genetic network architecture. Bioinformatics 17:405–414
4. Hoon MJL, de Imoto S, Nolan J, Miyano S (2004) Open source clustering software. Bioinformatics 20(9):1453–1454
5. Lukashin A, Futchs R (1999) Analysis of temporal gene expression profiles, clustering by simulated annealing and determining optimal number of clusters. Nat Genet 22:281–285
6. Xu Y, Olman V, Xu D (1999) Clustering gene expression data using a graph theoretic approach, an application of minimum spanning trees. Bioinformatics 17:309–318
7. Dembele D, Kastner P (2003) Fuzzy c-means method for clustering microarray data. Bioinformatics 19:973–980

8. de Souto MCP, R RBCP, Soares RGF, de Araujo DSA, Costa IG, Ludermir TB, Schliep A (2008) Ranking and selecting clustering algorithms using a meta-learning approach. In: Proceedings of IEEE international joint conference on neural networks. IEEE Computer Society, pp 3728–3734

9. Maulik U, Mukhopadhyay A, Bandyopadhyay S (2009) Combining pareto-optimal clusters using supervised learning for identifying co-expressed genes. BMC Bioinformatics 10:27

10. Maulik U, Sarkar A (2012) Efficient parallel algorithm for pixel classification in remote sensing imagery. GeoInformatica 16(2):391–407

11. Cordasco G, Scara V, Rosenberg AL (2007) Bounded-collision memory-mapping schemes for data structures with applications to parallel memories. IEEE Trans Parallel Distrib Syst 18 (7):973–982

12. Qin J, Lewis D, Noble W (2003) Kernel hierarchical gene clustering from microarray gene expression data. Bioinformatics 19:2097–2104

13. Eisen M, Spellman P, Brown P, Botstein D (1998) Cluster analysis and display of genome-wide expression patterns. Proc Natl Acad Sci USA 95:14863–14868

14. Pawlak Z (1982) Rough sets. Int J Comput Inf 11:341–356

15. Pawlak Z (1991) Rough sets, theoretical aspects of reasoning about data. Kluwer Academic Publishers, Dordrecht

16. Dixon B, Canade M (2008) Multispectral land use classification using neural networks and support vector machines: One or the other, or both? Int J Remote Sens 29(4):1185–1206

17. Mountrakis G, Im J, Ogole C (2011) Support vector machines in remote sensing: a review. ISPRS J Photogrammetry Remote Sens 66(3):247–259

18. Foody GM, Mathur A (2004) A Relative evaluation of multiclass image classification by support vector machines. IEEE Trans Geosci Remote Sens 42(6):1335–1343

19. Chiang JH, Hao PY (2004) Support vector learning mechanism for fuzzy rulebased modeling: A new approach. IEEE Trans Fuzzy Syst 12(1):1–12

20. Melgani F, Bruzzone L (2004) Classification of hyperspectral remote sensing images with support vector machines. IEEE Trans Geosci Remote Sens 42(8):1778–1790

21. Zhang H, Shi W, Liu K (2012) Fuzzy-topology integrated support vector machine for remotely sensed image classification. IEEE Trans Geosci Remote Sens 50(3):850–862

22. Guiyun L, Junfei C, Jiameng Z (2012) An integrated SVM and fuzzy AHP approach for selecting third party logistics providers. Electrotechnical Review (Electrical Review) ISSN 0033-2097, R. 88 NR 9b/2012

23. Liu Z, Yang H, Yang S (2009) Integration of multi-layer SVM classifier and multistage dynamic fuzzy judgment and its application in SCDA measurement. J Comput 4(11):1139–1144

24. Li Y, Yan L, Liu J (2009) Remote sensing image classification development in the past decade. Proceedings of SPIE 7494(74941D-1):338–343

25. Qiu ZG, Zhang C, Li L. Qiong, Hui XX, Zhang G (2008) High efficient classification on remote sensing images based on SVM. The International Archives of the Photogrammetry, Remote Sensing and Spatial Information Sciences XXXVII (B2)

26. Tsujinishi D, Abe S (2003) Fuzzy least squares support vector machines for multiclass problems. Neural Network 16(5/6):785–792

27. Warrender C, Forrest S (1999) Detecting intrusions using system calls: alternative data models. Proceedings of the IEEE Computer Society Symposium on Research in Security and Privacy, vol 6, issue 5, pp 133–145

28. Foody GM, Mathur A (2004) Toward intelligent training of supervised image classifications: directing training data acquisition for SVM classification. Remote Sens Environ 93(1–2):107–117

29. Mantero P, Gabriele M, Serpico SB (2005) Partially supervised classification of remote sensing images through SVM-based probability density estimation. IEEE Trans Geosci Remote Sens 43(3):559–570

30. Bruzzone L, Chi M, Marconcini M (2006) A novel transductive SVM for semisupervised classification of remote-sensing images. IEEE Trans Geosci Remote Sens 44(11):3363–3373

31. Foody GM, Mathur A (2006) The use of small training sets containing mixed pixels for accurate hard image classification: training on mixed spectral responses for classification by a SVM. Remote Sens Environ 103(2):179–189
32. Sahoo BC, Oommen T, Misra D, Newby G (2007) Using the one-dimensional s-transform as a discrimination tool in classification of hyperspectral images. Can J Remote Sens 33(6):551–560
33. Blanzieri E, Melgani F (2008) Nearest neighbor classification of remote sensing images with the maximal margin principle. IEEE Trans Geosci Remote Sens 46(6):1804–1811
34. Tuia D, Camps-Valls G (2009) Semisupervised remote sensing image classification with cluster kernels. IEEE Geosci Remote Sens Lett 6(2):224–228
35. Camps-Valls G, Mooij J, Scholkopf B (2010) Remote sensing feature selection by kernel dependence measures. IEEE Geosci Remote Sens Lett 7(3):587–591
36. Marconcini M, Camps-Valls G, Bruzzone L (2009) A composite semisupervised SVM for classification of hyperspectral images. IEEE Geosci Remote Sens Lett 6(2):234–238
37. Camps-Valls G, Gomez-Chova L, Munoz-Mari J, Rojo-Alvarez JL, MartinezRamon M (2008) Kernel-based framework for multitemporal and multisource remote sensing data classification and change detection. IEEE Trans Geosci Remote Sens 46(6):1822–1835
38. Camps-Valls G, Gomez-Chova L, Munoz-Mari J, Vila-Frances J, Calpe-Maravilla J (2006) Composite kernels for hyperspectral image classification. IEEE Geosci Remote Sens Lett 3 (1):93–97
39. Chi M, Feng R, Bruzzone L (2008) Classification of hyperspectral remote-sensing data with primal SVM for small-sized training dataset problem. Adv Space Res 41(11):1793–1799
40. Gómez-Chova L, Camps-Valls G, Muñoz-Marí J, Calpe J (2008) Semisupervised image classification with Laplacian support vector machines. IEEE Geosci Remote Sens Lett 5 (3):336–340
41. Castillo C, Chollett I, Klein E (2008) Enhanced duckweed detection using bootstrapped SVM classification on medium resolution RGB MODIS imagery. Int J Remote Sens 29(19):5595–5604
42. Wang WJ (1997) New similarity measures on fuzzy sets and on elements. Fuzzy Sets Syst 85:305–309
43. Demir B, Erturk S (2009) Clustering-based extraction of border training patterns for accurate SVM classification of hyperspectral images. IEEE Geosci Remote Sens Lett 6(4):840–844
44. Song X, Cherian G, Fan G (2005) Av-insensitive SVM approach for compliance monitoring of the conservation reserve program. IEEE Geosci Remote Sens Lett 2(2):99–103
45. Foody GM (2008) RVM-based multi-class classification of remotely sensed data. Int J Remote Sens 29(6):1817–1823
46. Ghoggali N, Melgani F, Bazi Y (2009) A multiobjective genetic SVM approach for classification problems with limited training samples. IEEE Trans Geosci Remote Sens 47 (6):1707–1718
47. Huang X, Zhang L (2010) Comparison of vector stacking, multi-SVMs fuzzy output, and multi-SVMs voting methods for multiscale VHR urban mapping. IEEE Geosci Remote Sens Lett 7(2):261–265
48. Su L (2009) Optimizing support vector machine learning for semi-arid vegetation mapping by using clustering analysis. ISPRS J Photogrammetry Remote Sens 64(4):407–413
49. Bazi Y, Melgani F (2006) Toward an optimal SVM classification system for hyperspectral remote sensing images. IEEE Trans Geosci Remote Sens 44(11):3374–3385
50. Pal M (2006) Support vector machine-based feature selection for land cover classification: a case study with DAIS hyperspectral data. Int J Remote Sens 27(14):2877–2894
51. Zhang R, Ma J (2009) Feature selection for hyperspectral data based on recursive support vector machines. Int J Remote Sens 30(14):3669–3677
52. Archibald R, Fann G (2007) Feature selection and classification of hyperspectral images with support vector machines. IEEE Geosci Remote Sens Lett 4(4):674–677
53. Zhang L, Huang X, Huang B, Li P (2006) A pixel shape index coupled with spectral information for classification of high spatial resolution remotely sensed imagery. IEEE Trans Geosci Remote Sens 44(10):2950–2961

54. Gawrys M, Sienkiewicz J (1994) Rsl–the rough set library version 2.0. ICS Research Report 27/94 Warsaw. Poland, Institute of Computer Science. W. U. of T
55. Dunn JC (1973) A fuzzy relative of the ISODATA process and its use in detecting compact well-separated clusters. J Cybern 3(3):32–57
56. David LD, Donald WB (1979) A cluster separation measure. IEEE Trans Pattern Anal Mach Intell PAMI 1(2):224–227
57. Rousseeuw PJ (1987) Silhouettes: a graphical aid to the interpretation and validation of cluster analysis. J Comput Appl Math 20:53–65
58. Young KY (2001) Validating clustering for gene expression dat. Bioinformatics 17:309–318
59. Hollander M, Wolfe D (1999) Nonparametric statistical methods, 2nd edn. Weily, USA
60. Sarkar A, Maulik U, (2009) Parallel point symmetry based clustering for gene microarray data, In: The proceedings of seventh international conference on advances in pattern recognition-2009 (ICAPR, 2009), Kolkata, IEEE Computer Society, Conference Publishing Services (CPS), pp 351–354
61. Sarkar A, Maulik U (2009) Parallel clustering technique using modified symmetry based distance. In: The proceedings of 1st international conference on computer, communication, control and information technology (C3IT 2009), MacMillan Publishers India Ltd, pp 611–618
62. Bandyopadhyay S (2005) Satellite image classification using genetically guided fuzzy clustering with spatial information. Int J Remote Sens 26(3):579–593
63. Smith III AR (1971) Two-dimensional formal languages and pattern recognition by cellular automata. In: IEEE conference record of 12th annual symposium on switching and automata theory
64. Wolfram S (1986) Cryptography with cellular automata. Lecture notes on computer Science 218:429–432
65. Wolfram S (1983) Statistical mechanics of cellular automata. Rev Mod Phys 55(3):601–644

A Study and Analysis of Hybrid Intelligent Techniques for Breast Cancer Detection Using Breast Thermograms

Usha Rani Gogoi, Mrinal Kanti Bhowmik, Debotosh Bhattacharjee,
Anjan Kumar Ghosh and Gautam Majumdar

Abstract The growing incidence and mortality rate of breast cancer draw the attention of the researchers to develop a technique for improving the survival rate of the cancer patients. Medical infrared thermography (MIT) with sensitivity 90 % has proved itself as a safe and promising method for early breast cancer detection. Moreover, an abnormal breast thermogram can signify breast pathology. The accurate classification and diagnosis of these breast thermograms is one of the major problem in decision making for treatments, which leads to the utilization of hybrid intelligent system in breast thermogram classification. Hybrid intelligent system plays a vital role in survival prediction of a breast cancer patient, and it is highly significant in decision making for treatments and medications. The primary objective of a hybrid intelligent system is to take the advantages of its constituent models and at the same time lessen their limitations. This chapter is an attempt to highlight the reliability of infrared breast thermography and hybrid intelligent system in breast cancer detection and diagnosis. A detailed overview of infrared breast thermography including its principles and role in early breast cancer detection is described here. Several research works are carried out by various researchers

U.R. Gogoi (✉) · M.K. Bhowmik · A.K. Ghosh
Department of Computer Science and Engineering, Tripura University
(A Central University), Suryamaninagar, 799022 Tripura, India
e-mail: ushagogoi.cse@gmail.com

M.K. Bhowmik
e-mail: mkb_cse@yahoo.co.in

A.K. Ghosh
e-mail: anjn@ieee.org

D. Bhattacharjee
Department of Computer Science and Engineering, Jadavpur University,
Kolkata 700032, India
e-mail: debotosh@ieee.org

G. Majumdar
Radiotherapy Department, Regional Cancer Center, Agartala Government
Medical College Agartala, 799006 Tripura, India
e-mail: drgmajumdar@yahoo.in

© Springer India 2016
S. Bhattacharyya et al. (eds.), *Hybrid Soft Computing Approaches*,
Studies in Computational Intelligence 611,
DOI 10.1007/978-81-322-2544-7_11

to identify the breast pathology from breast thermograms by using hybrid intelligent techniques which include extraction and analysis of several statistical features. A study of research works related to feature extraction and classification of breast thermograms using various types of hybrid classifiers is also included in this chapter.

Keywords Breast cancer · Digital infrared imaging · Infrared breast thermography · Breast asymmetry · Breast cancer detection

1 Introduction

The incidence of breast cancer has been increasing globally, and it is the most common among all cancers accounting for more than 1.6 % of deaths. Over the last few decades in India, the average age of developing breast cancer has shifted to 30–40 years [1]. According to the National Cancer Registry Program 2006–2008 (NCRP), breast cancer is the most common cancer in India accounting 25–32 % of all cancers in female [1]. This implies that one-fourth of all female cancer cases is breast cancer [1]. As per the report made by American Society of Clinical Oncology (ASCO) in 2009, the 5-year survival rate of breast cancer has increased from 79 % (in 1984–1986) to 89 % (1996–2004) in United States [1]. Due to the lack of breast cancer awareness and inadequate medical facilities, the survival rate of breast cancer patient is very poor in India comparative to United States. With the growing rate of breast cancer incidence, demand for developing new technologies and improving existing technologies for breast cancer prevention is increasing. Till now, no effective cure is there to prevent breast cancer. The probability of successful treatment and complete recovery of the patient entirely depends on the early detection and diagnosis of the breast cancer. If discovered early, breast cancer is a highly treatable disease, with 97 % chances of survival [2]. Medical imaging like mammography, breast ultrasound, infrared breast thermography, etc. has been considered as an essential tool for early detection, better diagnosis and effective treatment of breast cancer.

Among various breast imaging techniques, mammography is considered as the gold standard for breast cancer detection with a sensitivity of 80 %. However, in case of dense breast tissue in younger women, detection of suspicious lesions is very difficult from mammography. Compare to the older women of age above 50, breast cancers grow very faster in younger women under 40 years. The faster a malignant tumor grows the amount of infrared radiation it generates is also greater that can be recorded using a very sensitive thermal camera. Therefore, particularly for younger women under 40, thermography acts as a safe early risk marker of breast pathology. Moreover, mammography cannot detect a tumor until when it is of a certain size. Keyserlingk et al. [3] state that the minimum size of a cancer tumor to be detected in mammography is 1.68 cm while the average size of the tumor not getting detected in thermography is 1.28 cm which is much smaller. Thus, thermography can also detect those tumors or early changes that cannot be detected and missed by mammography. With the noninvasive, painless, noncontact, radiation-free and low-cost properties, infrared breast thermography is one of the best screening methods available today in medical science for the breast health. Thermography is capable of screening hard to reach areas like axilla and upper chest areas. After having the breast thermogram, sometimes due to the limitations of human perceptibility, the radiologists cannot accurately classify a breast thermogram, which may mislead the treatment options for the patient. In order to improve the survival rate of the breast cancer patient, early detection and decision-making process for initialization of medication and avoidance of aggressive therapies are primary requirements. Hence, accurate classification of the breast thermograms is also a vital component for medical decision-making and proper diagnosis. Compare to the conventional methods, the hybrid intelligent system provides a better classification accuracy which demonstrates that the hybrid intelligent system is proficient enough in undertaking breast thermogram classification.

This chapter provides an overview of breast cancer along with different breast imaging modalities for early detection of breast cancer. A special attention is given to the infrared thermography-based breast disease detection as it is a noninvasive, inexpensive method of breast imaging without using any radiation. This chapter is an attempt to highlight the feasibility and efficiency of hybrid intelligent system in infrared breast thermography-based breast cancer detection and diagnosis. The chapter is structured as follows. A brief outline including types, symptoms, and risk factors of breast cancer is given in Sect. 2. Section 3 illustrates the necessity of detecting breast cancer in early stage. The next section (Sect. 4) describes the importance of imaging modalities in breast cancer detection and diagnosis. Various imaging methods available for breast cancer detection are presented in Sect. 5. This section (Sect. 5) also includes the limitations of each breast imaging method. Section 6 presents the fundamental of digital infrared thermal imaging (DITI). Use of DITI in breast cancer early detection is described in Sect. 7. Section 8 presents all the necessary acquisition protocols for breast thermograms. A review work on image processing and hybrid intelligent system-based breast cancer detection from breast thermograms is presented in Sect. 9. Finally, Sect. 10 concludes this chapter.

2 Breast Cancer

Breast cancer begins in the cells of the breast tissue, either in the lobules that produce milk or in ducts that carry milk to the nipples. The cancerous cells of the breast continue to multiply to form a malignant tumor. A malignant tumor is a grouping of cancer cells that invade into nearby tissues and may invade to other parts of the body such as liver, lung, bone, and brain through the blood stream. Although it is rare, breast cancer can also develop in men. Breast cancer is categorized based on its origin and its level of invading. The entire female breast

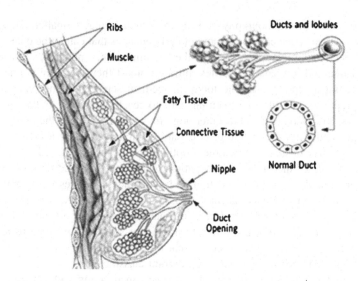

Fig. 1 Anatomy of the female breast [69]

anatomy is shown in Fig. 1. Based on the starting point of breast cancer, it is classified as ductal carcinoma (starts in milk duct) and lobular carcinoma (starts in breast lobule).

It is also classified as invasive and noninvasive (in situ) based on its level of spreading. Carcinoma in situ is a form of cancer where the tumor is confined to the region where it began. Invasive ductal carcinoma (IDC) is the most common breast cancer. It starts in a milk duct of the breast and then it breaks through the wall of the duct and invades into the fatty tissue, lymphatic vessels of the breast. Among all breast cancers, about 80 % is invasive ductal carcinoma. Invasive lobular carcinoma (ILC) is the second most frequent type of breast cancer after invasive ductal carcinoma. Sometimes, people are diagnosed when cancerous cells are totally inside a duct or lobule. It is called carcinoma in situ as no cancer cell has grown out from their original location. Comparative to invasive cancer, carcinoma in situ is easier to treat. The type of breast cancer known as 'Lobular Carcinoma in situ' is not a cancer but its presence indicates a higher risk of developing breast cancer in the future.

In the initial stage of breast cancer, it has no symptom. However, developing a tumor may be associated with the formation of a painless lump that persists after each menstrual cycle. Some other symptoms which may be noticed with the affected breast includes—Any changes in shape, texture, temperature and size of the breast; inflammation in the armpit; pain and tenderness in the breast; nipple discharge (sometimes may contain blood); dimpling of breast skin like orange peel; inverted nipple. When tumor grows larger, the patient may also notice bone pain, nausea, loss of appetite, weight loss, shortness of breath, muscle pain, etc.

2.1 Risk Factors of Breast Cancer

Though the breast cancer is the second deadliest disease after lung cancer, the exact cause of breast cancer is still unknown. The only known thing is that breast cancer is always caused by the damage to the cell's DNA (deoxyribonucleic acid). During the lifetime, the body's cells get reproduced and replaced in a controlled manner. When this control is lost, the cells start to divide more rapidly in an uncontrolled way and continue to form a lump or mass that causes breast cancer. The cells may spread to the lymph nodes or to other parts of the body. Several risk factors are associated with the breast cancer, but having a risk factor does not signify that women will acquire breast cancer while many risk factors increase the chances of having breast cancer. Some risk factors can be avoided or controlled, but some are there that can't be avoided. Some of these risk factors are described below:

- **Gender**: Breast cancer is more prevalent in female. Compared to female, less than 1 % men have breast cancer [4] i.e. the incidence rate is 100 times more common in female than in men.
- **Age**: The breast cancer developing risk increases with the increase of age. Women over 50 have a higher risk of getting breast cancer. Every woman within the age group 50 and 70 should undergo breast cancer screening program in every 3 years. 8 out of 10 women over 50 are diagnosed with breast cancer.
- **Family history**: Women, whose close relatives are suffering from either breast cancer or ovarian cancer before menopause, have a higher risk of developing breast cancer. Most breast cancer cases are not hereditary. However, two genes BRCA1 and BRCA2 (BReast CAncer genes 1 and 2) increases the risk of having breast cancer, and these genes can inherit from parents to the child. Presence of this gene indicates an 80 % likelihood of developing breast cancer. But, not having a close relative with breast cancer does not mean one would not get it.
- **Personal diagnosis of breast cancer**: A woman having breast cancer in one breast has greater chances of developing a new cancer in the other breast or same breast again.
- **Breast density**: The breast is made up of thousands of minute glands (lobules) which produce milk. Compare to the other breast tissue, the glandular cells contain a higher concentration of breast cells to make the breast denser, i.e., dense breast consists of more gland tissue and less fatty tissue. Women having dense breast have a greater chance of getting breast cancer. Also, identification of lump or any abnormal tissue in dense breast mammograms is very difficult. Young women have denser breast than the old women because, the amount of glandular tissue in the breast get decreased with age and replaced with fat.
- **Some benign breast problems**: Women having certain benign (noncancerous) breast problem or changes like developing of noncancerous lump or lobular carcinoma in situ have a greater chance of developing breast cancer.
- **Menstrual periods**: Women, who had periods early or before age of 12 or who entered menopause at a late age or after the age of 55, have a higher risk of

developing breast cancer. The ovaries where the eggs are stored produce oestrogen to regulate the periods. When the periods starts earlier and ends later, it results in over exposing of oestrogen for a longer period. The breast cancer cells are stimulated to grow by the hormone oestrogen.

- **Breast radiation in early life**: Women, who have undergone the radiation treatment to the chest area like X-rays, CT Scans (as treatment of another disease) when they are young have a significantly increased risk of developing breast cancer. The risk is much higher if the radiation is given when the breasts were still developing.

- **Have no child or having them in later life**: Women, having no children or who have the first baby after age 30, have a slightly higher risk of developing cancer. Getting pregnant many times or in younger age cut down the risk of breast cancer as the pregnancy interrupts the exposure to oestrogen. In order to prevent the oestrogen levels, the fat levels of the body must be maintained. Regular exercise or physical activity lowers the oestrogen level and thus reduce the risk of developing cancers.

- **Using oral contraceptives (birth control pills)**: Women who use oral contraceptive pills or any other birth control medication for a long time have higher chances of getting breast cancer. Once the pills are stopped, the risk also gets reduced.

- **Using hormone therapy after menopause**: Women who take hormone replacement therapy especially estrogen and progesterone after menopause, the possibility of developing breast cancer after 5 years of treatment get increased. The risk increases with the intake of hormone replacement therapy, but it becomes normal once the patient stop taking it.

- **Not breastfeeding**: Some research work shows that the breast cancer developing risk reduces with the breastfeeding if it lasts for 1 and half years to 2 years.

- **Alcohol**: The breast cancer developing risk increases with the amount of alcohol intake. Several breast cancers related study shows that one single drink may increase the risk.

- **Being overweight or obesity**: The breast cancer risk highly increases in women who become overweight or obsessed after menopause. It is due to the amount of oestrogen in the body, as obesity causes more oestrogen to produce.

- **Tobacco smoke**: Recent research activity suggests that women who started regularly smoking when they were young are 70 % more expected to develop breast cancer before the age of 50 than the nonsmokers.

- **Being tall**: The risk of having breast cancer is more in women who are taller than average than those who are shorter than average.

2.2 Incidence and Mortality Rate of Breast Cancer

Breast cancer is the second deadliest disease after lung cancer. More than a million women worldwide are identified with breast cancer every year, which accounts 23 % of all cases in female cancer [5]. Moreover, breast cancer is the leading cause of cancer-related death for women in both developed and developing countries. The influence of geographic variation on mortality rate is very less compared to the breast cancer incidence rate [6]. Breast cancer incidence rate is much higher in more developed countries than in the less developed countries, whereas the mortality rate is relatively much higher in less developed countries due to the lack of screening technology and inadequate medical facilities [7]. Being a developing country, the breast cancer incidence and mortality rate in India is increasing with growing migration of the rural population to the cities and due to changes in the lifestyles. According to the survey report made by the International Agency for Research on Cancer (IARC) and the specialized cancer agency of the World Health Organization (WHO), an estimated 70218 women died in India for the year 2012, due to breast cancer which is more than any other country in the world (second: china—47984 deaths and third: US—43909 deaths) though the number of newly diagnosed breast cancer cases was least in India [1]. In the year 2012, for every 5 or 6 women newly diagnosed with breast cancer, 1 woman died in US; for every 4 women newly diagnosed with breast cancer, 1 women died in China and in India for every 2 newly diagnosed cancer women, 1 lady died of it [1].

2.3 Importance of Breast Cancer Awareness for Early Detection of Breast Cancer

Breast cancer awareness is an attempt to make the people familiar with the deadliness of breast cancer through education about the symptoms and treatments. The awareness about the breast cancer symptoms will make the women to realize the importance of getting tested early and to visit a physician when he/she experiences any. This is also associated with a better curing option and long prognosis rate. There are several things like self-breast exam, regular exercise, having healthy diet, etc. that women can do to prevent the occurrence of this disease. Self-breast exam plays a significant role in breast cancer awareness through which a woman can find out the presence of cancer in early stage before moving to the doctor. In order to increase the breast cancer awareness of the people from rural areas, a Breast Cancer Awareness brochure in "BENGALI" language is designed in our Research Laboratory. The booklet provides all the necessary information about breast cancer including its symptoms, the way of performing breast-self-exams, various breast imaging modalities, breast cancer staging, and treatments. This Breast Cancer Awareness brochure will be published very soon in association with Regional

Cancer Centre, Agartala Government Medical College, Agartala, Tripura (West), Government of Tripura, India and Jadavpur University, Kolkata, India.

3 Necessity of Detecting Breast Cancer in Early Stage

With the growing rate of breast cancer incidence, demand for developing new technology and improving existing technology for breast cancer prevention is increasing. The probability of successful treatment and complete recovery of the patient entirely depends on the early detection and diagnosis of the breast cancer [8]. In the modern medical science, there are a large number of newly developed technologies for timely detection of breast cancer to save the lives of many women. The primary goal of screening exams for breast cancer is to detect cancer at a smaller size and at an earlier stage; otherwise, the cancer will extend to other parts of the body. If discovered early, breast cancer is a highly treatable disease, with 97 % chance of survival [2, 9] whereas Lahiri et al. [10], mentioned that early detection of breast cancer leads to 85 % survival chance. Thus, early detection of a breast tumor is the only means to reduce the mortality rate of breast cancer. Since the year 2008, the breast cancer incidence rate has been increased by 20 % while mortality rate has increased by 14 %. About 1.7 million women were newly diagnosed with breast cancer in the year 2012 [6].

4 Medical Imaging Methods for Detection of Breast Cancer

Medical imaging is an imperative diagnostic tool for early detection, better diagnosis, and effective treatment of breast cancer. From medical images, a doctor can evaluate the stage and extent of the cancer. The medical imaging is considered as a critical component of the nation's war on cancer. In Medical Science, there are a lot of medical imaging modalities like mammography, breast ultrasound, breast magnetic resonance imaging, breast-specific gamma imaging, molecular breast imaging, infrared thermography that decrease the mortality rate by playing a significant role in early detection of breast cancer. The incredible power of medical imaging allows researchers and physicians to observe not just within the body, but deep inside the chaos of cancer cells. In this role, imaging is used for:

- Screening, diagnosis, and staging of cancer;
- Guide cancer treatments;
- Finding out whether a treatment is working or not;
- Monitoring recurrence of cancer; and
- Facilitating medical research.

5 Various Medical Imaging Methods

As mentioned above the death rate of breast cancer can only be reduced, if the breast cancer get detected in early stage, i.e., the survival rate, complete recovery, and prognosis rate of breast cancer totally depends on the early detection and proper diagnosis of the breast cancer [8]. In the modern medical science, there are a large number of newly developed imaging modalities and techniques for timely detection of breast cancer to save the lives of many women. Also, the breast cancer mortality rate has been decreased as new imaging technologies have been introduced into today's medical system. Medical images play an important role in knowing the details of the human body for remedial or health science reasons. A succinct overview of the most widely used imaging methods for early breast cancer detection is given below:

1. **Mammography**: Mammography is essentially the only extensively used imaging modality for breast cancer screening. It is a low-dose x-ray of the breast. Breast mammography is of two types: screening mammography and digital mammography. A screening mammogram is suggested for women who have no symptoms of breast cancer. Screening mammography has long been considered as the "Gold standard" for breast cancer screening [11]. A diagnostic mammogram is used for evaluation of new abnormalities of patients having some symptoms of breast cancer. It is an invasive method which involves compression of breasts. Along with the cancerous tumor, breast mammography also identifies cysts, calcifications in the breast. Mammography can find out a cancer tumor in a curable stage. A physician recommends an annual screening mammography for all women over 40 years old since the sensitivity of mammography is very less in younger women with dense breasts [12]. Some samples of mammograms are shown in Fig. 2.

   Limitations of Mammography. Although mammography screening is presently considered as the most appropriate method for mass screening in asymptomatic women, it also has several limitations [13]. Some of them are described below:

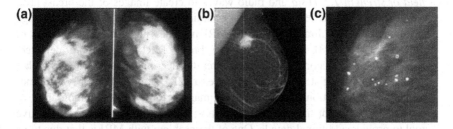

Fig. 2 a Dense breast mammogram where both dense breast tissues and tumors appear white while fatty tissue appears black [70]. **b** Mammogram, showing a breast tumor [71]. **c** Micro calcifications in breast mammogram [72]

- **Radiation risk of breast cancer**: Mammography possesses radiation risk of x-ray. Female breast tissue is highly sensitive to radiation, and this electromagnetic radiation triggers the factor that is responsible for cancerous growth. The radiation also raises the possibility of spreading or metastasizing an existing growth [13, 14].
- **False positives and false negatives**: While Mammography can detect breast cancer earlier, it also can generate false positives by detecting some abnormalities. But, the subsequent tests do not reveal the presence of any cancerous tumor [15]. It is called a 'false positive.' This is why women who have done the screening program run the risk of undergoing the tests that would not have been required if they had not been screened.

  The sensitivity of mammograms is about 90 % which indicates there is about a 10 % likelihood that a small tumor is present in the breast, but not detected. It is called a 'false negative' [15]. The screening mammography cannot detect every instance of breast cancer. One major limitation of mammography is that it cannot detect a cancerous tumor in the breast until when it is of certain size.
- **Dense breast**: Breast density differs widely among women. X-rays can easily pass through fat since fat is radiographically translucent whereas connective and epithelial tissue blocks x-rays to a greater extent as they are radiographically dense relative to fat. The detection of tumor is very difficult in mammography since both the tumor and dense breast tissue appear white in mammograms. That is why mammogram is not well suited for women having dense breast and fibrocystic breasts.
- **Risk of fracture**: During the process of mammogram 42 pounds of pressure is given to the breast that makes the compression of breast tissue [14]. This compression may increase the risk of rupture the encapsulation around the cancer tumor.
- **Age**: The accuracy, sensitivity, and specificity of mammography vary with age. Research shows that the mammogram sensitivity is higher for older women (age 60–69) at 85 % compared with younger women (<50 years) at 64 % [16]. It indicates that the mammography is less effective for a patient of younger age.

2. **Breast Magnetic Resonance Imaging (MRI)**: It is noninvasive imaging technique which does not involve any radioactivity and uses powerful *magnetic field of strength 1.5 Tesla* and radio waves to create images of the breast [17]. MRI can show smallest lesions/abnormalities which are not visible through mammography or ultrasound. Breast MRI provides highest quality images of breast anatomy [18]. In Fig. 3, some breast MRI samples are depicted. Breast MRI is not recommended for all breast cancer patients. But, due to higher sensitivity, radiologist recommends a breast MRI along with yearly mammograms. Some situations where another imaging tools like mammography or ultrasound could not find any abnormalities; the MRI is used as an adjunctive tool to provide additional details. One of the problem with MRI is that due to its high sensitivity it gives many false negatives for which it cannot be used alone as a standard tool for breast cancer detection. However, in some cases breast

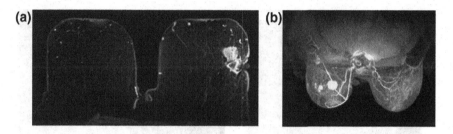

Fig. 3 **a** Breast MRI showing a lesion [73]. **b** Breast MRI of a dense breast [74]

MRI detects a potential and mammographically occult breast cancer threat very early. It is a very costly exam [18].

Limitations of breast MRI. The breast MRI is much more sensitive than the mammography and produces false positive results by detecting breast areas that do not have any cancer, and this leads to unnecessary biopsies [19]. Moreover, the breast MRI cannot detect the micro-calcifications that indicate a suspicious area. It cannot distinguish between cancerous and noncancerous abnormalities [19]. Compare to the other imaging modalities, breast MRI is very expensive exam, and pregnant women are not recommended to have breast MRI since a powerful magnet and a contrast agent is used in breast MRI.

3. **Breast Ultrasound (BUS)**: Breast Ultrasound is an important imaging technique that uses harmless high-frequency sound waves to detect and characterize tumors. The *7.5–12 MHz transducers* are usually used in BUS, which achieves adequate penetration in most of the women [20, 21]. BUS does not utilize ionizing radiation like mammography that makes it a preferred method for pregnant women [22]. It is a painless process as no compression is made on breast during the breast imaging. In ultrasound imaging, echoes reflected from normal, and abnormal tissues are captured by the computer to produce a 2D image called sonogram. BUS produces very sharp and high-contrast images of the breast. Ultrasound can show lumps that are filled with either solid mass or fluid. Radiologists examined these images to determine whether a mass is a solid tumor or just a fluid filled cyst [12, 22]. A cyst and a cancer tumor detected in breast ultrasound are shown in Fig. 4a, b, respectively. Ultrasound is often performed with mammography to identify the area of concern that requires further evaluation. Ultrasound is useful in finding very minute lesions that cannot be felt in a clinical exam, and it is also useful in guiding the needle during the biopsy [22]. Interpreting cancer tumor in mammogram of women with dense breast is difficult, for which breast ultrasound is considered as first diagnostic imaging method for women under the age of 35 [22]. Overall, ultrasound is a quite helpful investigative tool in the diagnosis of various breast cancer symptoms.

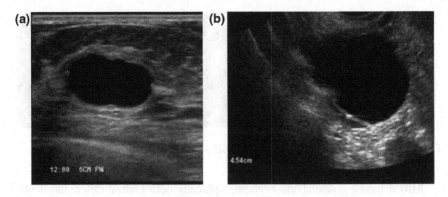

Fig. 4 **a** A simple breast cyst [75]. **b** Breast ultrasound, showing a cancer tumor [76]

Limitations of breast ultrasound. Comparative to the advantages and effectiveness of the breast ultrasound, it has limited drawbacks. One major disadvantage of breast ultrasound is that it cannot replace the mammography of women of age above 40 since like mammography breast ultrasounds are unable to identify calcifications in breasts which is also a sign of breast cancer in very early stage [23, 24]. In addition, ultrasound cannot screen many of cancers ,i.e., many cancer tumors are not visible in ultrasound. Therefore, in most of the cases breast ultrasound is followed by other diagnostic examinations like MRI, mammograms, etc. It cannot be alone used as standard screening tool for breast cancer [24].

4. **Nuclear medicine breast imaging**: It is a promising tool for screening and diagnosis of breast cancer in women with dense breast tissue. It is a noninvasive method of imaging. The FDA (Food and Drug Administration) approved short-term radioactive agent *Tc-99 sestamibi* is injected 5–10 min before the imaging procedure [25]. This radioactive tracer lights up (gamma rays) the cancerous area inside the breast. The breast cell absorbs this radioactive agent. But, the cancerous cells in the breast are found to absorb more of the agents than the normal breast cells, and these cancer cells can be imaged with special semiconductor-based γ-cameras [25]. Nuclear breast imaging covers all the imaging modalities including positron emission mammography to breast-specific gamma imaging. It is the safest diagnostic imaging exams available. Nuclear medicine imaging makes it possible to detect any abnormalities before progression of the disease. Figure 5 shows different types of nuclear medicine breast imaging. Some nuclear breast imaging techniques are:

   • **BSGI (Breast specific gamma imaging) and MBI (Molecular breast imaging)**: BSGI and MBI use a high-resolution gamma camera for imaging of the breast. During the image acquisition, a mild compression is made on breast. It can differentiate between the cancerous tissue and benign tissue of the breast. It is an ideal test to complement the mammography. The sensitivity of MBI is very high for small breast lesions detection. MBI has an

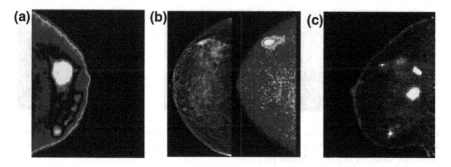

Fig. 5 **a** Detection of cancer tumor in BSGI of a dense breast [77]. **b** Shows the detection of 17 mm cancer tumor in MBI of a breast whose mammogram shows a negative result [78]. **c** Detection of the cancerous tumor in PEM of a breast [79]

overall sensitivity of 90 % in patient with suspected breast cancer. The sensitivity of MBI is 82 % for lesions less than 10 mm in size. Sensitivity is lowest for tumors less than 5 mm in size [25].

- **PEM (Positron Emission Mammography)**: In this imaging technique, a short-lived radioactive sugar-like substance is injected into the body. The substances get accumulated in the cancerous cells of the breast and radiate energy that can be captured using a complicated and advanced imaging camera. PEM can capture the shape; size (equal to pinpoint), and location of the breast tumor [26].

Limitations of Nuclear Medicine Breast Imaging. One of the essential components of nuclear medicine imaging is the use of radiotracer or radioisotopes which has a harmful impact on the health of the patient. Breast is one of the most radiosensitive organs in the body. Due to its radiation, pregnant women are not recommended for nuclear medicine breast imaging. Radioisotopes cause genetic mutation which is also a cause of breast cancer. Excessive use of nuclear medicine may cause the malfunctioning of an enzyme or protein. Also, nuclear medicine is very expensive and requires a huge amount of investment for which several medical institutions cannot afford it.

5. **Infrared Breast Thermography**: Since 1982, FDA (Food and Drug Administration) has approved IR imaging as an adjunct modality to mammography for breast cancer detection [27]. The underlying idea of breast thermography is that the temperature of the skin overlying a malignancy is higher than the skin overlying normal breast tissue which is caused by the increased rate of blood flow and metabolic activity to supplement the tumor's growth [28]. Due to ever increasing need for nutrients, cancerous tumors boost circulations to their cells by opening dormant vessels, and creating new ones. This process results in an increase in regional surface temperature of the breast. Ultra-sensitive medical infrared cameras and complicated computers are used in breast thermography to detect and produce high-resolution images of

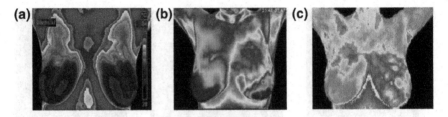

Fig. 6 **a** Healthy breast thermogram: the temperature patterns of two breasts are cool and almost identical [80]. **b** Shows invasive ductal carcinoma in left breast of the patient [81]. **c** Shows fibrocystic changes in the left breast [82]

temperature. Some sample breast thermograms are shown in Fig. 6. Thus, it detects and records the heat pattern of the breast surface. It has the potential to detect breast cancer 10 years earlier than the traditional golden method—mammography [9, 29].

Like the other imaging techniques, the digital infrared imaging does not identify the physical tumor. Instead, it detects the heat pattern produced by increased blood circulations and tumor-related metabolic changes. Identifying the minute variations in normal blood circulation activity, the infrared imaging can find the signs indicating a precancerous state of the breast or the presence of a tumor that is small enough to be detected by physical examination [9, 29]. In order to ensure that the thermographic examination is accurate, it is crucial to follow some simple instructions carefully. With its noninvasive, low-cost, non-radiation, noncontact basis, thermography has distinguished itself as the safe earliest detection technology for breast cancer. In the next section, the procedure for breast thermography is described elaborately.

6 Fundamentals of Digital Infrared Imaging (DII)

Any object whose surface temperature is above absolute 0 K (−273 °C) radiates infrared energy at a wavelength analogous to its surface temperature and its spectral emissivity. Thermographic camera is much sensitive to the radiation emitted by the human body [2, 30, 31]. The relationship between the energy radiated by an object and its temperature is defined by the Stefan–Boltzmann Law. According to the Stefan–Boltzmann Law, the radiation emitted by an object is directly proportional to the object's vicinity, emissivity, and the fourth power of its absolute temperature [31, 32]. Blackbody is considered as the hypothetical object that absorbs all incident radiation. Stefan–Boltzmann Law (Eq. (1)) states that the energy radiation from per unit area of a black body in per unit time is directly proportional to the fourth power of the absolute temperature of a black body. Thus, the total emissive power from a black body can be defined as:

$$E = \sigma T^4 \tag{1}$$

Here, E is the total emissive power (W/m^2), σ is the Stefan–Boltzmann Constant ($\sigma = 5.670373 \times 10^{-8}$ W/m^2 K^4), and T is the absolute temperature (K). According to the Stefan–Boltzmann's law, the emissivity (ε) of a blackbody is unity [10].

A body emits less energy than a black body when it does not absorb all the incident radiation. It is characterized by an emissivity, $\varepsilon < 1$ [10].

Thus, for objects which are not full radiator, the modified Stefan- Boltzmann equation can be defined as:

$$E = \varepsilon \sigma T^4 \tag{2}$$

where, ε is the emissivity of real surfaces or surfaces of non-full radiator. The emissivity of the human skin is more or less constant, and its value is 0.98 ± 0.01 for the wavelength range 2–14 μm [33, 34]. However, application of cosmetics, lotions, may change the emissivity of the human skin [32].

In the electromagnetic spectrum, the Infrared rays are found within the wavelengths of 0.75–1000 μm. The human skin emits infrared radiation in the range 2–20 μm. The entire IR range is subdivided into: near IR (NIR) having the spectral range between 0.75 and 1.4 μm; short-wave IR (SWIR) covers the wavelengths from 3 to 8 μm; long-wave IR (LWIR) covers the wavelength from 8 to 12 μm, and far IR (FIR) which covers all the wavelength beyond 12 μm. For medical IR imaging purposes, LWIR is the most vital IR spectral range. The clinical IR imaging depends upon the spectral transmission and reflection characteristics of tissue and blood. Penetration and reflection are maximum in the red end of the visible spectrum, where the radiation penetrates the superficial layers of skin and tissue up to 2.5 mm depth and is then reflected out again [32]. It is found that almost 90 % of the infrared energy radiated by the human body is in the range of longer wavelengths 6–14 μm [30].

7 Digital Infrared Imaging in Detection of Breast Cancer

The Congressionally Directed Medical Research Program has set some ideal characteristics for early breast cancer detection method which includes: Detection of early lesions, high sensitivity and high specificity, inexpensive, noninvasive, decrease mortality. All these requirements are met by the Infrared imaging [27]. Infrared breast thermography helps in early detection and monitoring of physiologic changes associated with breast pathology. Risk factors for the development of breast cancer can also be established from breast thermography. Digital Infrared imaging of the breast has achieved an average sensitivity and specificity of 90 %. Also, a persistent abnormal thermogram is 10 times more significant than the first-order family history of the disease. The IR image is the highest risk marker for screening the possibility of the presence of an undetected breast cancer or future development of breast cancer [30]. The aggressiveness of breast tumor is directly proportional to the thermo-vascular activity in the breast. Hence, infrared (IR) imaging can also be used as a prognostic indicator [30]. The patient's breast thermogram acts as thermal fingerprints of breast and any changes in this thermal fingerprint may indicate the presence of a breast disease like fibrocystic disease, cancerous tumor, vascular disease, Paget or an infection. Once abnormal heat pattern of the breast gets detected, necessary treatment can be taken to rule out the disease. Since, till now there are not any preventive for breast cancer, the only way to fight back with breast cancer is the early detection.

8 Acquisition of Breast Thermograms

Several factors are there that may modify the human body temperatures and create false findings in the thermogram for which thermograms need to be captured under strict protocols. Several components are there to be considered for characterizing thermal images as a potential tool for detecting breast cancer. It is crucial to follow these simple instructions carefully to ensure that the thermographic examination is accurate. All the requirements for preparing a patient for breast thermograms are described in subsequent sections.

8.1 Instructions Prior to Examinations

The thermographic procedure is performed for assisting the evaluation of the anomalous temperature patterns of the breasts that may or may not indicate the presence of a disease. Circulatory problems, previous injuries, can reduce the body surface temperature. Similarly, regular smoking can also decrease the body surface temperature. Again several activities like physical exertion, consumption of alcohol, and sunburn increase the skin surface temperature [2, 35, 36].

In order to get accurate results from thermal images without contamination of artifacts (anything unnatural that does not belong to normal human physiology), preparation of patients before screening is the utmost importance. The body temperature should be as normal as possible [37]. Hence, before capturing the temperature patterns of the patient body, some protocols are defined that must be followed by the patient to ensure valid test results. Application of cream, lotion, powder on chest, hormone replacement therapy (HRT), pregnancy and menstruation can also affect the breast surface temperature [2, 38]. Rigorous exercise, tight fitting cloth, sun-bathing, underarm shaving, utilization of deodorants, physical therapies, pain medication, smoking, intake of tea, coffee, alcohol, radiation treatment must be avoided prior to the examination [2, 31, 38–41]. Ng et al. [42] and

Acharya [43] considers the patients within the period of the 5th to the 12th day and after the 21st days of the commencement of the menstrual cycle.

8.2 Patient Intake Form

On arrival of the patient, the practitioner should provide a patient data form to the patient for filling up, which includes some necessary information regarding the symptoms and history of the patient for better understanding of the patient background. Some of these information may be the age and weight of the patient; family history of breast cancer; information about previous diagnosis of breast cancer; any treatments like biopsies or surgeries to the patient's breasts; previous breast screening; history of taking hormone treatment; experiencing any symptom of breast cancer etc. [44].

8.3 Patient Acclimation: Pre-imaging Equilibrium

After filling up the intake form, the patient needs to sit in a cool private room and should be informed with the testing procedure. The patients are instructed to undress from the waist up and to remove jewelry [30, 44]. After removing appropriate clothes, the patient is asked to sit and to leave the breasts exposed to air for 10-15 min so that the patient body can acclimate to the room temperature (to equilibrate to the atmosphere of the room) [30, 43, 45, 46]. This will create the "Thermal steady state." During this time of patient preparation, the patient must avoid folding or crossing of arms and legs or placing bare feet on a cold surface [2, 39]. Once acclimated, the patient will be asked to place her hands behind the head for taking the infrared images [39].

8.4 Environment of the Imaging Room

Thermograms are sensitive to environmental changes in temperature, humidity, and air flow for which infrared imaging must be captured under controlled environment [31]. It is essential that the "Infrared Imaging Room" itself is of adequate size to sustain a homogeneous temperature. The size of the examination room should be large enough to allow patients of different sizes to be positioned relatively equidistant from each wall. The room approximately of size 8 feet × 10 feet is sufficient to meet these requirements [47]. The nature of human physiology changes from different external environment as they produce thermal artifacts in the human body. The windows and doors should be adequately sealed to prevent direct airflow on the patient. The room must be free from drafts and sunlight [48]. The temperature of the examination room should be maintained such that the patient's physiology is not altered. The temperature range should be maintained at 18–23 °C [47, 49, 50]. The room temperature changes during the course of the examination must be kept within 1 °C in order to maintain a steady-state physiology. The humidity of the examination room must also be maintained at 60 ± 5 %, such that no air moisture is built upon the skin of the patient that can interact with radiant IR energy [47, 50]. During the time of examination, incandescent lighting should not be used due to the amount of radiation it produces.

8.5 Thermal Camera and Acquisition Systems

For a quality breast thermogram, thermal sensitivity and resolution are the two most important parameters of acquisition system [27]. Thermal sensors with good thermal sensitivity can detect a minute temperature difference. A slight temperature difference in breast thermogram may indicate a suspicious region. The resolution parameter is responsible for the number of colors in the computer display. The temperature transition is very smooth if the resolution of the thermal camera is better. Most of the infrared cameras used for breast imaging have a resolution of 320 × 240 pixels, and it is sufficient enough for informal screening of breast. Thermal camera having resolution of 640 × 480 and good sensitivity can provide more useful thermal and spatial details [27]. Zadeh et al. [51] had used the SDS D-series camera with thermal sensitivity 0.1 °C at 30 °C and resolution 160 × 120 pixels for collecting the breast thermograms of 200 patients at Hakim Sabzevari University in Sabzevar and with the cooperation of Sabzevar University of Medical Science. Arena et al. [52] had asked the patient to sit at a distance of approximately 5 feet away from the infrared camera. They had used an infrared camera having thermal sensitivity of 0.05 °C and resolution 320 × 240 pixels for capturing the breast thermogram of 238 normal patients, 67 newly discovered with cancer patients, and 46 patients who previously had a diagnosis of cancer. Acharya et al. [43] had used NEC-Avio Thermo TVS2000 MKIIST camera system for capturing 50 breast thermogram images, where 25 thermograms were of cancer patients and 25 thermograms were from healthy persons. They had collected their data from the Department of Diagnostic Radiology, Singapore General Hospital. Qi et al. [53] obtained the breast thermograms by using Inframetrics 600 M camera with thermal sensitivity of 0.050 K at Elliott Mastology Centre. Ng et al. [54] had collected the breast thermograms of 90 patients from Singapore General Hospital by using Avio TVS-2000 MkII ST infrared camera. Wishart et al. [55], collected the thermograms of 113 patients using a digital infrared breast scan called Sentinel BreastScan. A DITI system named as Sentinel BreastScan (Infrared Sciences Corp.) having thermal sensitivity 0.08 °C and resolution 320 × 240 pixels was used by Arora et al. [56] for capturing the breast thermograms of 92 patients.

8.6 Capturing Views of Breast Thermograms

All the thermography clinics or hospitals do not use a universal protocol for capturing of breast thermograms that makes it very difficult to follow a certain protocol for capturing. The accuracy of thermogram in detection of the breast abnormality entirely depends on the thermogram image resolution, thermal sensitivity and the number of views of breasts. Most of the FDA registered thermal systems are also not equipped with excellent image resolution and thermal resolution. Bharathi et al. [57] had mentioned that only three views of breast thermograms (Contra-lateral, Medio Lateral Oblique, and Axillary) may result in wrong diagnosis. They used 12 views of breast thermograms capturing at an angular interval of 30°. Campbell [58] suggested for taking the thermograms of the underside of the breast so that the cancer tumor in the lower portion of the breast should not get missed (if any). For capturing the breast underlying area, the patient needs to lie down on her back to cool down underside of the breast. During this process of cooling, the patient should keep her arms away from the breasts. Then the first breast thermogram is captured before capturing any other view of the breast. Figure 7a, shows a thermogram, where some abnormality is seen in the underside of the right breast. However, all the other thermograms of this patient were normal [58], which signifies the necessity of capturing the underside of the breast. This view of breast thermogram is known as supine view.

Kolarić et al. [59] had taken five views of breast thermograms for each patient: including frontal, right semi oblique, right oblique, left-semi oblique, and left oblique view. Five breast thermograms are usually acquired including frontal view, two lateral views and two oblique views in some thermography clinics [60, 61]. Figure 8a–c shows the frontal, right lateral, and left lateral views of breast thermogram.

Moreover, some other views including Bilateral Breast, Right Breast Close Up, Left Breast Close Up and areas of concern are also considered for breast thermography examination [58, 62]. Agostini et al. [63] captured only frontal view

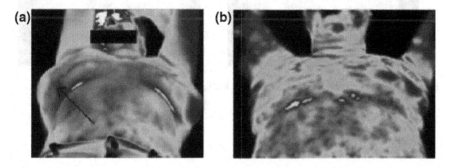

Fig. 7 **a** The arrow shows an abnormality in the underside of the right breast thermogram (supine view) [58]. **b** Normal supine view of breast [58]

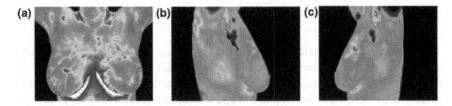

Fig. 8 **a** Anterior view or frontal breast thermogram [37]. **b** Right lateral view breast thermogram [83]. **c** Left lateral view breast thermogram [83]

images of the patient. During capturing, the patient was lying down on an examination table of 40° inclination with their arm up and resting their hands over head [63]. Kennedy et al. [14] asked the patients to stand about 10 feet away in front of the camera with raised arms resting over her head. They acquired three views of breasts including one anterior view and two lateral views of the breast. In breast thermography, capturing of the frontal view of breast thermogram is the utmost importance as it is the only view from which any abnormality or asymmetry between the left breast and right breast can be recognized. Figure 9a, b shows the left and right oblique views of breast thermogram. And Fig. 10 shows close views of left and right breast.

In thermography clinics only frontal breast thermogram is used to create the breast baseline which is the key for getting the benefits of infrared breast thermography in breast cancer detection.

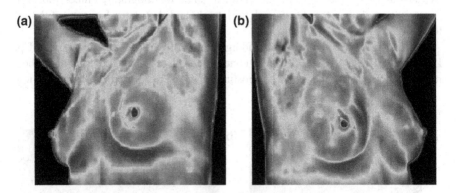

Fig. 9 **a** Left oblique breast thermogram [84]. **b** Right oblique breast thermogram [84]

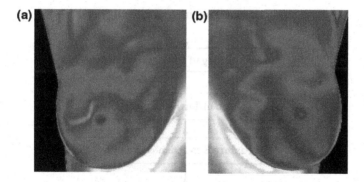

Fig. 10 **a** Close view of right breast; **b** Close view of left breast

9 Analysis of Breast Thermograms Using Hybridized Intelligent System for Abnormality Detection

Breast thermography is a promising technology for early breast cancer detection. Gautherie et al. [64] suggested that an abnormal thermogram is the single most reliable indicator of high risk of breast cancer in its early stage. With the availability of higher sensitive infrared cameras, application of thermography in breast cancer detection has drawn the interest of many researchers toward this domain. In normal breast thermograms, the thermal pattern in both the breasts is almost symmetrical while in case of cancerous breast, temperature asymmetry is observed. Based on this key idea several research work is going on to analyze the breast thermograms. A review on the research works related to the detection of breast cancer by extracting several statistical features from breast thermograms and application of intelligent systems for classification of breast thermograms into normal and abnormal is described in this section. The classification accuracy of hybrid intelligent system on medical data signifies that the hybrid intelligent system is very efficient in the task of detecting an abnormal thermogram. The intelligent and hybrid intelligent system-based computer aided diagnosis (CAD) system works as a promising tool for assisting and providing a "second opinion" to the radiologists or pathologists to produce an accurate and faster diagnosis results. The analysis of breast thermogram is a process of multiple steps including preprocessing of breast thermograms, background removal or extraction of region of interest, extraction of a set of features from each breast thermogram, asymmetry analysis and finally classification of breast thermograms into normal, abnormal, and benign breast. Figure 11 illustrates the general procedure for detecting breast cancer from breast thermograms.

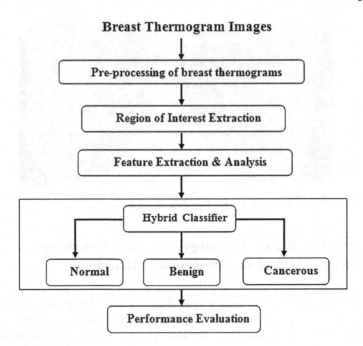

Fig. 11 Overview of the infrared thermography-based breast cancer detection system

9.1 Preprocessing of Breast Thermograms

The most important step in the detection of breast cancer from breast thermogram is the preprocessing of raw data. During capturing, the patients are sitting in a chair or standing, which results in a slight movement of the body. Also, the thermal images commonly exhibit a blurring effect that appears like defocusing. One of the most common techniques for removal of blur in images due to linear motion is the Wiener filter. Moreover, another most common preprocessing operation in breast thermogram is the conversion of the infrared thermal image into a gray scale image. Acharya et al. [43] and Borchartt et al. [11] had converted the breast thermograms into gray scale, after cropping the breast images. Kapoor and Prasad [65], had removed the background of the breast thermograms and resized them to remove the undesired body portion, before processing them.

9.2 Region of Interest Extraction

Along with the necessary information required for breast cancer detection, the breast thermogram also includes some unnecessary details that need to be discarded to improve the accuracy of the cancer detection system.

In the process of breast abnormality detection, identification of the background, and region of interest is a critical need of the system. Schaefer et al. [66] and Borchartt et al. [11] had manually segmented the frontal breast thermograms into left and right breast. In the method proposed by Borchartt, the segmented images are further refined manually to delete all contents not belonging to the breast. In [53, 67], Qi et al. used canny edge detector and Hough transformation to identify four dominant feature curves (left and right body boundaries and two lower boundaries of the breasts) for automatic segmentation of left and right breast.

9.3 Feature Extraction and Asymmetry Analysis

Feature extraction is the process of collecting a set of distinguishable image characteristic, which are most important for feature analysis and classification [42, 67]. It is the most significant part in the analysis of breast thermograms. In feature extraction, a series of statistical features are extracted from the region of interest. Schaefer et al. [66] had extracted total 38 statistical features including basic statistical features, moment features, histogram features, cross co-occurrence features, mutual information, and Fourier descriptors from each breast thermogram which describes the asymmetry between the left and right breasts. Qi et al. [53] had identified the asymmetry between left and right breasts by plotting the thermal histogram of the breast regions (left and right). Another method is there, presented by Qi et al. [67], where the asymmetry between left and right breast thermogram was measured by doing some feature extraction. Some high order statistics like mean, variance, skewness, kurtosis, correlation, entropy, and joint entropy were calculated as the components of the feature vector to quantify the distribution of different intensities in each breast. They suggested that the high order statistics (variance, skewness, and kurtosis) were most useful features to detect asymmetry while low-order statistics mean and entropy could not signify any asymmetry. In [43], Acharya et al. had extracted the statistical features like entropy, contrast and correlation from the gray level co-occurrence matrix (GLCM) to detect the presence of a cancerous tumor. Other features like gray level nonuniformity and run percentage were also calculated from the run-length matrix. After acquisition of breast thermograms, Acharya et al. [50] had converted the 2D thermograms into 1D data by using radon transform. From the transformed data, five higher order spectral features were extracted. The extracted features were—mean magnitude of the spectrum, entropy1, entropy2, entropy3, and phase entropy. For analyzing the asymmetry of the breasts, Zadeh et al. [51] had extracted diagnostic parameters including patient's age, mean, variance, kurtosis, skewness, entropy, difference between the two breasts and thermal pattern of the breasts from breast thermograms. Bharathi et al. [57] had extracted a series of statistical features and Haralick texture features from the breast thermograms. These extracted features before and after the cold stress were analyzed to identify any abnormality. For finding out the abnormal thermograms, Kapoor and Prasad [65] had extracted skewness, kurtosis,

entropy, joint entropy, energy, homogeneity and correlation from breast thermograms and analyzed them.

9.4 Classification of Breast Thermograms Using Intelligent Systems

The asymmetry analysis of breast thermograms is followed by the classification of
breast thermograms into normal and cancerous breasts. Fuzzy-rule based classifier
coupled with significant statistical features plays a vital role in improving the
survival rate of breast cancer patients. The life expectancy prediction made by
hybrid intelligent system is highly significant in decision making for treatments,
medication and therapies. Different classifiers are used in various research works.
Schaefer et al. [66] had employed a hybrid fuzzy rule-based classification system
for diagnosis where some genetic algorithms were applied to optimize the features
and parameters of fuzzy rules. With this classifier, they achieved a correct classification rate of about 80 %. Krawczyk et al. [68] had employed a hybrid multiple
classifier system for analyzing breast thermograms. Their multiple classifier system
was the hybridization of 3 different intelligence techniques: Neural Network
(NN) or Support Vector Machine (SVD) as base classifiers, a neural fuser to unite
the individual classifier, and a fuzzy measure. Using of this hybridized classifier for
evaluating 150 breast thermograms provided excellent classification accuracy. Qi
et al. [53] used unsupervised learning, and each pixel is relabeled to a certain
cluster. Finally, pixel distribution of each cluster was analyzed, and abnormalities
were determined. In another method proposed by Qi et al. [67], the asymmetry
identification was done using two methods: k-means clustering (unsupervised
learning) and k-nearest neighborhood (supervised learning) based on feature
extraction. They had used 6 normal and 18 cancer patient breast thermograms for
evaluating the performance of their method. Acharya et al. [43] employed the
Support Vector Machine (SVM) to have automatic classification of breast thermograms as normal and malignant breasts. For automatic classification of breast
thermograms, Borchartt et al. [11] had used free LibSVM classifier. The extracted
features were fed into the LibSVM software. The LibSVM classified the breast
thermograms into two classes: pathology and healthy. In the method proposed by
Zadeh et al. [51], a 3 layer Feed-Forward Neural Network with a sigmoidal activation function (logsig) in the middle layer had been used. Acharya et al. [50] had
used a feed–forward artificial neural network and SVM for classification of breast
thermograms. The extracted features were fed into the input of the feed-forward
neural network for classifying the breast thermograms. An Artificial Neural
Network had been used in the method proposed by Ng et al. [54] for the analysis of
breast thermograms. Kapoor and Prasad [65] had used a multilayer perceptron
neural network for classification of breast thermograms. The features extracted from
50 breast thermograms were fed into the neural network to train the system, and

remaining 10 breast thermograms were used for testing and validation of the classification system.

Different acquisition systems, used in various research works along with their specifications are listed in Table 1. Summary of different research works in breast cancer detection from breast thermograms is given in Table 2.

Table 1 Summary of acquisition system and number of patients considered in different research works

Authors	Acquisition system	Camera specifications	Number of patients/breast thermograms
Schaefer et al. [66]	Not specified	Not specified	146 Thermograms (29 malignant, 117 benign)
Ng et al. [54]	Avio TVS-2000 MkII ST	Not specified	90 Patients
Qi et al. [53]	Inframetrics 600 M camera	Thermal sensitivity 0.05 K	Not specified
Qi et al. [67]	Inframetrics 600 M camera	Thermal sensitivity 0.05 K	6 Normal, 18 cancerous
Acharya et al. [43]	NEC-avio thermo TVS2000 MKIIST	Not specified	25 Normal, 25 cancerous
Borchartt et al. [11]	Not specified	Resolution 320 × 240	24 Cancerous, 4 normal
Acharya et al. [50]	Thermo TVS2000 MkIIST Avio short wavelength system	Not specified	25 Normal, 25 cancerous
Zadeh et al. [51]	SDS D-series;	Thermal sensitivity 0.1 °C at 30 °C, resolution 160 × 120	200 Patients
Bharathi et al. [57]	MAMRIT (mammary rotational infrared thermographic system)	Not specified	8 Normal, 8 abnormal
Arena et al. [52]	Not specified	Thermal sensitivity 0.05°, resolution 320 × 240	238 Normal, 67 newly discovered and 46 who previously diagnosed with cancer
Wishart et al. [55]	Sentinel breast scan	Not specified	113 Patients
Arora et al. [56]	Sentinel breast scan (infrared sciences corp.)	Thermal sensitivity 0.08 °C, resolution 320 × 240	92 Patients
Kapoor et al. [65]	Infrared thermal imager by irisys	Resolution 320 × 240	60 Patients

Table 2 Summary of statistical features and classifiers used by different researchers

Authors	Feature extracted	Classifier	Accuracy
Schaefer et al. [66]	38 features (basic statistical features, moments, histogram features, cross co-occurrence matrix features, mutual information and Fourier descriptors)	Hybrid fuzzy rule-based classification	Accuracy: 80 %
Qi et al. [53]	No feature extraction	K-means clustering	Not provided
Qi et al. [67]	Mean, variance, skewness, kurtosis, correlation, entropy and joint entropy	K-nearest neighborhood	Not provided
Acharya et al. [43]	First 4 moments, entropy, contrast and correlation, gray level non-uniformity and run percentage	Support vector machine	Accuracy: 88.10 %
			Sensitivity: 90.48 %
			Specificity: 85.71 %
Borchartt et al. [11]	Range of temperature, mean temperature, standard deviation, and quantization of higher tone in eight level posterization	Free LibSVM	Accuracy: 85.71 %
			Sensitivity: 95.83 %
			Specificity: 25.00 %
Acharya et al. [50]	Mean-magnitude of the spectrum, entropy1, entropy2, entropy3, and phase entropy	ANN	Accuracy: 90 %
		SVM	Accuracy: 80 %
Zadeh et al. [51]	Patient's age, mean, variance, kurtosis, skewness, entropy, difference between the two breasts and thermal pattern of the breasts	3 Layer feed-forward neural network	Accuracy: 70 %
			Sensitivity: 50 %
			Specificity: 75 %
Bharathi et al. [57]	Mean, variance, skewness, kurtosis, angular second moment (ASM), contrast, correlation, sum of square, inverse difference moment, sum entropy, sum average, sum variance, entropy, difference variance, information measure of correlation1 and information measure of correlation2, difference entropy	Support vector machine	Not provided
Kapoor et al. [65]	Skewness, kurtosis, entropy, joint-entropy, energy, homogeneity, and correlation	Artificial neural network	Accuracy: 80 %

10 Conclusion

Despite the advances in treatments, the breast cancer remains the second leading cause of the cancer-induced death after lung cancer. There is no diagnostic tool that is capable of significantly reducing the breast cancer mortality. Only possible way of saving lives is the early detection of breast cancer. Most of the breast imaging modalities, although promising but too expensive for routine use. Compare to the other imaging modalities, breast thermography has several advantages that benefits women of all ages. A solid tumor which is small enough to be detected by any of the available diagnostic tool can be detected by infrared thermography due to its higher temperature compared with the surrounding tissue. This facilitates the early detection of the tumor before it invades to the surrounding region. Thermography is a noninvasive, radiation-free, painless, inexpensive imaging modality that can be used as a complementary method to other screening methodologies. Breast mammography alone has a sensitivity of 84 % while used along with thermography its sensitivity has increased to 95 %. Thus, thermography is very adequate for both asymptomatic and symptomatic patients for routine checkups. It also helps the physicians to decide the proper treatment for the symptomatic patients.

Acknowledgments The work presented here is being conducted in the Bio-Medical Infrared Image Processing Laboratory (B-MIRD), Department of Computer Science and Engineering, Tripura University (A Central University), Suryamaninagar-799022, Tripura(W). The research work is supported by the Grant No. BT/533/NE/TBP/2013, Dated 03/03/2014 from the Department of Biotechnology (DBT), Government of India. The first author would like to thank Prof. Barin Kumar De, Department of Physics, Tripura University (A Central University) for his kind support to carry out this work. The second author also would like to thank Prof. Siddhartha Majumder, Advisor, Medical Education, Government of Tripura for his valuable advices to carry out this project.

References

1. Breast Cancer India: Pink Indian statistics. http://www.breastcancerindia.net/bc/statistics/stat_global.htm
2. Ng EYK (2009) A review of thermography as promising non-invasive detection modality for breast tumor. Int J Therm Sci 48(5):849–859
3. Keyserlingk JR, Ahlgren PD, Yu E, Belliveau N, Yassa M (2000) Functional infrared imaging of the breast. IEEE Eng Med Biol Mag 19(3):30–41
4. Anderson WF, Jatoi I, Tse J, Rosenberg PS (2009) Male breast cancer: a population-based comparison with female breast cancer. J Clin Oncol 28(2):232–239
5. Gallardo-Caballero R, García-Orellana CJ, García-Manso A, González-Velasco HM, Macías-Macías M (2012) Independent component analysis to detect clustered microcalcification breast cancers. Sci World J 2012:6
6. International Agency for Research on Cancer. http://www.iarc.fr/en/media-centre/pr/2013/pdfs/pr223_E.pdf
7. World Cancer Research Fund International: Comparing more & less Developed Countries. http://www.wcrf.org/int/cancer-facts-figures/comparing-more-less-developed-countries

8. Bozek J, Mustra M, Delac K, Grgic M (2009) A survey of image processing algorithms in digital mammography. In: Grgic M, Delac K, Ghanbari M (eds) Recent advances in multimedia signal processing and communications, 231:631–657. Springer, Berlin Heidelberg

9. Gautherie M (1983) Thermobiological assessment of benign and malignant breast diseases. Am J Obstet Gynecol 147(8):861–869

10. Lahiri BB, Bagavathiappan S, Jayakumar T, Philip J (2012) Medical applications of infrared thermography: a review. Infrared Phys Technol 55(4):221–235

11. Borchartt TB, Resmini R, Conci A, Martins A, Silva AC, Diniz EM, Paiva A, Lima RCF (2011) Thermal feature analysis to aid on breast disease diagnosis. In: Proceedings of 21st Brazilian congress of mechanical engineering—COBEM2011, Natal, RN, Brazil

12. Kuhl CK, Schrading S, Leutner CC, Morakkabati-Spitz N, Wardelmann E, Fimmers R, Kuhn W, Schild HH (2005) Mammography, breast ultrasound, and magnetic resonance imaging for surveillance of women at high familial risk for breast cancer. J Clin Oncol 23 (33):8469–8476

13. Kobrunner SH, Hacker A, Sedlacek S (2011) Advantages and disadvantages of mammography screening. Breast Care (Basel, Switzerland) 6(3):199–207

14. Kennedy D, Lee T, Seely D (2009) A comparative review of thermography as a breast screening technique. Integr Cancer Ther 8(1):9–16

15. Malterud K (1986) Advantages and disadvantages of mammography screening of healthy women. A critical evaluation. Tidsskr Nor Laegeforen 106(19–21):1608–1610, 1615

16. Ng EYK, Sudarshan NM (2001) Numerical computation as a tool to aid thermographic interpretation. J Med Eng Technol 25(2):53–60

17. Price J (2012) Handbook of Breast MRI. Cambridge University Press, Cambridge

18. Sardanelli F, Boetes C, Borisch B, Decker T, Federico M, Gilbert FJ, Helbich T, Heywang-Kobrunner SH, Kaiser WA, Kerin MJ, Mansel RE, Marotti L, Martincich L, Mauriac L, Meijers-Heijboer H, Orecchia R, Panizza P, Ponti A, Purushotham AD, Regitnig P, Del Turco MR, Thibault F, Wilson R (2010) Magnetic resonance imaging of the breast: recommendations from the EUSOMA working group. Eur J Cancer 46(8):1296–1316

19. Breast Cancer MRI—Magnetic Resonance Imaging. http://www.imaginis.com/mri-scan/magnetic-resonance-breast-imaging-mri-mr-2#limitations-to-an-mri-exam-of-the-breast

20. Cardenosa G (2004) Breast Imaging. Lippincott Williams & Wilkins, Greensboro

21. Stavros AT (2004) Breast Ultrasound. Lippincott Williams & Wilkins, Philadelphia

22. Gokhale S (2009) Ultrasound characterization of breast masses. Indian J. Radiol Imaging 19 (3):242–247

23. RadiologyInfo.org. Produced by Radiological Society of North America (RSNA) and American College of Radiology (ACR).Ultrasound Breast. http://www.radiologyinfo.org/en/info.cfm?pg=breastus#part_one

24. Madjar H (2002) Advantages and limitations of breast ultrasound. Gynakol Obstet Rundsch 42 (4):185–190

25. O'Connor MK, Rhodes D, Hruska C (2009) Molecular breast imaging. Expert Rev Anticancer Ther 9(8):1073–1080

26. Positron Emission Mammography. http://www.inlandimaging.com/breast/pem

27. Diakides NA, Diakides M, Lupo JC, Paul JL, Balcerak R (2007) Advances in medical infrared imaging. In: Diakides Nicholas A, Bronzino Joseph D (eds) Medical infrared imaging: 1-1-1-13. Taylor and Francis, CRC Press, Boca Raton

28. Wang J, Chang KJ, Chen CY, Chien KL, Tsai YS, Wu YM, Teng YC, Shih TTF (2010) Evaluation of the diagnostic performance of infrared imaging of the breast: a preliminary study. J BioMed Eng Online 9:3 (2010)

29. Keyserlingk JR, Ahlgren PD, Yu E, Belliveau N (1998) Infrared imaging of breast: Initial reappraisal using high-resolution digital technology in 100 successive cases of stage I and II breast cancer. Breast J 4(4):245–251

30. Amalu WC, Hobbins WB, Head JF, Elliot RL (2007) Infrared imaging of the breast: a review. In: Diakides Nicholas A, Bronzino Joseph D (eds) Medical infrared imaging: 9-1-9-22. Taylor and Francis, CRC Press, Boca Raton

31. Borchartt TB, Conci A, Lima RCF, Resmini R, Sanchez A (2013) Breast thermography from an image processing viewpoint: a survey. Sign Proc (in press) 93(10):2785–2803
32. Jones CH (1988) Physical aspects of infrared imaging. In: Webb S (ed) The physics of medical imaging, pp 488–508. Taylor & Francis, New York
33. Jones BF (1998) A reappraisal of the use of infrared thermal image analysis in medicine. IEEE Trans Med Imaging 17(6):1019–1027
34. Steketee J (1973) Spectral emissivity of the skin and pericardium. Phys Med Biol 18:686–694
35. Staiger PK, White JM (1988) Conditioned alcohol-like and alcohol-opposite responses in humans. Psychopharmacology 95(1):87–91
36. Melnizky P, Ammer K (2000) Einfluss von Alkohol und Rauchen auf die Haut-temperature des Gesichts, der Hände und der Kniegelenke. ThermologyInternational 10(4):191–195
37. Oregon Natural Medicine. http://www.oregonnaturalmedicine.com/breast-thermography/preparation-for-scan
38. Kapoor P, Prasad SVAV (2010) Image processing for early diagnosis of breast cancer using infrared images. Paper presented at the 2nd international conference on computer and automation engineering (ICCAE) 3(1):564–566, 26–28
39. Ring EFJ, Ammer K (2000) The Technique of infrared imaging in medicine. Thermol Int 10(1):7–14
40. Ring EFJ, Ammer K (2012) Infrared thermal imaging in medicine. Physiol Meas 33(3):R33–R46
41. Carlo AD (1995) Thermography and the possibilities for its applications in clinical and experimental dermatology. Clin Dermatol 13(4):329–336
42. Ng EYK, Ung LN, Ng FC, Sim LS (2001) Statistical analysis of healthy and malignant breast thermography. J Med Eng Technol 25(6):253–263
43. Acharya UR, Ng EYK, Tan JH, Sree SV (2010) Thermography based breast cancer detection using texture features and support vector machine. J Med Syst 36(3):1503–1510
44. American College of Clinical Thermology (ACCT) Breast Screening Procedure. http://www.thermologyonline.org/Breast/breast_thermography_procedure.htm
45. Kontos M, Wilson R, Fentiman I (2011) Digital infrared thermal imaging (DITI) of breast lesions: sensitivity and specificity of detection of primary breast cancers. Clin Radiol 66(6):536–539
46. Frize M, Herry C, Scales N (2003) Processing thermal images to detect breast cancer and assess pain. Paper presented at the 4th international IEEE EMBS special topic conference, pp 234–237, 24–26 April 2003
47. International Academy of Clinical Thermology. http://www.iactorg.org/professionals/thermog-guidelines.html
48. Herry CL, Frize M, Goubran RA (2008) Search for abnormal thermal patterns in clinical thermal infrared imaging. IEEE Int Workshop Med Meas Appl 61–65:9–10 (Ottawa)
49. A Review of Breast Thermography. http://www.iact-org.org/articles/articles-review-btherm.html
50. Acharya UR, Ng EYK, Sree SV, Chua CK, Chattopadhyay S (2014) Higher order spectra analysis of breast thermogram for the automated identification of breast cancer. J Expert Syst 31(1):37–47
51. Zadeh HG, Haddadnia J, Hashemian M, Hassanpour K (2012) Diagnosis of breast cancer using a combination of genetic algorithm and artificial neural network in medical infrared thermal imaging. Iran J Med Phys 9(4):265–274
52. Arena F, Barone C, Di Cicco T (2003) Use of digital infrared imaging in enhanced breast cancer detection and monitoring of the clinical response to treatment. In: Proceedings of the 25th annual international conference on engineering in medicine and biology society (EMBS), vol 2, pp 1129–1132
53. Qi H, Head JF (2001) Asymmetry analysis using automatic segmentation and classification for breast cancer detection in thermograms. In: Proceedings of the 23rd annual international conference of the IEEE EMBS, vol 3. Turkey, pp 2866–2869

54. Ng EYK, Kee EC (2008) Advanced integrated technique in breast cancer thermography. J Med Eng Technol 32(2):103–114
55. Wishart GC, Campisi M, Boswell M, Chapman D, Shackleton V, Iddles S, Hallett A, Britton PD (2010) The accuracy of digital infrared imaging for breast cancer detection in women undergoing breast biopsy. Eur J Cancer Surg 36:535–540
56. Arora N, Martins D, Ruggerio D, Tousimis E, Swistel AJ, Osborne MP, Simmons RM (2008) Effectiveness of a noninvasive digital infrared thermal imaging system in the detection of breast cancer. Am J Surg 196(4):523–526
57. Bharathi GB, Francis SV, Sasikala M, Sandeep, JD (2014) Feature analysis for abnormality detection in breast thermogram sequences subject to cold stress. In: Proceedings of the national conference on man machine interaction (NCMMI), pp 15–21
58. Beware of Poor Breast Thermograms, ILSI Thermography Service. http://www. doctormedesign.com/HTMLcontent/Beware%20of%20Poor%20Thermograms.htm
59. Kolarić D, Herceg Z, Nola IA, Ramljak V, Kulis T, Holjevac JK, Deutsch JA, Antonini S (2013) Thermography–a feasible method for screening breast cancer? Coll Antropol 37 (2):583–588
60. OHIO INFRARED HEALTH, Radiation-Free breast imaging to aid in the earliest detection of breast disease. http://www.ohioinfraredhealth.com/our-services.htm
61. Thermography Center of Memphis.Radiation free screening provides early warning signs. http://www.memphisthermography.com/breast-health.html
62. Thermography Service. http://breastthermogram.wordpress.com/thermography-services/
63. Agostini V, Knaflitz M, Molinari F (2008) Motion artifact reduction in breast dynamic infrared imaging. IEEE Trans Biomed Eng 56(3):903–906
64. Gautherie M, Gros CM (1980) Breast thermography and cancer risk prediction. Cancer 45 (1):51–56
65. Kapoor P, Prasad SVAV, Patni S (2012) Automatic analysis of breast thermograms for tumor detection based on biostatistical feature extraction and ANN. Int J Emerg Trends Eng Dev 7 (2):245–255
66. Schaefer G, Nakashima T, Zavisek M (2008) Analysis of breast thermo-grams based on statistical image features and hybrid fuzzy classification. In: Bebis G, Boyle R, Parvin B, Koracin D, Remagnino P, Porikli F, Peters J, Klosowski J, Arns L, Chun Y, Rhyne T, Monroe L (eds) Advances in visual computing: lecture notes in computer science, vol 5358. Springer, Las Vegas, pp 753–762
67. Qi H, Kuruganti PT, Snyder WE (2008) Detecting breast cancer from thermal infrared images by asymmetry analysis. In: Diakides NA, Bronzino JD (eds) Medical infrared imaging. pp 11.1–11.14
68. Krawczyk B, Schaefer G (2014) A hybrid classifier committee for analysing asymmetry features in breast thermograms. J Appl Soft Comput 20:112–118
69. BreastCancerTreatment.in. http://www.breastcancertreatment.in/breast_anatomy.htm
70. HRT doubles breast density. http://www.fhcrc.org/en/news/center-news/2004/11/HRT.html
71. Breast Cancer Awareness. http://www.sheknows.com/sheknows-cares/breast-cancer-up-close-personal-photo-gallery/breast-cancer-up-close-and-personal/mammogram-showing-a-breast-tumor
72. Hayashi Y, Okuyama F (2010) New approach to breast tumor detection based on fluorescence x-ray analysis. German medical Science (GMS), vol 8
73. Breast Cancer MRI—Magnetic Resonance Imaging. http://www.imaginis.com/mri-scan/magnetic-resonance-breast-imaging-mri-mr-3
74. http://lubbockonline.com/sites/default/files/imagecache/superphoto/12724180.jpg
75. Ultrasound of the Breast—Pathology. http://www.ultrasoundpaedia.com/pathology-breast/
76. Breast Cancer Ultrasound. http://medicalpicturesinfo.com/breast-cancer-ultrasound/
77. Dilon Breast Cameras. http://www.diagimaging.com/dilon%20breast%20camera.htm
78. Molecular breast imaging more effective than mammography at detecting cancer in high-risk women with dense breasts. http://mcnewsblog.wordpress.com/2008/09/03/breast-cancer-molecular-breast-imaging-mammography/

79. The role of positron emission mammography in breast cancer imaging and management. http://www.appliedradiology.com/articles/the-role-of-positron-emission-mammography-in-breast-cancer-imaging-and-management
80. Breast Thermography. http://www.breastthermography.com/breast_thermography_proc.htm
81. Articles of health. http://articlesofhealth.blogspot.in/2012/04/reversal-of-breast-cancer-invasive.html
82. American College of Clinical Thermology. Available at: http://www.thermologyonline.org/breast/breast_thermography_what.htm
83. Innerimage. Clinical Thermography. http://www.myinnerimage.com/who-should-have-this-test.php
84. The Breast Thermography Journal (Digital Infrared Analysis). http://advancedbreastthermography.blogspot.in/2013/02/birads-iii-to-be-or-not-to-be.html

Neuro-Fuzzy Hybridized Model for Seasonal Rainfall Forecasting: A Case Study in Stock Index Forecasting

Pritpal Singh

Abstract The ensemble of statistics and mathematics has increased the accuracy of forecasting the Indian summer monsoon rainfall (ISMR) up to some extent. But due to the nonlinear nature of the ISMR, its forecasting accuracy is still below the satisfactory level. Mathematical and statistical models require complex computing power. Now a day, artificial neural networks (ANNs)-based models are used to forecast the ISMR. Various experiments signify that alone ANN cannot deal with the dynamic nature of the ISMR. So, in this chapter, we present a novel model based on the ensemble of ANN and fuzzy time series (FTS). This model is referred to as "Neuro-Fuzzy hybridized model for time series forecasting". The ISMR data set from the period 1901–1990 for the monsoon season (mean of June, July, August, and September) is considered for the experimental purpose. The forecasted results obtained for the training (1901–1960) and testing (1961–1990) data sets are then compared with existing models. The results clearly exhibit the superiority of our model over the considered existing models. The applicability of the proposed model has also been examined in the stock index data set.

Keywords Neuro-fuzzy hybridized model · Seasonal rainfall forecasting · Stock index forecasting · Fuzzy time series

1 Introduction

The Indian economy is based on agriculture and its agricultural products, and crop yield is heavily dependent on the summer monsoon (June–September) rainfall. Therefore, any decrease or increase in annual rainfall will always have a severe impact on the agricultural sector in India. About 65 % of the total cultivated land in

P. Singh (✉)
Department of Computer Science & Engineering, Thapar University,
Patiala 147004, Punjab, India
e-mail: pritpal.singh@thapar.edu

© Springer India 2016   361
S. Bhattacharyya et al. (eds.), *Hybrid Soft Computing Approaches*,
Studies in Computational Intelligence 611,
DOI 10.1007/978-81-322-2544-7_12

India is under the influence of rain fed agriculture system [1]. Therefore, prior knowledge of the monsoon behavior (during which the maximum rainfall occurs in a concentrated period) will help the Indian farmers and the government to take advantage of the monsoon season. This knowledge can very useful in reducing the damage of crops during the less rainfall in the monsoon season. Therefore, forecasting the monsoon is a major scientific issue in the field of monsoon meteorology. The ensemble of statistics and mathematics has increased the accuracy of forecasting of the ISMR up to some extent. But the nonlinear nature of the ISMR, its forecasting accuracy is still below the satisfactory level.

In 2002, Indian Meteorological Department (IMD) fails to predict the deficit of rainfall during the summer monsoon season, which led to considerable concern in the meteorological community [2]. In 2004, drought has been again observed in the country with a deficit of more than 13 % rainfall [3], which could not be predicted by any statistical or dynamical model.

Forecasting of the ISMR started more than 100 years ago [4–7]. Mathematical and statistical models require complex computing power [8–11]. Therefore, many researchers have tried to apply ANN for ISMR forecasting. In literature, several types of neural networks are available, but usually Feed-forward neural network and Back-propagation neural network are used in the ISMR forecasting [12–21].

Another soft computing approach, in which historical values from time series are used to forecast the future values, is fuzzy time series (FTS). Forecasting using the FTS is applied in several areas including forecasting university enrollments, sales, road accidents, and financial forecasting. In a conventional time series, the recorded values of a special dynamic process are represented by crisp numerical values. However, in a FTS model, the recorded values of a special dynamic process are represented by linguistic values. Based on the FTS concept, first forecasting model was introduced by Song and Chissom [22, 23]. They presented the FTS model by means of fuzzy relational equations involving max-min composition operation and applied the model to forecast the enrollments in the University of Alabama. In 1996, Chen [24] used simplified arithmetic operations avoiding the complicated max-min operations and their method produced better results. Later, many studies provided some improvements to the existing methods in terms of effective lengths of intervals [25–33], fuzzification [27, 34, 35], fuzzy Logical relationships [33, 36–43] and defuzzification techniques [44, 45]. Recent advancement in the FTS forecasting model can be found in these articles [27, 46–53].

The literature review reveals that ANN and fuzzy logic theory have great capability in solving many real-life problems, especially, when it comes to the process of complex decision making. So, it is beneficial to fuse ANN and fuzzy set techniques together by substituting the demerits of one technique by the merits of another techniques. These two techniques can be fused as:

- Application of ANN for designing fuzzy logic-based systems, and
- Application of fuzzy logic for designing ANN-based systems.

Many researchers tried to design hybridize-based models to solve complex decision making problems, such as rough-fuzzy hybridization scheme for case generation [54], genetic fuzzy systems and ANN for stock price forecasting [55], genetic algorithm and rough set theory for stock price forecasting [56], hybridization of ANN and FTS for TAIEX forecasting [28, 57], neuro-fuzzy model for domestic debt forecasting [58], FTS and particle swarm optimization for TAIFEX forecasting [59, 60], ANN and genetic algorithm for dwelling fire occurrence prediction [61], cumulative probability distribution approach, rough set rule induction and FTS for stock market forecasting [62], fuzzy-neuro approach for sales forecasting of printed circuit board industry [63], enrollments forecasting based on FTS and particle swarm optimization [64, 65].

So far our knowledge is concerned no works have been reported in the literature that involves hybridization of ANN and FTS concept for involving rainfall prediction. So, in this work, we have designed a model based on these two techniques that can be employed for predicting rainfall in the summer monsoon season for the north-eastern region of India, where heavy rainfall is observed during this season. To show the effectiveness of the proposed model, it is also applied on the stock index data set.

This chapter is organized as follows. In Sect. 2, various problems corresponding to FTS model and effective measures to resolve these problems have been discussed. In Sect. 3, some basic concepts of fuzzy set theory have been explained with an overview of FTS. Application of ANN for designing the proposed model has been discussed in Sect. 4. A brief description of the data set is provided in Sect. 5. The architecture of the proposed model is presented in Sect. 6. In Sect. 7, the proposed neuro-fuzzy hybridized model has been presented. In Sect. 8, statistical parameters for analyzing the performance of the model have been discussed. The performance of the model has been assessed, and presented in the results Sect. 9. The application of the proposed model on the stock index data set is discussed in Sect. 10. The conclusions and future works are discussed in Sect. 11.

2 Problem Definitions

From review of literature, it is obvious that accuracy of the FTS model depend on four major factors: (a) Lengths of intervals, (b) Handling of repeated fuzzy sets, (c) Handling of trend associated with fuzzy sets, and (d) Defuzzication operation. In this study, we aim to deal with the problems associated with these four factors based on the hybridization of ANN and FTS. The main purpose of hybridized these two techniques together are explained as follows:

Factor 1 (*Lengths of intervals*). For fuzzification of time series data set, determination of lengths of intervals of the historical time series data set is very important. In case of most of the FTS models [22–24, 34, 38], the lengths of the intervals were kept the same. No specific reason is mentioned for using the fixed lengths of intervals. Huarng [25] shows that the lengths of intervals always affect the results of forecasting. So, for creating the effective lengths of intervals, an ANN based technique is used in this model.

Factor 2 (*Handling of repeated fuzzy sets*). In existing FTS models, each fuzzy set is given equal importance, which is not effective to solve real-time problems. Because, each fuzzy set represents various uncertainty involved in the domain. There are two possible ways to determine the weights as: (i) Assign weights based on human interpretation, and (ii) Assign weights based on occurrence of fuzzy sets in the fuzzified time series data set. Assignment of weights based on human-knowledge is not an acceptable solution for real-world problems as human-interpretation varies from one to another. Moreover, human-interpretation is still an issue which is not understood by the computational scientists. Therefore, second way is considered, where all the fuzzy sets are given importance based on their frequency of occurrences. For example, if the fuzzy set $A_i (i = 0, 1, 2, .., n)$ occurs two times in the fuzzified time series data set, then assign its weight as 2 in terms of percentage (%). Sometimes, the nature of event is very dynamic. Therefore, it is advantageous to assign weight in %, because it will help to capture the variation in value in the particular time instance.

Factor 3 (*Handling of trend associated with fuzzy sets*). In the FTS modeling approach, fuzzified time series values are further used to establish the fuzzy logical relations (FLRs). These FLRs are represented in the form of $A_i \rightarrow A_j$ $(i, j = 0, 1, 2, .., n)$, where condition attribute "A_i" and decision attribute "A_j" represent two different fuzzy sets for two different time instances "$t - 1$" and "t", respectively. In these FLRs, the decision attribute can exhibit different trends with respect to the condition attribute, which can be determined based on the following three conditions:

Condition 1 If $i < j$, then the trend of A_j will be upward ($\uparrow$)
Condition 1 If $i > j$, then the trend of A_j will be downward ($\downarrow$)
Condition 1 If $i = j$, then the trend of A_j will be unchanged ($=$)

Here, each i and j ψ represent the indices of the fuzzy sets A_i and A_j, respectively. Previously, trends represented by these decision attributes are not considered by the researchers during final decisions making. Therefore, to enhance the predictive skill of our model, these three trends (*i.e.*, $\uparrow, \uparrow, =$) are employed at the time of simulating and forecasting the results.

Factor 4 (*Defuzzification operation*). Song and Chissom [22] adopted the following method to forecast enrollments of the University of Alabama:

$$Y(t) = Y(t - 1) \circ R, \tag{1}$$

where $Y(t - 1)$ is the fuzzified enrollment of year "$t - 1$", $Y(t)$ is the forecasted enrollment of year "t" represented by fuzzy set, "$\circ$" is the max-min composition operator, and "R" is the union of fuzzy relations. This method takes much time to compute the union of fuzzy relations R, especially when the number of fuzzy relations is more in Eq. 1 [66, 67]. In 1996, Chen [24] used simplified arithmetic operations for defuzzification operation by avoiding this complicated max-min operations, and their method produced better results than the Song and Chissom models [22, 23]. Most of the existing FTS models have used Chen's defuzzification method [24] to acquire the forecasting results. However, forecasting accuracies of these models are not good enough, so it is required to adopt a better method that can be employed for the defuzzification operation. So, to resolve this problem, we introduce a new "Frequency-Based Defuzzification Technique" for the defuzzification operation.

3 Fuzzy Sets and the FTS—A Brief Overview

In 1965, Zadeh [68] introduced the fuzzy sets theory involving continuous set membership for processing data in presence of uncertainty. He also presented fuzzy arithmetic theory and its application [69–71]. Here, we briefly reviewed some concepts of FTS from [22, 23].

Definition 1 (*Fuzzy Set*) [68] A fuzzy set is a class with varying degrees of membership in the set. Let U be the universe of discourse, which is discrete and finite, then fuzzy set A can be defined as follows:

$$A = \{\mu_A(x_1)/x_1 + \mu_A(x_2)/x_2 + \cdots\} = \sum_i \mu_A(x_i)/x_i, \qquad (2)$$

where μ_A is the membership function of A, $\mu_A : U \rightarrow [0, 1]$, and $\mu_A(x_i)$ is the degree of membership of the element x_i in the fuzzy set A. Here, the symbol "+" indicates the operation of union and the symbol "/" indicates the separator rather than the commonly used summation and division in algebra, respectively.

When U is continuous and in finite, then the fuzzy set A of U can be defined as:

$$A = \left\{ \int \mu_A(x_i)/x_i \right\}, \forall x_i \in U, \qquad (3)$$

where the integral sign "$\int$" stands for the union of the fuzzy singletons, $\mu_A(x_i)/x_i$.

FTS concept was proposed in [22, 23] and the main difference between the traditional time series and the FTS is that the values of the former are crisp numerical values, while the values of the latter are fuzzy sets. The crisp numerical values can be represented by real numbers whereas in fuzzy sets, the values of observations are represented by linguistic values. The definition of FTS is briefly reviewed as follows:

Definition 2 (*FTS*) [22, 23] Let $Y(t)(t = 0, 1, 2, \ldots)$ be a subset of R and the universe of discourse on which fuzzy sets $\mu_i(t)(i = 1, 2, \ldots)$ are defined and let $F(t)$ be a collection of $\mu_i(t)(i = 1, 2, \ldots)$. Then, $F(t)$ is called a FTS on $Y(t)(t = 0, 1, 2, \ldots)$.

Definition 3 *(Fuzzy logical relationship)* [22–24] Assume that $F(t-1) = A_i$ and $F(t) = A_j$. The relationship between $F(t)$ and $F(t-1)$ is referred as a FLR, which can be represented as:

$$A_i \rightarrow A_j, \tag{4}$$

where A_i and A_j refer to the previous state and current state of the FLR, respectively.

Definition 4 *(Fuzzy logical relationship group)* [22–24] Assume the following FLRs:

$$A_i \rightarrow A_{k1},$$

$$A_i \rightarrow A_{k2},$$

$$A_i \rightarrow A_{km}$$

Chen [24] suggested that the FLRs having the same previous state are grouped into a fuzzy logical relationship group (FLRG). So, based on the Chen's model, these FLRs can be grouped into the FLRG as:

$$A_i \rightarrow A_{k1}, A_{k2}, \ldots, A_{km}$$

4 ANN and Its Application for Creation of Intervals

ANN is a computational model that is inspired by the human brain [72, 73]. ANN is composed of large number of interconnected nodes or neurons, which usually operate in parallel, and are configured in regular architectures. Researchers employ ANN in various forecasting problems such as electric load forecasting [74], short-term precipitation forecasting [75], credit ratings forecasting [76], tourism demand forecasting [77] etc., due to its capability to discover complex nonlinear relationships [78, 79] in the observations.

Data clustering is a popular approach for automatically finding classes, concepts, or groups of patterns [80]. Time series data are pervasive across all human endeavors, and their clustering is one of the most fundamental applications of data mining [81, 82]. In the literature, many data clustering algorithms [83–85] have been proposed, but their applications are limited to the extraction of patterns that represent points in multidimensional spaces of fixed dimensionality [86]. In our proposed model, a distance-based clustering algorithm, i.e., the Self-organizing feature maps (SOFM) is employed for determining the intervals of the historical time series data set by clustering them into different groups. The SOFM is developed by Kohonen [87], which is a class of neural networks with neurons arranged in a low dimensional (often two-dimensional) structure, and trained by an iterative unsupervised or self-organizing procedure [63, 88]. The SOFM converts the patterns of arbitrary dimensionality into response of one-dimensional or two-dimensional arrays of neurons, i.e., it converts a wide pattern space into a feature space. The neural network performing such a mapping is called feature map. The training process of the SOFM can be found in [63, 73]. In this work, MATLAB 7.2 [89] is used for implementing the SOFM to determine the intervals of time series data set.

Based on the SOFM, the historical time series data set is partitioned into different lengths of intervals. These intervals are presented in Sect. 7.

5 Description of Data Set

The north-eastern region of India refers to the easternmost region of India consisting of the neighboring seven states as: Arunachal Pradesh, Assam, Nagaland, Meghalaya, Manipur, Tripura, and Mizoram. All these states are shown in Fig. 1.

In this region, the monsoon rainfall starts in the month of June and ends in the month of September. July and August fall in the mid of the monsoon season. For modeling purpose, time series data set of all these individual months are collected from the period 1901–1990 [90]. The seasonal rainfall data set is prepared by taking the mean (June, July, August and September) of all these time series data set [17]. This data set is shown in Table 1. For simulating purpose, this data set is divided into the training set from the period 1901–1960, and the testing set from the period 1961–1990. In this study, forecasting of the seasonal rainfall is main objective.

6 Architecture of the Model

Most of the existing FTS models as discussed earlier use the following six common steps to deal with the forecasting problems:

Step 1. Divide the universe of discourse into intervals
Step 2. Define linguistic terms for each of the interval
Step 3. Fuzzify the time series data set
Step 4. Establish the FLRs based on Definition 4
Step 5. Construct the FLRGs based on Definition 5
Step 6. Defuzzify and compute the forecasted values

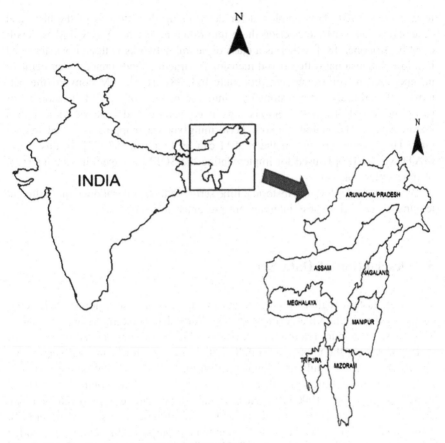

Fig. 1 Map showing the north-eastern region of India

Table 1 A sample of mean seasonal rainfall data set (in mm) for the north-eastern region of India [90]

Year	Rainfall
1901	1361.70
1902	1525.70
1903	1406.20
1904	1329.40
1905	1551.90
...	...
1989	1544.40
1990	1311.80

In this study, an improved FTS forecasting model is proposed, which is based on the hybridization of ANN and FTS concept. Therefore, the above six steps are modified, which is presented in Fig. 2. The proposed model is referred to as "Neuro-Fuzzy hybridized model for time series forecasting".

7 Proposed Neuro-Fuzzy Hybridized Model

We apply the proposed model to forecast the seasonal rainfall values in the north-eastern region of India from the period 1901–1990. This model is initially trained with the data set for the period 1901–1960. Each step of the training process is explained next.

Step 1. Define the universe of discourse U of the historical time series data set, and partition it into different length of intervals.

[**Explanation**] Define the universe of discourse U of the training data set. Assume that $U = [M_{min}, M_{max}]$, where M_{min} and M_{max} are the minimum and maximum values of the time series data set. In this data set, the universe of discourse is: $U = [1229.4, 1793.2]$. Now, by applying the SOFM NN, divide the universe of discourse U into different lengths of intervals as $a_0, a_1, \ldots, a_n$. Now, assign the data to their corresponding intervals. Centroid of each interval is recorded by taking the mean of upper bound and lower bound of the interval. The resulting intervals and centroids for the considered data set are shown in Table 2.

Step 2. Define linguistic terms for each of the interval. Assume that the historical time series data set is distributed among n intervals (i.e., $a_0, a_1, \ldots, a_n$). Therefore, define n linguistic variables $A_0, A_1, \ldots, Aj$, which can be represented by fuzzy sets, as shown below:

$$A_0 = 1/a_0 + 0.5/a_1 + 0/a_2 + \cdots + 0/a_{n-2} + 0/a_{n-1} + 0/a_n,$$
$$A_1 = 0.5/a_0 + 1/a_1 + 0.5/a_2 + \cdots + 0/a_{n-2} + 0/a_{n-1} + 0/a_n,$$
$$A_2 = 0/a_0 + 0.5/a_1 + 1/a_2 + \cdots + 0/a_{n-2} + 0/a_{n-1} + 0/a_n,$$
$$\vdots$$
$$A_j = 0/a_0 + 0/a_1 + 0/a_2 + \cdots + 0/a_{n-2} + 0.5/a_{n-1} + 1/a_n.$$

For ease of computation, the degree of membership values of fuzzy set $A_j (j = 0, 1, \ldots, n)$ are considered as either 0, 0.5, or 1. Here, the maximum degree of membership of fuzzy set A_i occurs at interval a_i and $0 \leq i \leq n$.

[**Explanation**] We define 43 linguistic variables $A_0, A_1, \ldots, A_{42}$ for the rainfall data set, because total 43 intervals are generated. All these defined linguistic variables are shown as follow:

$$A_0 = 1/a_0 + 0.5/a_1 + 0/a_2 + \cdots + 0/a_{40} + 0/a_{41} + 0/a_{42},$$
$$A_1 = 0.5/a_0 + 1/a_1 + 0.5/a_2 + \cdots + 0/a_{40} + 0/a_{41} + 0/a_{42},$$
$$A_2 = 0/a_0 + 0.5/a_1 + 1/a_2 + \cdots + 0/a_{40} + 0/a_{41} + 0/a_{42},$$
$$\vdots$$
$$A_{42} = 0/a_0 + 0/a_1 + 0/a_2 + \cdots + 0/a_{40} + 0.5/a_{41} + 1/a_{42}.$$

Here, the maximum degree of membership of the fuzzy set Ai occurs at interval a_i, and $0 \leq i \leq 42$.

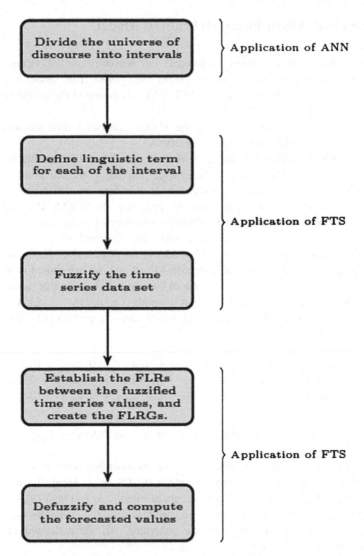

Fig. 2 Neuro-Fuzzy hybridized model for time series forecasting

For each interval, the centroid is calculated by taking the average of lower bound and upper bound of the interval.

Step 3. Fuzzify the time series data set. If one year's datum belongs to the interval a_i, then datum is fuzzified into A_i, where $0 \leq i \leq n$.

[Explanation] If any datum of rainfall data set belongs to the interval a_i, where $0 \leq i \leq 42$, then it is fuzzified into A_i. For example, the actual rainfall value for year 1958 is 1259.10, where 1259.10 belong to the interval a_2. Hence, the rainfall value

Table 2 A sample of intervals produced by the SOFM NN for the training data set with their corresponding centroids

Interval	Centroid
$a_0 = [1229.40, 1229.40]$	1229.40
$a_1 = [1241.10, 1241.10]$	1241.10
$a_2 = [1259.10; 1259.10]$	1259.10
$a_3 = [1262.10, 1262.10]$	1262.10
$a_4 = [1286.50, 1286.50]$	1286.50
…	…
$a_{41} = [1626.80, 1626.80]$	1626.80
$a_{42} = [1793.20; 1793.20]$	1793.20

for year 1958 is fuzzified into fuzzy set A_2. The fuzzified rainfall values for the training data set are shown in Table 3.

Based on the frequency of occurrence of each fuzzified values, their corresponding weights are assigned. For example, fuzzy set A_{15} has occurred three times in the fuzzified time series data set. Therefore, we have assigned its weight as 3 in terms of percentage (%). All these corresponding weights for the fuzzified time series data set in terms of % are also shown in Table 3.

Step 4. Establish the FLRs between the fuzzified time series values, and create the FLRGs.

[Explanation] Based on Definition 4, we can establish FLRs between two consecutive fuzzified rainfall values. For example, in Table 3, fuzzified rainfall values for Years 1901 and 1902 are A_{10} and A_{36}, respectively. So, we can establish a FLR between A_{10} and A_{36} as: $A_{10} \rightarrow A_{36}$. In this way, we have obtained the FLRs for the fuzzified rainfall values, which are presented in Table 4.

Based on Definition 5, the FLRs can be grouped into a FLRG. For example, in Table 4, there are two FLRs with same previous state, $A_6 \rightarrow A_{39}$ and $A_6 \rightarrow A_7$. Therefore, these FLRs can be grouped into the FLRG, $A_6 \rightarrow A_7, A_{39}$. A sample of FLRGs for the remaining FLRs is presented in Table 5.

The FLRGs shown in Table 5 are in the form of $A_i \rightarrow A_{j0}, A_{j1}, ..., A_{jp}$; where left-hand side and right-hand side of the expression represent previous state and current state, respectively. Now, each fuzzy set associated with the current state of the FLRG is classified into different trends as per the conditions provided in **Factor 3**. For example, consider the "FLRG 16: $A15 \rightarrow A_6, A_{21}, A_{22}$". Here, fuzzified rainfall values of the current state $A_6, A_{21,}$ and A_{22} can be represented with the following trends as:

$$A_{15} \rightarrow A_6(\downarrow), A_{21}(\uparrow), A_{22}(\uparrow)$$

In this way, we have recorded all the trends of fuzzified rainfall values for the training data set for the future consideration. If the same fuzzy set appears more than once in the current state of the FLRG, it is included only once in the rule.

Step 5. Defuzzify and compute the forecasted values from the fuzzified time series data set.

Table 3 A sample of fuzzified rainfall values for the training data set with their corresponding centroids and weights

Year	Rainfall	Fuzzified rainfall	Centroid	Weight (in %)
1901	1361.70	A_{10}	1361.70	1
1902	1525.70	A_{36}	1525.70	1
1903	1406.20	A_{15}	1406.40	3
1904	1329.40	A_6	1329.30	2
...	...	...	...	...
1959	1229.40	A_0	1229.40	1
1960	1530.30	A_{37}	1530.30	2

Table 4 A sample of FLRs for the rainfall data set

FLR
$A_{10} \rightarrow A_{36}$
$A_{36} \rightarrow A_{15}$
$A_{15} \rightarrow A_6$
$A_6 \rightarrow A_{39}$
$A_{39} \rightarrow A_{34}$
...
$A_2 \rightarrow A_0$
$A_0 \rightarrow A_{37}$

Table 5 A sample of FLRGs

FLRG
FLRG 1: $A_0 \rightarrow A_{37}$
FLRG 2: $A_1 \rightarrow A_{22}$
FLRG 3: $A_2 \rightarrow A_0$
FLRG 4: $A_3 \rightarrow A_2$
FLRG 5: $A_4 \rightarrow A_{25}$
...
FLRG 42: $A_{41} \rightarrow A_{26}$
FLRG 43: $A_{42} \rightarrow A_6$

To defuzzify the fuzzified time series data set and to obtain the forecasted values, we have proposed the "Frequency-Based Defuzzification Technique". This defuzzification technique employs the FLRGs obtained in Step 4 to get the forecasted values. The proposed defuzzification technique is designed in such a way that it can deal with the trend exhibited by different fuzzy sets of the corresponding FLRGs.

Based on the application of technique, it is categorized as: **Principle 1** and **Principle 2**. The **Principle 1** is given as follows:

Principle 1: For forecasting year $Y(t)$, the fuzzified value for the year $Y(t-1)$ is required, where "t" is the current year which we want to forecast. The Principle 1 is applicable only if there are more than one fuzzified values available in the current state. The steps under **Principle 1** are explained next.

Step 1. Obtain the fuzzified rainfall value for the year $Y(t-1)$ as $A_i(i = 0, 1, 2, \ldots, n)$.

Step 2. Obtain the FLRG whose previous state is A_i and the current state is $A_{j0}, A_{j1}, \ldots, A_{jp}$, i.e., the FLRG is in the form of $A_i \to A_{j0}, A_{j1}, \ldots, A_{jp}$. The fuzzy sets associated with the current state of the FLRG may exhibit miscellaneous trends (i.e., ↑, ↓ and =), because it contains more than one fuzzified values in the current state.

Step 3. Find the interval where the maximum membership value of the fuzzy set A_i (previous state) occurs, and let this interval be $a_i(i = 0, 1, 2, \ldots, n)$.

Step 4. Compute the variation in rainfall for the fuzzy set A_i as:

$$V_r = \left[\frac{C_i W_i}{100} \right] \tag{6}$$

Here, C_i and W_i represent the corresponding centroid of the interval and weight (in %) for the fuzzy set A_i respectively, where $i = 0, 1, 2, \ldots, n$.

Step 5. Find the intervals where the maximum membership values of the fuzzy sets $A_{j0}, A_{j1}, \ldots, A_{jp}$ (current state) occur. Let these intervals be $a_{j0}, a_{j1}, \ldots, a_{jp}$. All these intervals have the corresponding centroids $C_{j0}, C_{j1}, \ldots, C_{jp}$.

Step 6. Calculate the average of the centroids $C_{j0}, C_{j1}, \cdots, C_{jp}$ as:

$$A_{\text{centroid}} = \left[\frac{C_{j0} + C_{j1} + \cdots + C_{jp}}{p} \right] \tag{7}$$

Here, p represents the total number of fuzzy sets associated with the current state of the FLRG.

Step 7. Compute the trend value as:

$$T_{\text{value}} = \left[\frac{(C_{j0} * V_r) + (C_{j1} * V_r) + \ldots + (C_{jp} * V_r)}{p} \right] \tag{8}$$

In Eq. 8, if any fuzzified rainfall value in the current state exhibits downward trend (↓), then the symbol "*" represents "addition" operation; otherwise it represents "subtraction" operation for the upward trend (↑).

In the considered FLRG "$A_i \to A_{j0}, A_{j1}, \ldots, A_{jp}$", if any fuzzified rainfall value in the current state exhibits unchanged trend with respect to the previous state fuzzified rainfall value (i.e., $i = j$), then find the interval where the maximum membership value of the fuzzy set A_j occurs. Let this interval be a_j, which has the centroid C_j. This centroid C_j can directly be added to the numerator of left-hand side of Eq. 8, without addition or subtraction of V_r value from C_j.

Step 8. Compute the forecasted rainfall value as:

$$F_{\text{forecast}} = \left[\frac{A_{\text{centroid}} + T_{\text{value}}}{2} \right] \tag{9}$$

Principle 2: This principle is applicable only if there is only one fuzzified rainfall value in the current state. The steps under **Principle 2** are given as follows:

Step 1. Obtain the fuzzified rainfall value for the year $Y(t-1)$ as $A_i(i = 0, 1, 2, \ldots, n)$.

Step 2. Obtain the FLRG whose previous state is A_i and the current state is A_j, i.e., the FLRG is in the form of $A_i \rightarrow A_j$. The fuzzy sets associated with the current state of the FLRG may exhibit either upward ($\uparrow$), downward ($\downarrow$), or unchanged ($=$) trend, because it contains only one fuzzified value in the current state.

Step 3. Find the interval where the maximum membership value of the fuzzy set A_i (previous state) occurs, and let this interval be $a_i(i = 0, 1, 2, \ldots, n)$.

Step 4. Compute the variation in rainfall for the fuzzy set A_i as:

$$V_r = \left[\frac{C_i W_i}{100} \right] \tag{10}$$

Here, C_i and W_i represent the corresponding centroid of the interval and weight (in %) for the fuzzy set A_i, respectively, where $i = 0, 1, 2, \ldots, n$.

Step 5. Compute the trend value as:

$$T_{\text{value}} = [A_j \ast V_r] \tag{11}$$

In Eq. 11, if any fuzzified rainfall value in the current state exhibits downward trend ($\downarrow$), then the symbol "$\ast$" represents "addition" operation; otherwise it represents "subtraction" operation for the upward trend ($\uparrow$).

In the considered FLRG "$A_i \rightarrow A_j$", if any fuzzified rainfall value in the current state exhibits unchanged trend with respect to the previous state fuzzified rainfall value (i.e., $i = j$), then find the interval where the maximum membership value of the fuzzy set A_j occurs. Let this interval be a_j, which has the centroid C_j. This centroid C_j can directly be added to the left-hand side of Eq. 11, without addition or subtraction of V_r value from C_j.

Step 6. Find the interval where the maximum membership value of the fuzzy set A_j (current state) occurs. Let this interval be C_j.

Step 7. Compute the forecasted rainfall value as:

$$F_{\text{forecast}} = \left[\frac{C_j + T_{\text{value}}}{2} \right] \tag{12}$$

Based on the proposed model, here two examples have been presented to compute the forecasted values of the summer monsoon rainfall.

Example 1 Suppose we want to forecast the rainfall for the year $Y(1927)$, then we must have the fuzzified rainfall value for the previous year $Y(1926)$. The fuzzified rainfall value for the year $Y(1926)$ is obtained from the Table 3, which is A_{22}. Now, obtain the FLRG whose previous state is A_{22}. In Table 5, this FLRG is in the form of $A_{22} \rightarrow A_{12}, A_{31}$ (i.e., FLRG 23). Here, **Principle 1** is applicable, because there are more than one fuzzified rainfall values (A_{12} and A_{31}) available in the current state.

In this FLRG, the fuzzified rainfall values in the current state exhibit two different trends as: $A_{22} \rightarrow A_{12}$ ($\downarrow$), A_{31} ($\uparrow$). Now, obtain the interval where the maximum membership value of the fuzzy set A_{22} occurs from Table 2, which is a_{22}. This interval a_{22} has the corresponding centroid 1440.60 ($=C_{22}$). Here, the fuzzy set A_{22} has the weight 3 % ($=W_{22}$). Based on Eq. 6, compute the variation in rainfall for the fuzzy set A_{22} as:

$$V_r = \left[\frac{1440.60 \times 3}{100} \right] = 43.218$$

Now, find the intervals where the maximum membership values of the fuzzy sets A_{12} and A_{31} occur from Table 2, which are a_{12} and a_{31}, respectively. All these intervals a_{12} and a_{31} have the corresponding centroids 1372.80 ($=C_{12}$) and 1484.10 ($=C_{31}$), respectively. Based on Eq. 7, compute the average of the centroids as:

$$A_{centroid} = \left[\frac{1372.80 + 1484.10}{2} \right] = 1428.45$$

Now, compute the trend value based on Eq. 8 as:

$$T_{value} = \left[\frac{1372.80 + 43.218}{2} \right] + \left[\frac{1484.10 - 43.218}{2} \right] = 1428.45$$

Here, "addition $(+)$" operation is done for downward trend, whereas "sub-traction $(-)$" operation is done for upward trend. Now, calculate the forecasted rainfall value for the $Y(1927)$ based on Eq. 9 as:

$$F_{forecast} = \left[\frac{1428.45 + 1428.45}{2} \right] = 1428.45$$

Example 2 Suppose we want to forecast the rainfall for the year $Y(1915)$, then we must have the fuzzified rainfall for the previous year $Y(1914)$. The fuzzified rainfall value for the year $Y(1914)$ is obtained from Table 3, which is A_4. Obtain the FLRG whose previous state is A_4. In Table 5, this FLRG is in the form of $A_4 \rightarrow A_{25}$, (i.e., FLRG 5). Here, **Principle 2** is applicable, because there is only one fuzzified rainfall value (A_{25}) available in the current state.

In this FLRG, the fuzzified rainfall value in the current state exhibits only a single trend as: $A_4 \rightarrow A_{25}(\uparrow)$. Now, obtain the interval where the maximum membership value of the fuzzy set A_4 occurs from Table 2, which is a_4. This interval a_4 has the corresponding centroid 1286.50 ($=C_4$). Here, the fuzzy set A_4 has the weight 1 % ($=W_4$). Based on Eq. 10, compute the variation in rainfall for the fuzzy set A_4 as:

$$V_r = \left[\frac{1286.50 \times 1}{100}\right] = 12.865$$

Now, find the interval where the maximum membership value of the fuzzy set A_{25} occurs from Table 2, which is a_{25}. This interval a_{25} has the corresponding centroid 1458.50 ($=C_{25}$).

Now, compute the trend value based on Eq. 11 as:

$$T_{\text{value}} = [1458.50 - 12.865] = 1445.635$$

Here, "subtraction $(-)$" operation is done for the upward trend. Now, calculate the forecasted rainfall value for the year Y(1915) based on Eq. 12 as:

$$F_{\text{forecast}} = \left[\frac{1458.50 + 1445.635}{2}\right] = 1452.07$$

8 Performance Analyses

The performance of the model is evaluated with the help of root mean square error (RMSE), average forecasting error rate (AFER), evaluation parameter (δ_r), correlation coefficient (CC), coefficient of determination (CC^2) and tracking signal (TS). Definitions of all these parameters can be found in article [101].

9 Empirical Analysis

The proposed model is trained and tested using the data set mentioned in Sect. 5. While training and testing the model, 43 and 20 number of intervals is used, respectively. Results for the training and testing data set are presented in Table 6 in terms of various performance analysis parameters (as discussed in Sect. 8). The computed value of δ_r is much less than 1 as shown in Table 6. The CC between actual and forecasted rainfall is close to one, which indicates the efficiency of the model. The CC^2 values for the training and test data set exhibit the strong linear association between actual and forecasted rainfall values. The TS values are plotted for the forecasted rainfall values, which are depicted in Fig. 3. In Fig. 3, it is observed that TS values for each of the forecasted value lies between the ranges ± 4, which indicate that the model is working correctly.

During simulation of the SOFM NN, a number of experiments were carried out to set additional parameters [73], viz., initial weight, learning rate, epochs, learning

Table 6 Experimental results of the training and testing data set

Statistical parameters	Experimental result (training data set)	Experimental result (testing data set)
RMSE	59.42	67.80
AFER	2.41 %	3.49 %
δ_r	0.4	0.4
CC	0.85	0.86
CC^2	0.72	0.74
M_{ad}	34.55	48.64
R_{sfe}	12.56	5.42
TS	0.36	0.11

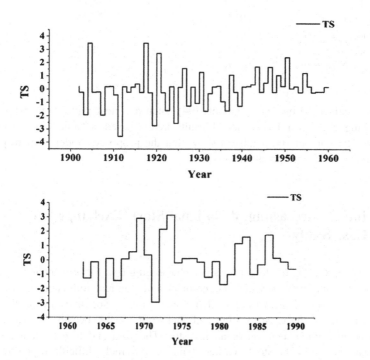

Fig. 3 TS curves for the training and testing data set (*top* to *bottom*)

radius, and activation function to obtain the optimal results, and we have chosen the ones that exhibit the best behavior in terms of accuracy. The determined optimal values of all these parameters are given in Table 7.

To assess the performance of the proposed model, it is compared with the existing FTS model proposed by Yu [93]. This model is chosen for comparative study, because its forecasting accuracy is better than the existing FTS models [24, 37]. Furthermore, the performance of the proposed model is also compared with

Table 7 Additional parameters and their values during the training and testing processes of neural network

Serial number	Additional parameter	Input value
1	Initial weight	0.3
2	Learning rate	0.5
3	Epochs	1000
4	Learning radius	3
5	Activation function	Sigmoid

Table 8 AFER for the training and test data set based the existing FTS model and linear models

Model	AFER (training data set)	AFER (testing data set)
Yu [93]	3.62 %	4.96 %
Linear regression	4.91 %	7.82 %
Quadratic regression	4.91 %	7.82 %
Compound regression	4.91 %	7.82 %
Exponential regression	4.91 %	7.82 %

various statistical time series models [94], such as linear regression, exponential smoothing, and so on. Experimental results are discussed in Table 8 in terms of the AFER. The comparative analyzes show that the proposed model is more precise than exiting FTS and linear models.

10 Index Forecasting of National Stock Exchange—A Case Study

Accurate forecasting of stock prices is a challenging task for brokers and investors. Various factors such as social, economical and political influence on the stock market, and control the variation of the stock prices. So, investing on the stock market is one of the most exciting and challenging tasks, because lose and profit depends on a good decision of an investor. The gain and lose explain the non-stationary nature of the stock market, which is extremely difficult to predict accurately in advance without an expert system.

In this section, we apply the proposed method for forecasting the daily index of the National Stock Exchange (NSE). The NSE is stock exchange situated at Mumbai, India. It is in the top 20 largest stock exchanges in the world by market capitalization and largest in India by daily turnover and number of trades, for both equities and derivative trading. The NSE has a market capitalization of around US $one trillion and over 1,652 listings as of July, 2012. The NSE's key index is

Table 9 Index price data set of the NSE from the period 8/1/2012 to 9/28/2012

Date (mm/dd/yyyy)	Actual index price (in rupee)
8/1/2012	5240.50
8/2/2012	5227.75
8/3/2012	5215.70
8/6/2012	5282.55
8/7/2012	5336.70
...	...
...	...
9/18/2012	5600.05
9/20/2012	5554.25
9/21/2012	5691.15
9/24/2012	5669.60
9/25/2012	5673.90
9/27/2012	5649.50
9/28/2012	5703.30

Table 10 Experimental results of the stock index data set

Statistical parameters	Experimental result
RMSE	249.61
AFER	0.67 %
δ_r	0.25
CC	0.96
CC^2	0.92
M_{ad}	36.24
R_{sfe}	−4.84
TS	−0.13

known as the NSE NIFTY (National Stock Exchange 50), an index of 50 major stocks weighted by market capitalization (http://www.nse-india.com).

For application purpose, the NSE's index data set (NIFTY) for the period 1/08/2012 to 28/09/2012 is collected from Yahoo Finance (http://in.finance.yahoo.com). This data set is shown in Table 9.

The consistency of the proposed model in forecasting the stock index has been analyzed on the basis of statistical parameters as discussed in Sect. 8. Experimental results are listed in Table 10. All these statistical analyzes signify the robustness of the proposed model for forecasting the stock index. From Fig. 4, it is obvious that the forecasted values for the stock index lie between the ranges ±4, which indicates that the proposed model is able to predict the stock index with reasonable range of errors.

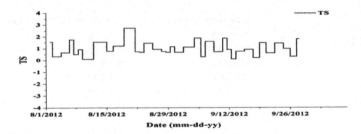

Fig. 4 TS curve for the stock index data set

11 Conclusions and the Way Ahead

In the field of forecasting, many researchers agree that none of the existing methods is best individually to handle the real time problem [95–97] due to complexity of their underneath architecture. Both theoretical as well as experimental results indicate that ensembling of different models together can be very effective and efficient way to improve the quality of forecasting [98, 99], as a single model may not be able to capture various types of patterns effectively [100].

This study presents a novel approach combining ANN and FTS for building a time series forecasting expert system. The main contributions of this chapter are presented as follows:

- First, the author shows that the forecasted accuracy of the FTS model can be enhanced by resolving the problems associated with different factors such as determination of lengths of intervals, handling of repeated fuzzy sets, handling of trends associated with fuzzy sets, and defuzzification operation. In this work, unique strength of ANN (SOFM NN) for determination of effective lengths of intervals, assignment of weights for repeated fuzzy sets based on their frequency of occurrences, determination of trends for fuzzy sets based on the proposed conditions, and defuzzification operation based on the proposed technique have been utilized for improving the forecasted accuracy of the model.
- Second, the author shows that the seasonal rainfall forecasting problem can be solved effectively by the proposed Neuro-Fuzzy hybridized model.
- Third, the author demonstrates that the proposed model is superior to existing FTS model for prediction of rainfall.
- Fourth, the author validates the proposed model using various statistical parameters, and shows that the presented model has the ability to predict the nonlinear behavior of the summer monsoon rainfall much more accurately compared to various linear models.
- Fifth, the author shows the applicability of the proposed model for forecasting the stock index. Experimental results show that the proposed model can effectively be utilized as an expert decision making system in stock as well as financial time series forecasting.

A significant drawback of the FTS forecasting model is that increase in the number of intervals of time series data set increases the accuracy rate of forecasting, but decreases the fuzziness of time series data set. Therefore, in this study, results are obtained with minimum number of intervals.

The proposed model has the limitation that it can applicable only in one-factor time series data set. Therefore, the author has tried to make the proposed model more generalize so that it can applicable in different kinds of one-factor time series data set, and can be employed in various domains flexibly. Work is underway to apply the proposed model on other domains in the following way:

- Apply the proposed model on different regions of seasonal rainfall data set (one-factor), and check its accuracy and performance with different size of intervals, and
- To test the performance of the model for different types of financial, stocks, insurance and marketing data set (one-factor).

Hence, this study implies that the approaches that have been adopted in the proposed model can be applied to improve the accuracy and performance of the FTS forecasting model.

References

1. Swaminathan MS (1998) Padma Bhusan Prof. P. Koteswaram First Memorial Lecture-23rd March 1998. In: Climate and Sustainable Food Security, vol 28, Vayu Mandal, pp 3–10
2. Gadgil S, Srinivasan J, Nanjundiah RS, Kumar KK, Munot AA, Kumar KR (2002) On forecasting the Indian summer monsoon: the intriguing season of 2002. Curr Sci 83(4):394–403
3. Gadgil S, Rajeevan M, Nanjundiah R (2005) Monsoon prediction-why yet another failure? Curr Sci 88(9):1389–1400
4. Blanford HF (1884) On the connection of the Himalayan snow with dry winds and seasons of droughts in India. In: Proceedings of the Royal Society of London, pp 3–22
5. Walker GT (1908) Correlation in seasonal variation of climate. Mem Ind Meteorol Dept 20:117–124
6. Walker GT (1910) On the meteorological evidence for supposed changes of climate in India. Indian Meteorol Memoirs 21:1–21
7. Walker GT (1933) Seasonal weather and its prediction. Brit Assoc Adv Sci 103:25–44
8. Krishna KK, Soman MK, Kumar KR (1995) Seasonal forecasting of Indian summer monsoon rainfall: a review. Weather 50:449–467
9. Mooley DA, Parthasarathy B (1984) Fluctuations in All-India summer monsoon rainfall during 1871-1978. Clim Change 6:287–301
10. Satyan V (1988) Is there an attractor for the Indian summer monsoon? In: Proc Ind Acad Sci (Earth Planet Sci), vol 97, pp 49–52
11. Basu S, Andharia HI (1992) The chaotic time-series of Indian monsoon rainfall and its prediction. In: Proc Ind Acad Sci (Earth Planet Sci), vol 101, pp 27–34
12. Goswami P, Srividya (1996) A novel neural network design for long range prediction of rainfall pattern. Current Science 70:447–457
13. Navone HD, Ceccatto HA (1994) Predicting Indian monsoon rainfall:a neural network approach. Clim Dyn 10:305–312

14. Shukla J, Mooley DA (1987) Empirical prediction of the summer monsoon rainfall over India. Mon Weather Rev 115:695–703
15. Hastenrath S (1988) Prediction of Indian monsoon rainfall: further exploration. J Clim 1:298–304
16. Guhathakurta P, Rajeevan M, Thapliyal V (1999) Long range forecasting Indian summer monsoon rainfall by a hybrid principal component neural network model. Meteorol Atmos Phys 71:255–266
17. Sahai AK, Soman MK, Satyan V (2000) All India summer monsoon rainfall prediction using an artificial neural network. Clim Dyn 16:291–302
18. Guhathakurta P (2006) Long-range monsoon rainfall prediction of 2005 for the districts and sub-division Kerala with artificial neural network. Curr Sci 90:773–779
19. Chakraverty S, Gupta P (2007) Comparison of neural network configurations in the long-range forecast of southwest monsoon rainfall over India. Neural Comput Appl 17:187–192
20. Rajeevan M, Pai DS, Dikshit SK, Kelkar RR (2004) IMD's new operational models for long-range forecast of southwest monsoon rainfall over India and their verification for 2003. Curr Sci 86:4220–4431
21. Goswami P, Kumar P (1997) Experimental annual forecast of all-India mean summer monsoon rainfall for 1997 using a neural network model. Curr Sci 72:781–782
22. Song Q, Chissom BS (1993) Forecasting enrollments with fuzzy time series—part I. Fuzzy Sets Syst 54(1):1–9
23. Song Q, Chissom BS (1994) Forecasting enrollments with fuzzy time series—part II. Fuzzy Sets Syst 62(1):1–8
24. Chen SM (1996) Forecasting enrollments based on fuzzy time series. Fuzzy Sets Syst 81:311–319
25. Huarng K (2001) Effective lengths of intervals to improve forecasting in fuzzy time series. Fuzzy Sets Syst 123:387–394
26. Li ST, Chen YP (2004) Natural partitioning-based forecasting model for fuzzy time-series
27. Cheng C, Chang J, Yeh C (2006) Entropy-based and trapezoid fuzzification-based fuzzy time series approaches for forecasting IT project cost. Technol Forecast Soc Chang 73:524–542
28. Huarng K, Yu THK (2006) The application of neural networks to forecast fuzzy time series. Physica A 363(2):481–491
29. Cheng CH, Cheng GW, Wang JW (2008) Multi-attribute fuzzy time series method based on fuzzy clustering. Expert Syst Appl 34:1235–1242
30. Kai C, Ping FF, Gang CW (2010) A novel forecasting model of fuzzy time series based on k-means clustering. In: 2010 second International Workshop on Education Technology and Computer Science, China, pp 223–225
31. Liu HT, Wei ML (2010) An improved fuzzy forecasting method for seasonal time series. Expert Syst Appl 37(9):6310–6318
32. Singh P, Borah B (2011) An efficient method for forecasting using fuzzy time series. Machine Intelligence. Tezpur University, Assam (India), Narosa, India, pp 67–75
33. Chen SM, Chen CD (2011) Handling forecasting problems based on high-order fuzzy logical relationships. Expert Syst Appl 38(4):3857–3864
34. Hwang JR, Chen SM, Lee CH (1998) Handling forecasting problems using fuzzy time series. Fuzzy Sets Syst 100:217–228
35. Sah M, Degtiarev K (2005) Forecasting enrollment model based on first-order fuzzy time series. Preceedings World Acad Sci Eng Technol. 1:132–135
36. Tsai CC, Wu SJ (2000) Forecasting enrolments with high-order fuzzy time series. In: 19th International Conference of the North American, pp 196–200
37. Chen SM (2002) Forecasting enrollments based on high-order fuzzy time series. Cybern Syst 33(1):1–16
38. Huarng K (2001) Heuristic models of fuzzy time series for forecasting. Fuzzy Sets Syst 123:369–386
39. Chang J, Lee Y, Liao S, Cheng C (2007) Cardinality-based fuzzy time series for forecasting enrollments. New Trends Appl Artif Intell Japan 4570:735–744

40. Chen TL, Cheng CH, Teoh HJ (2008) High-order fuzzy time-series based on multi-period adaptation model for forecasting stock markets. Physica A 387(4):876–888
41. Aladag CH, Basaran MA, Egrioglu E, Yolcu U, Uslu VR (2009) Forecasting in high order fuzzy times series by using neural networks to define fuzzy relations. Expert Syst Appl 36 (3):4228–4231
42. Aladag CH, Yolcu U, Egrioglu E (2010) A high order fuzzy time series forecasting model based on adaptive expectation and artificial neural networks. Math Comput Simul. 81 (4):875–882
43. Singh SR (2009) A computational method of forecasting based on high-order fuzzy time series. Expert Syst Appl 36(7):10551–10559
44. Lee HS, Chou MT (2004) Fuzzy forecasting based on fuzzy time series. Int J Comput Math 81 (7):781–789
45. Qiu W, Liu X, Li H (2011) A generalized method for forecasting based on fuzzy time series. Expert Systems with Applications 38(8):10,446–10,453
46. Chen TL, Cheng CH, Teoh HJ (2007) Fuzzy time-series based on fibonacci sequence for stock price forecasting. Physica A 380:377–390
47. Singh S (2007) A robust method of forecasting based on fuzzy time series. Appl Math Comput 188(1):472–484
48. Li ST, Cheng YC (2007) Deterministic fuzzy time series model for forecasting enrollments. Comput Math Appl 53(12):1904–1920
49. Wang JW, Liu JW (2010) Weighted fuzzy time series forecasting model. Proceedings of the Second international conference on Intelligent information and database systems: Part I. Springer-Verlag, Berlin, Heidelberg, pp 408–415
50. Chou HL, Chen JS, Cheng CH, Teoh HJ (2010) Forecasting Tourism Demand Based on Improved Fuzzy Time Series Model. In: Nguyen N, Le M, Swiatek J (eds) Intelligent Information and Database Systems. Lecture Notes in Computer Science, Springer, Berlin, Heidelberg
51. Shah M (2012) Fuzzy based trend mapping and forecasting for time series data. Expert Syst Appl 39(7):6351–6358
52. Gangwar SS, Kumar S (2012) Partitions based computational method for high-order fuzzy time series forecasting. Expert Syst Appl 39(15):12158–12164
53. Li ST, Kuo SC, Cheng YC, Chen CC (2011) A vector forecasting model for fuzzy time series. Appl Soft Comput 11(3):3125–3134
54. Pal SK, Mitra P (2004) Case generation using rough sets with fuzzy representation. IEEE Trans Knowl Data Eng 16(3):292–300
55. Hadavandi E, Shavandi H, Ghanbari A (2010) Integration of genetic fuzzy systems and artificial neural networks for stock price forecasting. Knowl-Based Syst 23(8):800–808
56. Cheng CH, Chen TL, Wei LY (2010) A hybrid model based on rough sets theory and genetic algorithms for stock price forecasting. Inf Sci 180(9):1610–1629
57. Yu THK, Huarng KH (2008) A bivariate fuzzy time series model to forecast the TAIEX. Expert Syst Appl 34(4):2945–2952
58. Keles A, Kolcak M, Keles A (2008) The adaptive neuro-fuzzy model for forecasting the domestic debt. Knowl-Based Syst 21(8):951–957
59. Park JI, Lee DJ, Song CK, Chun MG (2010) TAIFEX and KOSPI 200 forecasting based on two-factors high-order fuzzy time series and particle swarm optimization. Expert Syst Appl 37 (2):959–967
60. Kuo IH, Horng SJ, Chen YH, Run RS, Kao TW, Chen RJ, Lai JL, Lin TL (2010) Forecasting TAIFEX based on fuzzy time series and particle swarm optimization. Expert Syst Appl 37(2):1494–1502
61. Yang L, Dawson C, Brown M, Gell M (2006) Neural network and GA approaches for dwelling fire occurrence prediction. Knowl-Based Syst 19(4):213–219
62. Teoh HJ, Cheng CH, Chu HH, Chen JS (2008) Fuzzy time series model based on probabilistic approach and rough set rule induction for empirical research in stock markets. Data Knowl Eng 67(1):103–117

63. Chang YC, Chen SM (2009) Temperature Prediction Based on Fuzzy Clustering and Fuzzy Rules Interpolation Techniques. Proceedings of the 2009 IEEE International Conference on Systems, Man, and Cybernetics. San Antonio, TX, USA, pp 3444–3449

64. Kuo IH, Horng SJ, Kao TW, Lin TL, Lee CL, Pan Y (2009) An improved method for forecasting enrollments based on fuzzy time series and particle swarm optimization. Expert Syst Appl 36(3):6108–6117

65. Huang YL, Horng SJ, He M, Fan P, Kao TW, Khan MK, Lai JL, Kuo IH (2011) A hybrid forecasting model for enrollments based on aggregated fuzzy time series and particle swarm optimization. Expert Syst Appl 38(7):8014–8023

66. Chen SM, Hwang JR (2000) Temperature Prediction Using Fuzzy Time Series. IEEE Trans syst man cyberns 30:263–275

67. Huarng KH, Yu THK, Hsu YW (2007) A multivariate heuristic model for fuzzy time-series forecasting. IEEE Trans Syst Man Cybern B Cybern 37:836–846

68. Zadeh LA (1965) Fuzzy sets. Inf Control 8(3):338–353

69. Zadeh LA (1975) The concept of a linguistic variable and its application to approximate reasoning. Inform Sci 8:199–249

70. Zadeh LA (1973) Outline of a new approach to the analysis of complex system and decision process. IEEE Tran Syst Man Cybern 3:28–44

71. Zadeh LA (1971) Similarity relations and fuzzy orderings. Inf Sci 3:177–200

72. Bose NK, Liang P (1998) Neural Network Fundamentals with Graphs, Algorithms, and Applications. Tata McGraw-Hill, New Delhi, India

73. Sivanandam SN, Deepa SN (2007) Principles of Soft Computing. Wiley India (P) Ltd., New Delhi

74. Taylor JW, Buizza R (2002) Neural Network Load Forecasting with Weather Ensemble Predictions. IEEE Trans on Power Syst 17:626–632

75. Kuligowski RJ, Barros AP (1998) Experiments in short-term precipitation forecasting using artificial neural networks. Mon Weather Rev 126:470–482

76. Kumar K, Bhattacharya S (2006) Artificial neural network vs. linear discriminant analysis in credit ratings forecast: A comparative study of prediction performances. Rev Account Finance 5(3):216–227

77. Law R (2000) Back-propagation learning in improving the accuracy of neural network-based tourism demand forecasting. Tour Manag 21(4):331–340

78. Indro DC, Jiang CX, Patuwo BE, Zhang GP (1999) Predicting mutual fund performance using artificial neural networks. Omega 27(3):373–380

79. Donaldson RG, Kamstra M (1996) Forecast combining with neural networks. J Forecast 15 (1):49–61

80. Gondek D, Hofmann T (2007) Non-redundant data clustering. Knowl Inf Syst 12:1–24

81. Rakthanmanon T, Keogh E, Lonardi S, Evans S (2012) MDL-based time series clustering. Knowl Inf Syst 33(2):371–399

82. Keogh E, Lin J (2005) Clustering of time-series subsequences is meaningless: implications for previous and future research. Knowl Inf Syst 8(2):154–177

83. Cazarez-Castro NR, Aguilar LT, Castillo O (2012) Designing Type-1 and Type-2 Fuzzy Logic Controllers via Fuzzy Lyapunov Synthesis for nonsmooth mechanical systems. Eng Appl Artif Intell 25(5):971–979

84. Ordonez C (2003) Clustering binary data streams with K-means. Proceedings of the 8th ACM SIGMOD workshop on Research issues in data mining and knowledge discovery. ACM Press, New York, USA, pp 12–19

85. Wu X, Kumar V, Quinlan JR, Ghosh J, Yang Q, Motoda H, McLachlan G, Ng A, Liu B, Yu P, Zhou ZH, Steinbach M, Hand D, Steinberg D (2008) Top 10 algorithms in data mining. Knowl Inf Syst 14:1–37

86. Xiong Y, Yeung DY (2002) Mixtures of ARMA models for Model-Based Time Series Clustering. IEEE International Conference on Data Mining. Los Alamitos, USA, pp 717–720

87. Kohonen T (1990) The self organizing maps. In: Proc. IEEE, vol 78, pp 1464–1480

88. Liao TW (2005) Clustering of time series data-a survey. Pattern Recogn 38(11):1857–1874

89. MATLAB (2006) Version 7.2 (R2006). http://www.mathworks.com/
90. Pathasarathy B, Munot AA, Kothawale DR (1994) All india monthly and seasonal rainfall series: 1871-1993. Theoritical and Applied Climatolgy 49:217–224
91. Singh P, Borah B (2013) An e_cient time series forecasting model based on fuzzy time series. Eng Appl Artif Intell 26:2443–2457
92. Yu HK (2005) Weighted fuzzy time series models for TAIEX forecasting. Physica A 349(3–4):609–624
93. PASW (2012) PASW Statistics 18. http://ww.spss.com.hk/statistics/
94. Jenkins GM (1982) Some practical aspects of forecasting in organisations. J Forecasting 1:3–21
95. Makridakis S, Anderson A, Carbone R, Fildes R, Hibdon M, Lewandowski R, Newton J, Parzen E, Winkler R (1982) The Accuracy of Extrapolation (Time Series) Methods: Results of a Forecasting Competition. J Forecast 1(2):111–153
96. Chatfield C (1988) What is the 'best' method of forecasting? J Appl Statist 15:19–39
97. Newbold P, Granger CWJ (1974) Experience with Forecasting Univariate Time Series and the Combination of Forecasts. J Royal Stat Soc Series A Gen 137(2):131–165
98. Makridakis S (1989) Why combining works? Int J Forecast 5(4):601–603
99. Zhang GP (2003) Time series forecasting using a hybrid ARIMA and neural network model. Neurocomputing 50:159–175

Hybridization of 2D-3D Images
for Human Face Recognition

Suranjan Ganguly, Debotosh Bhattacharjee and Mita Nasipuri

Abstract Now-a-days face recognition is more realistic biometric approach for biometric based system for human authentication purpose. It has been aimed by the researchers and scientists over some decades to provide more reliable and secure environment. Although face recognition techniques have gained significant level of success, it is still having some challenging tasks due to the presence of facial pose, expression as well as illumination variations. With the trends of decrease in the cost of cameras, increase in the technological aspects and availability of processing power, face recognition task has now gained most of the researchers' attention in handling this complex task of computer vision. The human face images can be acquired by different methodologies, such as: from video sequences, from various sensors like optical, thermal and 3D etc. The variations of face images have also motivated the researchers to design the intelligent system for feature estimation purpose. In this chapter, an overview of hybrid techniques with its application in the domain of 3D face registration and recognition is discussed. Authors have also proposed a new 3D face recognition scheme from 2D and 3D hybrid face images using two supervised classifiers. The authors have also reported the contribution of their research work by considering all the related and recent works with proposed methodology. The investigation is accomplished on Frav3D database and achieved maximum 95.17 % accurate face recognition rate.

Keywords Face recognition · Range images · Optical images · Hybridization · ANN · K-NN

S. Ganguly (✉) · D. Bhattacharjee · M. Nasipuri
Department of Computer Science and Engineering, Jadavpur University, Kolkata, India
e-mail: suranjanganguly@gmail.com

D. Bhattacharjee
e-mail: debotoshb@hotmail.com

M. Nasipuri
e-mail: mnasipuri@cse.jdvu.ac.in

© Springer India 2016
S. Bhattacharyya et al. (eds.), *Hybrid Soft Computing Approaches*,
Studies in Computational Intelligence 611,
DOI 10.1007/978-81-322-2544-7_13

1 Introduction

Biometrics is the distinctive biological and behavioral human characteristics that can be used to uniquely identify a person or to claim the verification of an individual automatically. The word 'Biometric' is the combination of two ancient Greek words, i.e. 'bios' that is meant for 'life' and 'metron' which is used to mean 'measure' [1, 2]. Now, from these two meanings, 'Biometric' can be termed as 'life measuring'. Biometrics based recognition techniques are often used as a security measurement that can be deployed to control the security at secure buildings, airports or border crossings or for recording employee attendance, criminal identification purpose etc. But today, in modern life the use of biometrics has gradually spread to other applications also. The 'identification' and 'verification' are the two different approaches to recognize an individual. The 'identification' task is composed by 'one-to-many' comparison technique. Here, the probe images are tested by all the gallery images where as 'verification' task is little simple which is basically 'one-to-one' comparison. This chapter is briefly focused on face based human recognition. Other than human face, there are also different types of biometric measurements. The variations of the biometrics can be abstracted into two groups, namely: behavioral and biological. The biometrics measurements like 'Face', 'Fingerprint', 'Hand', 'Irish', 'DNA' etc. are in biological group, whereas 'Signature', 'Voice', 'Gait', 'Keystroke' etc. are within behavioral based biometric measurement group.

Recognizing the faces is some 'life measuring' technique that can be done usually without any effort. Along with this, several other advantages are also discussed below.

- The sensors can capture the face of any individual in public area by surveillance cameras or closed circuit television (CCTV).
- It can also be captured from some distance away.
- Face images can also be captured without any physical contact.

There are other characteristics of face images that are described in Fig. 1, for which human faces are highly acknowledged over other biometric traits.

Considering all these key features of the face images, human face can be processed for several directions. It is listed in Fig. 2.

Human face images are digitized for different computation purpose especially for face detection, face reconstruction, face recognition (FR) etc. In case of face detection, the location and approximation of face size can be determined. During face localization mechanism, it overlooks the other details from image, like: wall, tree, building or any kind of object other than face. Face reconstruction is the process by which missing facial information or different facial components (such as eyes, lips, nose, hair style, ear etc.) a face image can be constructed. It may be widely used in forensic or criminal identification purpose. Besides these, FR is another domain of human face processing. It is a very established biometric based research area. In this categorization of face image, it allows to recognize the individuals automatically. Though visual, thermal, and sketch based FRs are the well

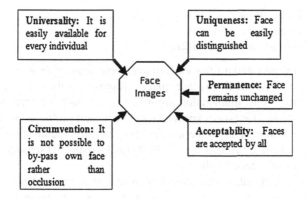

Fig. 1 The advantages of face images as a biometric measurement

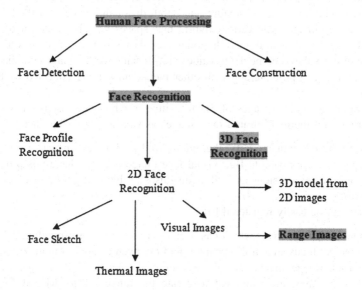

Fig. 2 Different approaches of human face processing

known approaches, 3D FR is also the foremost research interest of many scientists. There are mainly two advancements for 3D face based recognition scheme. One is 3D face model from 2D images and another one is range image based FR. Within the context of this chapter, authors have only focused on hybridization of 2D and 3D (especially 2.5D range image) face image based recognition techniques.

With all the said pros, there are still some challenges involved that makes the FR task more complex.

- Different illumination challenges
- Huge pose or viewing variation

- Large facial expression
- Face deformation due to age
- The disguises as well as occlusions

There are several solutions that exist to overcome some of these limitations. One solution is to use thermal IR [3] face images. It preserves the heat emissions from individuals' face. Thus, illumination problem [3] can easily be dealt with. But for determining solution for pose, expression and occlusion problems 3D face images might be better choice. Due to the inherent property, the illumination challenge is automatically solved in this new era of FR. Actually, using 3D sensors, face images are independent of external light source.

2D images are the flat arrangement of pixels or texture whereas 3D images consist of three distinct point clouds along X, Y and Z axes. From 3D face image, 2.5D range face image [4] is created. It is actually a normalized depth image. The motivation for hybridization of 2D and 3D face image is to design a new hybrid face space where both facial texture and depth data can be accomplished. Hence, from two face image, nine various hybrid face spaces have been designed.

A 3D face image with three orthogonal axes shown in Fig. 3 is used to describe the role of Z's value (i.e. the depth value) in X-Y plane. A 3D image contains [4, 5] X, Y, Z, three data points along with other meta-data that are collected by the 3D sensors.

There are some advantages of 3D face images which are discussed below by comparing with thermal infrared face images as well as 2D visible face images.

- Due to the inherent property, thermal as well as 3D face images are free from any kind of illumination. But, visual face image does not have this property.
- Variation of expressions as well as disguises problem might be resolved using 3D images.
- Poses can be easily registered [6].

Range face image (RFI) which has been hydrided with corresponding 2D image is the principal focus area of this chapter. RFI has been created from the original 3D face images. Range images are also sometime known as 2.5D images [4]. The 'depth' or Z-values are examined to create the range image. RFI of Fig. 3 is illustrated in Fig. 4.

In general, any FR system consists of five steps which are strongly bonded to each other. In Fig. 5, these steps are explained.

With the available face datasets, it is required to process the central (the core) computation by an intelligent system. The central processing may include different subsections, namely: selection of feature attributes and classification, for automatic FR system. These two core parts are very much correlated to each other. The intelligence system which may be involved with FR system might be 'Precious Model' [6] (traditionally named as Hard Computing) or 'Approximate Model (also known as Soft Computing). Though there are some limitations exist for soft computing techniques, Approximate Reasoning, Functional Approximation etc. are chosen maximum.

Fig. 3 Visualization of 3D
face image

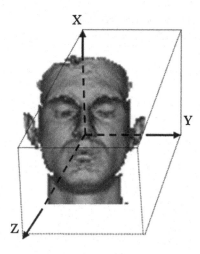

Fig. 4 Range face image
(RFI) of Fig. 3

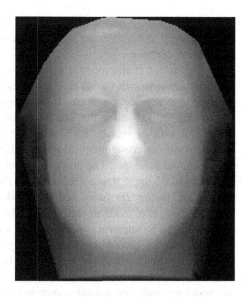

2 Literature Survey on Application of Hybrid Techniques for 3D Face Registration and Recognition

With the limitations of 2D face images [7] (both thermal as well as visual), 3D face images has gained much attention of researchers. As discussed earlier, the major advantages of 3D face images are that it is independent of illumination and also the rotated face images can be registered to frontal pose leading to better, accurate and robust recognition. With the advancement of various techniques, hybridization of these techniques has also been implemented for automatic image registration as

Fig. 5 General approaches
for face recognition

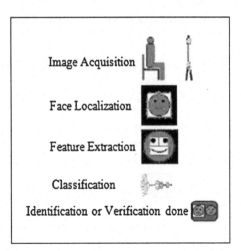

well as recognition purpose. Considering the context of this chapter, biological object namely face images are selected as a research topic for registration as well as recognition purpose.

In [8], Hutton et al. proposed a model named as 'Dense Surface Model (DSM)'. It is a hybrid approach for 3D face registration. The proposed hybridization method is done by alliance of iterative closest-point (ICP) and active shape model (ASM) fitting. This algorithm has been tested on 21 test example set and it was successful with an average RMS error 3.0 mm. In [9], author has proposed a 3D model based on evolutionary algorithm (EA) and skull-face overlay mechanism using EA and fuzzy set. The fitness function that has been proposed by the authors is the mean of the distances between the facial images and the projected cranial landmarks. In the 2D fuzzy set, landmarks are the fuzzy points. Due to such consideration, the more uncertainty of the landmark makes the broader fuzzy region. In [5] Santamaría et al. proposed an image registration method which is hybridization of global and local strategies. It is also named as 'sequential hybridization approach'. In this sequential hybridization method, a global search is first carried out by considering the advantage of the EAs or EAs based search methodology and then a local search is carried out for fine tuning. It is tested on human skull dataset.

In addition, a series of research work on 3D FR is also performed. In this section, their brief introduction in this area is also reported. Lee et al. [10] proposed a 3D FR algorithm that used to classify the feature set by fuzzy neural network. The feature set is derived from the eigenface images from maximum and minimum curvature maps derived from the 3D face images. In [11], Lin et al. proposed a hybrid intelligent technique that is used to recognize 2D as well as 3D face images. A neural network alliancing with hybrid Taguchi-PSO (particle swarm optimization) i.e. HTPSO is used to classify the obtained feature vector. A series of operations have been performed to obtain the feature vector. Features from Gabor wavelet transform is combined with surface feature vector by PCA for final derivation of feature set. Thakare et al. [12] has introduced a fuzzy rule with neural network or fuzzy neural network (FNN) for pose and illumination invariant 3D faces recognition purpose. It is tested on CASIA 3D face database. Their proposed FNN is having four layers namely: Input layer, Fuzzification layer, Inference layer and Defuzzification layer. This hybrid mechanism has been tested on different dataset like: varying lighting effects on frontal images, varying lighting effects on non-frontal images, varying rotation and lighting effects on frontal and non-frontal images.

3 Hybridization of 2D-3D Face Images for Recognition

Hybridization of multimodal face images is the process by which the dissimilarities of both the modalities are crossed to produce more significant face space or feature space. In case of multimodal face image based recognition task, like 3D (especially 2.5D) and 2D during hybridization process, a new space is created. The intensity values from 2D texture and depth values, shape details from 3D face images have been hydrided or fused in this investigation. In the domain of FR, hybridization process can be grouped into two sections, such as matching and feature estimation by hybridization techniques. Furthermore, hybrid matching technique can also be implemented either in feature level or in holistic approach or both the approach can be combined. In this case two or more classifiers with a fixed architecture are used. Again, feature estimation can be derived from hybrid face space or feature vector. Creation of face space signifies the importance of multimodality and creates a new type of face image that might be useful for better recognition purpose. Whereas, hybridization of multiple feature set from single images emphasize the selection of optimal feature space.

Here, authors have followed pixel level hybridization process to derive new face space for recognition purpose of 3D face images. For the hybrid approach two different ratios are considered. Among these ratios, one is complement of other and vice versa. In Fig. 6, this process is illustrated.

The constants 'α' and 'β' are the weighted values to accomplish the hybrid face space for 3D FR purpose. The mathematical computation that has been followed is shown in Eq. 1.

$$H(x,y) = (R(x,y) \times \alpha) + (O(x,y) \times \beta) \qquad (1)$$

In the equation, $H(x, y)$ is the hybrid face space where as $R(x, y)$ and $O(x, y)$ are the 2.5D range and 2D optical face images respectively. The 'α' value lies within the range of 0–1 and $\beta = 1 - \alpha$.

The significance of hybrid face space over optical and RFI is characterized in Fig. 7. The depth values from RFI are fused over intensity values from optical face

Fig. 6 The hybrid face space

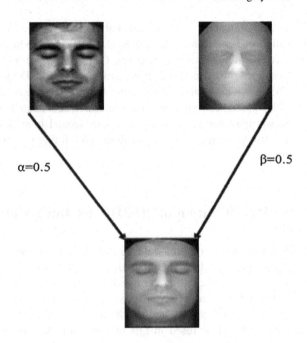

α=0.5   β=0.5

images. Thus, with different weighted values, different variations in hybrid face space are captured.

It is investigated that the qualities of 3D face images (i.e. to preserve the sharpness, depth values) along with the reflectance characteristics [13] of human face image (i.e. intensity data) are all blended in hybrid face images. The absolute difference between optical image and corresponding hybrid image is shown in Fig. 7c. The difference is used to serve this phenomenon. The bendness, sharpness i.e. the curvature points [7, 14, 15] are associated with the hybrid face image. On the other hand in Fig. 7f, significance of depth values than intensity values is displayed. In Fig. 8, the varieties of hybrid face images are shown for a randomly selected subject from Frav3D database [16].

3.1 Domain Knowledge Preparation

In this section authors have prepared a literature survey to demonstrate the current FR methodologies based on hybrid 2D-3D multimodal approach. Along with the study, a discussion of the proposed methodology by comparing with others is summarized here.

In [17] researchers have proposed a hybrid or fused feature vector for recognizing face images from FRGCv2.0 face database. In that paper authors have created a feature vector from 27 fiducial points or land marks. Each point is

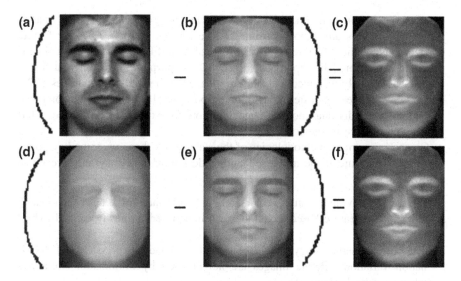

Fig. 7 Significance of hybrid face image. **a** Optical image. **b** Hybrid image. **c** Absolute difference map between (a) and (b). **d** Range image. **e** Hybrid image. **f** Absolute difference map between (d) and (e)

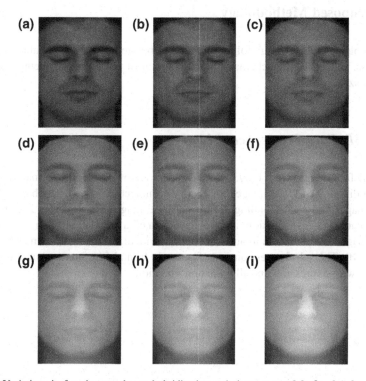

Fig. 8 Variations in face images due to hybridization technique. **a** $\alpha = 0.9$, $\beta = 0.1$, **b** $\alpha = 0.8$, $\beta = 0.2$, **c** $\alpha = 0.7$, $\beta = 0.3$, **d** $\alpha = 0.6$, $\beta = 0.4$, **e** $\alpha = 0.5$, $\beta = 0.5$, **f** $\alpha = 0.4$, $\beta = 0.6$, **g** $\alpha = 0.3$, $\beta = 0.7$, **h** $\alpha = 0.2$, $\beta = 0.8$, **i** $\alpha = 0.1$, $\beta = 0.9$

convolved with 40 Gabor kernels to extract feature points around land marks. Again, the outliers are removed from 3D face images to apply PCA for feature extraction. Now, these feature vectors are appended for FR purpose. Mian et al. [18] proposed a multimodality approach for 2D-3D FR purpose. Two different fusion approaches (namely weighted and sum) are made for the similarity matrices of 2D and 3D face images for better recognition purpose. In [19] authors have combined the normalized individual modal's matching score for experimental purpose. It has been tested on face images of 198 persons.

Comparing various methodologies of the state-of-art, there are some key points came up, which are to be addressed during designing a new system, are given below.

- Instead of individual feature vectors, a new hybrid face space has been created.
- Holistic and feature based techniques can be applied on newly created hybrid face space.
- The current methodology should be capable to handle expression and illumination variations. Range images are not affected by illumination variations whereas optical images do.

4 Proposed Methodology

The implemented algorithm follows three steps before it is used for final recognition purpose. In Fig. 9, the schematic block diagram of the proposed methodology is emphasized.

4.1 3D Face Image Acquisition

The 3D face images from Frav3D face database are of VRML file format [4]. There are 16 different 3D face images for each 106 subject in Frav3D database. The 3D face images and corresponding 2D optical face images are shown in Fig. 10.

To carry out investigation, authors have created a sub-dataset with eight frontal face images out of sixteen images for each individual, which are shown as bold line in Fig. 10. The sub-dataset consists of not only frontal face images but contains of images with expression and illumination variations.

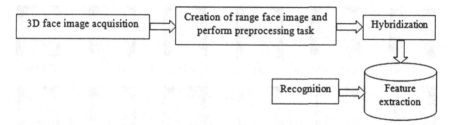

Fig. 9 Block diagram of the proposed algorithm

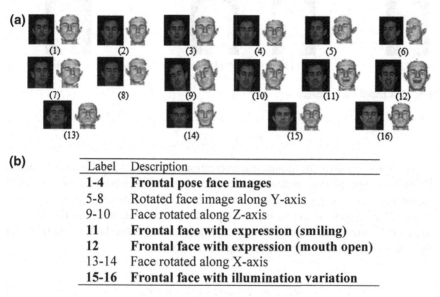

Fig. 10 Detail description of the considered database. **a** Optical and corresponding 3D face images from Frav3D database. **b** Description of the face dataset

4.2 Pre-Processing

Preprocessing is an essential step between image acquisition and recognition process. We have generated the range images from 3D face images. The generated RFI are shown in Fig. 11.

The range images are created following the equation (as shown in Eq. 2). The depth value at any position (x, y) is the function of X, Y. The 'X' and 'Y' are the 2D co-ordinate space where 'Z' value (i.e. depth values are preserved).

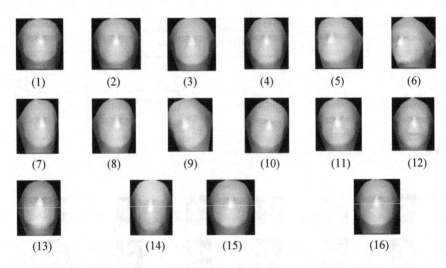

(1)   (2)   (3)   (4)   (5)   (6)

(7)   (8)   (9)   (10)   (11)   (12)

(13)   (14)   (15)   (16)

Fig. 11 Created range images

$$R(x, y) = f(X, Y) \tag{2}$$

The very next step after range image creation is to extract the Region of Interest (RoI) from optical face image. Optical face images are having more unwanted region compared to created range images. For isolating these regions, the color human face images are converted to grey scale images following the Eq. 3. Now, these gray images are converted to binary image using a threshold value 0.15 and then face region is extracted as the largest component.

$$G = R * C_r + G * C_g + B * C_b \tag{3}$$

$$\text{where } C_r = 0.2989; C_g = 0.5870; C_b = 0.1140$$

The stages which are followed to extract the RoI are detailed in Fig. 12.

Authors have developed a method by which only the RoI from the optimal face region can be extracted by omitting the outer regions namely ear, neck, hair etc. At first, cropped 2D and 2.5D range images of both the modalities are resized. Hence, the two different modalities have been normalized. Then, depth based 'pronasal' landmark is localized [7] from range image and a fixed window from all the frontal images of both the modalities are extracted for further feature extraction and recognition purpose. The outcome of this technique is highlighted in Fig. 13. Though it is required to resize the input images, the proposed method is independent of image scaling.

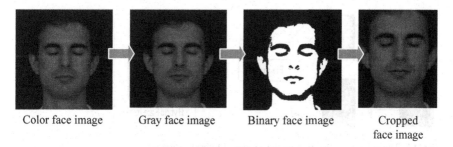

Color face image   Gray face image   Binary face image   Cropped
                                                                               face image

Fig. 12 Illustration of different steps for extraction of RoI

Fig. 13 The optimal face
region from RoI

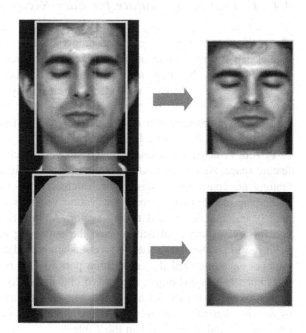

4.3 Hybridization Technique

Authors have applied a weighted pixel level hybridization to 2D and 3D face imaging system to establish the effectiveness of the hybrid face image. The weights are varied from 10 to 90 % instead of 0–100 %. Selecting 0 and 100 % weight will form either pure range image or optical image like input images. Thus, no hybridization will be followed.

More specifically, 50–50 % weighted from both the optical and range images have been acknowledged for this analysis. The same priority is also provided for depth and intensity values. Thus a new hybrid face is created and it has already shown in Fig. 6. The hybridization process of 2D optical and 2.5D RFI is described in Fig. 14.

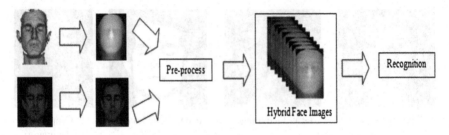

Fig. 14 The schematic diagram of the proposed hybrid technique

4.4 Estimation of Feature for Face Recognition

The estimation process to evaluate proper distinct characteristic is very much crucial step towards classification task. Selection and reduction of feature space may lead to the 'ugly duckling' story. The estimated good features may lead to poor recognition rate whereas neglected features may be very fruitful during classification.

It has been observed that in literature researchers have used ANN [20, 21] and K-NN [22, 23] followed by PCA [24, 25].

PCA or Principal Component Analysis is very much well accepted to reduce the feature space. Among several eigenvalues, top five eigenvalues and corresponding features are considered to design the feature vector for recognition purpose.

For the recognition purpose, the feature vector is distributed among two sets namely training vector and testing vector. Among two partitioned sets, one set contains even number of features and the remaining odd number of features are belonged to another set. These feature sets are named as 'U' and 'V' respectively. Authors have analyzed the recognition performance by two-fold cross validation technique. In this technique, supervised classifiers [26, 27] (ANN and K-NN) are trained with 'U' feature set and tested with 'V' feature set. The same is followed again with alternative training-testing feature set. The variations of recognition rate are reported and summarized in the Table 1.

The five layer feed forward back propagation neural network and K-Nearest Neighbor classifiers are used to demonstrate the potentiality of the selection of hybrid face space and finally the feature vector. Out of five layers of the neural network, first hidden layer contains 100 neurons, second hidden layer contains 50 neurons; third hidden layer contains 10 neurons and input layer contains the neurons with exact number of features. Last layer contain number of classes. Similarly, in case of K-NN, K is used to denote the number of possible subjects for this research work.

Belghini et al. [28] have proposed a recognition scheme with the recognition rate 93 % from Frav3D database. Here, from global depth information, Gaussian Hermite Moments (GHM) feature has been computed for recognition reason. Ganguly et al. [7] also proposed curvature based 3D FR technique, tested on Frav3D database. They have considered two curvature pairs Mean-Maximum and

Table 1 Performance of the hybrid face images

Classifier (training set + testing set)	Recognition rate (%)
ANN (U+V)	**95.17**
ANN (V+U)	94.97
K-NN (U+V)	**91**
K-NN (V+U)	90.57

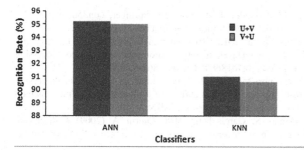

Mean-Gaussian as individual feature vectors that are classified by ANN. The recognition rate that has been proposed by them is 86.51 %. In compared to these methodologies, authors have designed a new hybrid face space from where more detailed information can be accumulated. Both the reflectance characteristics and depth information has been accomplished in this hybridization method. Hence, more detailed data from hybrid face space has been used for more accurate recognition purpose.

5 Conclusion and Future Scope

FR (especially 3D FR) is a very complex and challenging task in the domain of computer vision and pattern recognition. Now-days it has also gained much attention of researchers. Though many scientists have conducted numerous experiments but the ideal goal is still beyond the scope. The goal of this research was to take a step forward for automation of 3D face image recognition using a hybrid procedure. It is invariant of expression as well as illumination.

However, the present approach has some limitations. The extreme pose changes along yaw, pitch and roll are not considered here. As a part of our future work, we shall develop more robust method for the objective of face registration as well as recognition. Thus, a new era is emerging towards hybrid technology for more accuracy in automation purpose.

Acknowledgments Authors are thankful to a project supported by DeitY (Letter No.: 12(12)/ 2012-ESD), MCIT, Govt. of India, at Department of Computer Science and Engineering, Jadavpur University, India for providing the necessary infrastructure for this work.

References

1. Jain AK, Flynn P, Ross AA (2007) Handbook of biometrics. Springer, London
2. Toth B (2005) Biometric liveness detection, Information Security Bulletin
3. Seal A, Bhattacharjee D, Nasipuri M, Basu DKr (2014) Thermal face recognition for biometric security system, in the book Research Developments in Biometrics and Video Processing Techniques, IGI Global, pp 1–24 doi:10.4018/978-1-4666-4868-5.ch001
4. Ganguly S, Bhattacharjee D, Nasipuri M (2014) 2.5D Face Images: Acquisition, Processing and Application, in ICC 2014-Computer Networks and Security, International Conference on Communication and Computing (ICC-2014), pp 36–44, ISBN: 978-93-5107-244-7
5. Santamaría J, Cordón O, Damas S, García-Torres JM, Quirin A (2009) Performance evaluation of memetic approaches in 3D reconstruction of forensic objects, Soft Computing, doi:10.1007/s00500-008-0351-7
6. Spreeuwers L (2011) Fast and accurate 3D face recognition. Int J. Comput Vis 93:389–414. doi:10.1007/s11263-011-0426-2
7. Ganguly S, Bhattacharjee D, Nasipuri M (2014) 3D Face recognition from range images based on curvature analysis. ICTACT J Image Video Process 4(3)
8. Hutton TJ, Buxton BF, Hammond P (2003) Automated registration of 3D faces using dense surface models
9. Cordón O (2011) An automatic method for forensic identification based on soft computing techniques, soft computing for forensic identification, EUSFLAT—LFA Ainx-Les-Bains (France) 19–22 July, URL: http://sci2s.ugr.es/Tutorials_Talks/files/Plenary-Talk-Eusflat-LFA-2011-Cordon.pdf
10. Lee YH, Han CW, Kim TS (2008) 3D facial recognition with soft computing. Digital Human Model LNAI 4650:194–205
11. Lin CJ, Wang JG, Chen SM (Feb 2011) 2D/3D Face recognition using neural network based on hybrid taguchi-particle swarm optimization. Int J Innovative Comput Inform Control 7(2)
12. Thakare NM, Thakare VM (June 2012) Supervised hybrid methodology for pose and illumination invariant 3D face recognition, Int. J. Comput. Appl. (0975–8887) 47(25)
13. Gonzalez RC, Woods RE (2007) Digital Image Processing, 3rd edn. Aug 31
14. Ganguly S (2014) Curvature Based 3D Face Recognition. Jadavpur University, India
15. Szeptycki P, Ardabilian M, Chen L (2009) A coarse-to-fine curvature analysis-based rotation invariant 3D faces landmarking, URL: http://liris.cnrs.fr/Documents/Liris-4503.pdf
16. Conde C, Rodríguez-Aragón LJ, Cabello E (2006) Automatic 3D face feature points extraction with spin images, ICIAR 2006. LNCS 4142:317–328
17. Arca S, Lanzarotti R, Lipori G (2007) Face recognition based on 2D and 3D features. knowledge-based intelligent information and engineering systems, Lect. Notes Comput. Sci. 4692:455–462.
18. Mian AS, Bennamoun M, Owens R (2007) An efficient multimodal 2D-3D hybrid approach to Automatic face recognition. IEEE Trans Pattern Anal Mach Intell 29(11):1927–1943
19. Chang KI, Bowyer KW, Flynn PJ (2005) An evaluation of multimodal 2D + 3D face biometrics. IEEE Trans Pattern Anal Mach Intell 27(4):619–624
20. Lin, L (1996) Neural Fuzzy Systems, Prentice Hall International
21. Haddadnia J, Faez K, Ahmadi M (2003) An efficient human face recognition system using pseudo zernike moment invariant and radial basis function neural network. Int J Pattern Recognit Artif Intell 17(1):41–62
22. Deokar S (Apr 20 2009) Weighted K-Nearest-Neighbor

23. Hiremath PS, Manjunatha Hiremath (2014) RADON transform and PCA based 3D face recognition using KNN and SVM. Int J Comput Appl (0975 – 8887) Recent Adv Info Technol
24. Turk M, Pentland A (1991) Eigenfaces for recognition. J Cogn Neurosci 3(1):71–86
25. Turk MA, Pentland AP (1991) Face recognition using eigenfaces. In: Proceedings of the IEEE Computer Society Conference on Computer Vision and Pattern Recognition, pp 586–591 June 1991
26. Theodoridis S, Koutroumbas K (2008) Pattern Recognition, 4th edn. Nov 3
27. Mitchell TM Machine Learning, McGraw-Hill (1997) Higher Education
28. Belghini N, Zarghili A, Kharroubi J (2012) 3D face recognition using gaussian hermite moments, Special Issue of Int J Comput Appl (0975–8887) on Software Engineering, Databases and Expert Systems—SEDEXS, pp 1–4 Sept 2012

Neutrosophic Trust Evaluation Model in B2C E-Commerce

Swati Aggarwal and Anurag Bishnoi

Abstract Trust is an important term in the context of E-Commerce. It lies at the top in the present trend of E-Commerce. Trust lists the diverse prospects of trustworthiness that exist between the merchant and customer, inducing a better customer liking and Business-to-Customer (B2C) E-Commerce. Considering the imprecise nature of E-Commerce trust, various researchers proposed different trust models and integrated them with fuzzy logic to handle inherent uncertainty. Conventional models of trust are based on subjective logic which falls short in mapping the real-time environment of E-commerce that deals with tentative behavioural values. Though fuzzy logic representation of the facts is a way to deal with improbability but it fails to capture the indeterminacy and false values given by respondents during survey. Authors in this chapter have attempted to target the indeterminacy involved while capturing the perception of respondents during survey for any website. Quite recently, neutrosophic logic (NL) has been proposed by Florentine Smarandache that gives a mathematical model for representing uncertainty, vagueness, ambiguity, imprecision, incompleteness, inconsistency, redundancy and contradictions. All the factors stated are very integral to human thinking, as it is very rare that we tend to conclude/judge in definite environments, imprecision of human systems could be due to the imperfection of knowledge that the human receives (observation) from the external world. Imperfection leads to a doubt about the value of a variable, a decision to be taken or a conclusion to be drawn for the actual system. This chapter suggests computation of perceived trust value by integrating a neutrosophic logic with the proposed fuzzy based trust model that considers all the chief features which affect the trust in E-Commerce.

Keywords Fuzzy logic · B2C E-Commerce · Trust · Neutrosophic logic

S. Aggarwal (✉)
COE, NSIT Dwarka, New Delhi, India
e-mail: swati1178@gmail.com

A. Bishnoi
Cognizant, Chennai, India

© Springer India 2016
S. Bhattacharyya et al. (eds.), *Hybrid Soft Computing Approaches*,
Studies in Computational Intelligence 611,
DOI 10.1007/978-81-322-2544-7_14

405

1 Introduction

Presently, E-Commerce is growing very rapidly and it is a convenient way for the vendors to reach the broader consumer market. In this competitive business environment, E-Commerce has shown tremendous growth and competition. Though the merchant and the consumer do not have face-to-face interaction in the E-Commerce environment, they still create their mutual trust. So, good relationships can be maintained in the business but a risk factor is involved from the consumer's perspective; for example, we can see only an image of the item but we do not have any actual idea of item [1, 2]. Thus a consumer may be in an unclear thought process that whether or not he should opt for online shopping. Thus loyalty involves as a major deciding factor if a vendor wants the consumer to buy something from his website. Low loyalty can certainly become an obstruction in the path of growing business. The rate of online shopping is low in comparison to the exponential growth in the number of Internet users; because of low consumer trust, as suggested by Hoffman et al. [3].

In extreme cases, business may come to an end because of the absence of trust [4, 5]. Vendor's motive is only to enhance the business and this can be possible by developing the consumer trust.

Trust of consumers may be influenced by several factors:

1. Unusual behaviour of the website during payment.
2. Request for more personal information demand during registration on website.
3. Repudiation by the merchant for exchange even if the product is encountered defective.
4. Brand value of the organization etc.

Although loyalty has such a colossal importance, still there is not much research work that has been done in the detection of antecedents and consequents of online shopping. Available research papers [6] focus on small-scale models that lack the important attributes such as security and privacy. There exist uncertainties in E-Commerce data and fuzzy logic [7] is an essential tool to deal with such type of data and is expressed in linguistic form which makes it easy to understand the level of trust. Authors in this chapter have given a new model that considers all the important facets that govern E-Commerce transaction and then finally computing a trust value using neutrosophic logic (NL) for handling the implicit indeterminacy involved. This chapter explains various sections as follows: Sect. 2 describes the related work in the field of trust evaluation in E-Commerce; Sect. 3 explains the proposed model, basics of neutrosophic logic and a detailed algorithm of working is given in Sect. 4; Sect. 5 describes the integration of Neutrosophic logic with the proposed model; Sects. 6 and 7 discusses the experimental results and conclusions, respectively.

2 Related Work

Loyalty is a term which has an important role in the field of E-Commerce. Vendor is always curious to know how many users use his website and what their perceived trust is. So trust evaluation has been in great demand from a long time and the number of models [8–10] for trust calculation have been proposed by various researchers. This section discusses the prominent models in this domain. Based on trusting beliefs and trusting intentions, McKnight et al. [11] develops a trust model that depends on three factors: structural assurances, reputation of the vendor and website quality. These factors help in trust building of a consumer only in the initial phase of business relation; but to keep this relationship longer, there is a need for considering some more factors which are missing in this model.

According to Zuhang's trust model [12] trust depends on the price of a product and credit history of the customer. Though it is a pragmatic model for trust calculation because trust optimization is also considered in it, the author here fails to resolve the contour discovery issue. This model also needs a broker to handle the task of user verification which makes its working complex. Akhter [13] proposed a model according to which trust of a consumer depends on security, familiarity and website design. These again are only limited number of factors for trust calculation and author has not explained anything about these terms. Nilashi et al. [14] has extended the model proposed by Akhter et al. [13], though the author listed all the attributes on which security, familiarity and website design depends but this research falls short in context of not explaining the clustering technique that is integrated with the rule base. The idea of making clusters for the rules does not even seem beneficial on performance of trust calculation.

According to Qin and Tian [15], the trust in E-Commerce is based on the psychology status of consumers that depends on several factors. According to theauthor, trust directly depends upon environments of technique and society, the viewpoint of consumer and vendor and the indirect trust depends upon reputation, recommendation and similarity. The author has analyzed the factors influencing trust and established an integrated model for direct and indirect trust in B2C E-Commerce.

The research paper [16] addresses the trust affecting factors: existence, affiliation, policy and fulfilment. This is a superior system but still it is not cost effective which is the requirement according to the consumer's perception [17, 18].

Ludwig et al. [19] have discussed the evaluation of trust based on user interaction with the vendor and reputation services. Author has shown a comparison of proposed approach with weighted approach for trust calculation and concludes that the proposed fuzzy based trust model performs better in case of either positive or negative deception.

Nafi et al. [20] has evaluated the trust value using the model proposed by Nefti et al. [16]. In this model, author has calculated trust based on certainty and average rating corresponding to each attribute in the module. Here, probabilistic behaviour of trust over expected is also shown but in this methodology there is no fuzziness.

However, extended version by Nafi et al. [21], includes the fuzzy approach and it can also be used in cloud computing.

Quite recently, many researchers have conducted different types of study dealing with different aspects of trust in e-commerce websites [22–28].

3 Proposed Model

Bishnoi and Aggarwal [29] proposed model that aims to evaluate the trust in E-Commerce and it considers all the prominent factors that affect the trust of a consumer. The final trust is composed of two types of trust, i.e. indirect trust and indirect trust as shown in Fig. 1.

Indirect trust is evaluated from recommendation by others, policies of the organization and website design which has its own importance for attraction of consumers. Recommendations further rely on three attributes:

Words of mouth from family, friends or through advertisements.

Trust certification by a third party which is more trustable from a consumer's perspective [30].

Feedbacks for a website or its products by other customers: Feedback about a product or the website has a great affect on human tendency to purchase or not. If the feedback is in favour, it sets the mind that the item/product is good; otherwise, an opposite reaction may be seen from the consumer's perspective.

Policy of E-Commerce Organization is dependent on following three attributes: Security of the website [31, 32].

Privacy policies for the information of its consumers [33].

Services for the satisfaction of its customers such as tracking of order, product exchange, cash on delivery etc.

Website design [34] is also a crucial factor which further depends on three things:

1. Interactive GUI of the website.
2. Categories: various classes should be there depending upon the type of the item like an E-Commerce website of books should classify its books according to topics like science, poetry, religion, art, computers etc.
3. Effective navigation between web pages for easily finding the products or for understanding of the website [35].

Direct trust is evaluated using two factors, i.e. the past experience of the consumer with the organization and price of the product. Past experience [36, 37] is further categorized into three attributes on which it depends. These are:

Behaviour of the customer support (vendor) during communication with them. They should provide a complete assistance to their customers and should have a problem-solving attitude from the consumers' perspective.

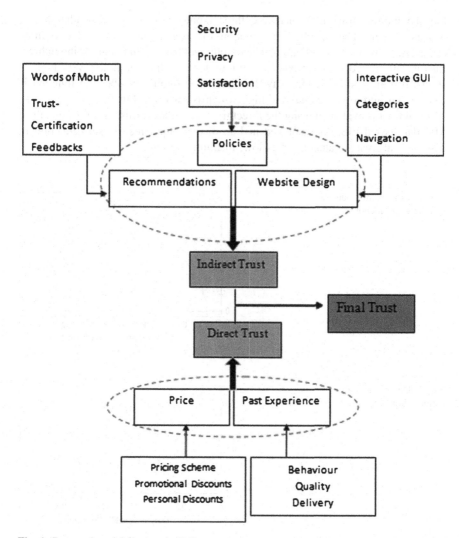

Fig. 1 Proposed model for trust in E-Commerce

Quality of the product: Product quality should be good for achieving the faith of customers. Some customers believe in buying only quality product even at high price of the product.

Delivery of the product: It is the tendency of customers to anticipate the delivery of the ordered item as soon as possible. So delivery service should be fast.

This model is completely based on fuzzy logic. Authors [29] are using three Mamdani-based fuzzy inference systems (FIS), first for the evaluation of indirect trust(FIS_indirect_trust), second for direct trust(FIS_direct_trust) calculation and third for computation of final trust(FIS_final_trust) for the E-Commerce website.

For the indirect trust FIS, there are three inputs: policy, recommendation and website design. The membership function for each input is divided into three categories: low, medium and high. The output membership function for the indirect trust is categorized into five types: very low, low, medium, high and very high as shown in Fig. 2. In the same way the direct trust output membership function is categorized into low, medium and high which is shown in Fig. 3.

Final trust is evaluated using the direct trust and indirect trust as inputs in a third Mamdani-based FIS. As displayed in Fig. 4, the output membership function in case of final trust is categorized as low, medium and high.

Fig. 2 Indirect trust output membership function

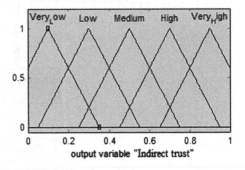

Fig. 3 Direct trust output membership function

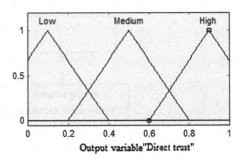

Fig. 4 Final trust membership function

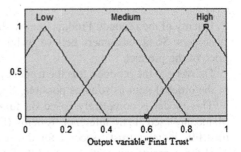

3.1 Basics of Neutrosophic Logic

Quite recently, neutrosophic logic was proposed by Florentine Smarandache which is based on the non-standard analysis that was given by Abraham Robinson in 1960s [38]. Neutrosophic logic was developed by Smarandanche to represent the mathematical model of uncertainty, vagueness, ambiguity, imprecision, incompleteness, inconsistency, redundancy and contradiction [39]. Neutrosophic logic is a logic in which each proposition is estimated to have the percentage of truth in a subset T, the percentage of indeterminacy in a subset I and the percentage of falsity in a subset F, where T, I and F are standard or non-standard real subsets of $]^-0, 1^+[$ [40],

with sup $T = t\_sup$, inf $T = t\_inf$, sup $I = i\_sup$, inf $I = i\_inf$, sup $F = f\_sup$, inf $F = f\_inf$, and $n\_sup = t\_sup + i\_sup + f\_sup$, $n\_inf = t\_inf + i\_inf + f\_inf$.

The sets T, I and F are not necessarily intervals, but may be any real sub-unitary subsets: discrete or continuous; single-element, finite, or (countably or uncountably) infinite; union or intersection of various subsets; etc. They may also overlap [41]. Statically T, I, F are subsets. Here, a subset of truth (or indeterminacy, or falsity), instead of a number, is used because in many cases we are not able to exactly determine the percentages of truth and of falsity but approximate them: for example, a proposition is between 30 and 40 % true and between 60 and 70 % false, even worst: between 30 and 40 % or 45–50 % true (according to various analyzers), and 60 % or between 66 and 80 % false. Neutrosophic logic suggests that neutrosophic probability (using subsets—not numbers—as components) should be used for better representation as it is a more natural and justified estimation [40].

All the factors stated by neutrosophic logic are very integral to human thinking, as it is very rare that we tend to conclude/judge in definite environments, imprecision of human systems could be due to the imperfection of knowledge that human receives (observation) from the external world [42]. For example: for a given proposition "Movie ABC would be a blockbuster movie", human brain certainly in this situation cannot generate precise answers in terms of yes or no, as indeterminacy is the sector of unawareness of a proposition's value, between truth and falsehood; undoubtedly neutrosophic components best fits in the modelling of simulation of human brain reasoning.

Definition Neutrosophic Set [40]: Let X be a space of points (objects), with a generic element in X denoted by x.

A neutrosophic set A in X is characterized by a truth-membership function T_A, a indeterminacy-membership function I_A and a falsity-membership function F_A. $T_A(x)$, $I_A(x)$ and $F_A(x)$ are real standard or non-standard subsets of $]^-0, 1^+[$. That is

$$T_A : X \rightarrow]^-0, 1^+[$$
$$I_A : X \rightarrow]^-0, 1^+[$$
$$F_A : X \rightarrow]^-0, 1^+[$$

There is no restriction on the sum of $T_A(x)$, $I_A(x)$ and $F_A(x)$, so

$$^-0 \leq \sup T_A(x) + \sup I_A(x) + \sup F_A(x) \leq 3^+$$

Also as neutrosophy allows the provision of reflecting the dynamics of things and ideas [41]; the proposition "Movie ABC would be a blockbuster movie" does not mean fixed value components structure; the truth value of the proposition may change from place to place. For example: proposition "Movie ABC would be a blockbuster movie" may yield neutrosophic components 0 % true, 0 % indeterminate and 100 % false in north region of the country and may yield $(1, 0, 0)$ in south region.

Neutrosophy also allows change in values with respect to the observer [41]. For example: proposition "Movie ABC would be a blockbuster movie" may yield neutrosophic components $(t = 0.60, i = 0.30, f = 0.20)$ if observed by any film critic then results would differ; like $(t = 0.30, i = 0.15, f = 0.80)$ if analyzed by other critic.

4 Working of Model

The proposed model in Fig. 1 is simulated over Mamdani-based fuzzy logic controller which is composed of rule base and fuzzy inference system (FIS). Fuzzy inference system [43] is actually a mapping system that maps the given input into to an output with the help of fuzzy logic. Two main concepts of fuzzy inference system are linguistic variables and rule base. The value of linguistic variable is usually a word or sentence that is understandable to humans. Fuzzy inference system contains rules and only some particular rules are fired out based on the values of input. Rules are defined in if-then form which is composed from linguistic variables. Crisp value of inputs such as $x_i \in U_i$ where $\{i = 1, 2, 3,..., n\}$ are fuzzified into linguistic values $F1, F2, ..., Fn$ where Fi is defined over universe of discourse $U = \{U1 * U2 * \cdots * Un\}$. If output linguistic variables are $G1, G2, G3...Gn$ then the rules of rule base are represented in general form as:

$$R_{(j)} : \text{IF } x_1 \in F1_j \text{ AND} ... x_n \in Fn_j \text{ THEN } y \in G_j. \tag{1}$$

Each input and output may have any type of membership function such as triangular, gaussian or sigmoidal function. Here in this work authors have used triangular membership function. Based upon the model and experts' suggestions, rules of the FIS are designed. The antecedents of rule base are connected using AND, NOT or OR operators. Authors have introduced 27 rules for indirect trust calculation (Table 1) and direct trust calculation (Table 2) which are designed on the basis of inputs and output trust perception of 20 expert users in online shopping.

For all the rules, antecedents are connected using only AND operator. As for the calculation of final trust, there are two inputs (direct and indirect trust) each with three membership functions; so, again nine rules are defined for final trust based upon expert's decision. Table 3 shows the rules for final trust.

Table 1 Indirect trust rules

Rule number	Recommendations linguistic value	Web design linguistic value	Policies linguistic value	Indirect trust linguistic value
1	High	High	High	Very high
2	High	Medium	High	Very high
3	High	Low	High	High
4	High	High	Low	High
–				
–				
–				
25	Medium	High	Medium	Medium
26	Medium	Medium	Medium	Low
27	Medium	Low	Medium	Low

Table 2 Direct trust rules

Rule number	Past experience linguistic value	Price linguistic value	Direct trust linguistic value
1	High	Low	High
2	High	Medium	High
–	–	–	–
–	–	–	–
8	Low	Medium	Low
9	Low	High	Low

Table 3 Final trust rules

Rule number	Indirect trust linguistic value	Direct trust linguistic value	Final trust linguistic value
1	High	Low	High
2	High	Medium	High
–	–	–	–
–	–	–	–
8	Low	Medium	Low
9	Low	High	Low

Rule formation mapping surface can be visualized that reflects the variation of output depending upon the change in input. In Fig. 5, it can be seen that with increase in value of policy and web design the value of indirect trust increases. The yellow portion of the mapping surface shows very high value of indirect trust for high value of both policy and web design. In case of direct trust mapping surface as shown in Fig. 6, it can be seen that with rise in value of price the output, i.e. direct trust value decreases. Trust is inversely proportional to price value because if a

Fig. 5 Indirect trust mapping
surface (FIS_indirect_trust)

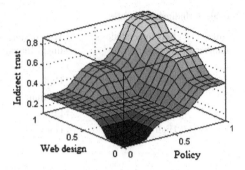

Fig. 6 Direct trust surface
(FIS_direct_trust)

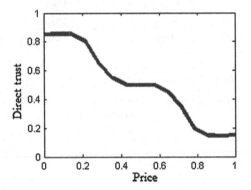

consumer finds the same item at lower price on some other website then he or she
would prefer to buy it in lower price.

As final trust value depends upon indirect and direct trust values, same priority
has been given to both the types. Its mapping surface is symmetric as shown in
Fig. 7. Yellow portion shows high value of final trust for high value of input
variables and in the same way blue portion shows low value of final trust for low
value of input variables.

Fig. 7 Final trust surface
(FIS_final_trust)

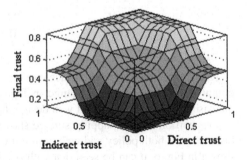

5 Integration of Neutrosophic Logic with Model

To allow effective integration of the neutrosophy with our model, respondents were asked to answer the questionnaire based on neutrosophy approach. This means the respondents were given the flexibility of recording their opinions in three ways: true (when they are sure about a certain aspect), false (when they disagree) and not sure (when they are not sure of, meaning indeterministic). Integrating neutrosophy in data collection plays a very important role. Normally, the questionnaires are designed such that they capture the agreement/disagreement, do not take into account the indeterminacy of the respondent for certain aspect, which is quite natural. For the aspects where respondent expresses indeterminacy, there are equal chances of agreement/disagreement. Hence, it is very important to record the indeterminacy too, thus the authors have integrated neutrosophy with the proposed model. The survey data is then collected and summarized based upon the views of all users. After that data values are normalized and new value for each of the attribute is calculated as given in Table 6. The indirect and direct trust value is calculated using FIS for each. Using value of indirect and direct trust final value of trust can be calculated using third FIS.

5.1 Steps for Working with the Model Are as Follows

1. Design the questionnaire based on neutrosophy approach.
   Neutrosophy is integrated with the proposed model by capturing the agreement/disagreement/indecisiveness of the respondent/customer during the data collection phase. To facilitate this survey, questionnaire is designed like-wise (Fig. 8).
2. Conduct the survey and collect data.
   Respondents opinions are captured along three dimensions of neutrosophy: truthness, falsity and indeterminacy.
3. Summarize the data from Step 2 for all the users (like Table 4).
   This step consolidates the respondents view along three dimensions of truthness, falsity and indeterminacy for all the parameters. Like for certain aspect security (s), how many respondents agreed: truthness recorded (s_t), disagreed: falseness recorded (s_f) or are indecisive: indeterminacy recorded (s_i), where $s_t + s_f + s_i =$ total number of repondents.
4. Normalize the data generated from Step 3 in the range from 0 to 1 (like Table 6).
   Neutrosophic components (t, i, f) are real standard or non-standard subsets of $]^-0, 1^+[$. In this step, normalization of the values is done: where each value (like s_t, s_f, s_i) is divided by the total number of respondents, so we get the values between the range of 0 and 1; which is desirable for neutrosophic components. This gives a complete three-dimensional breakdown (along truthness, falsity and

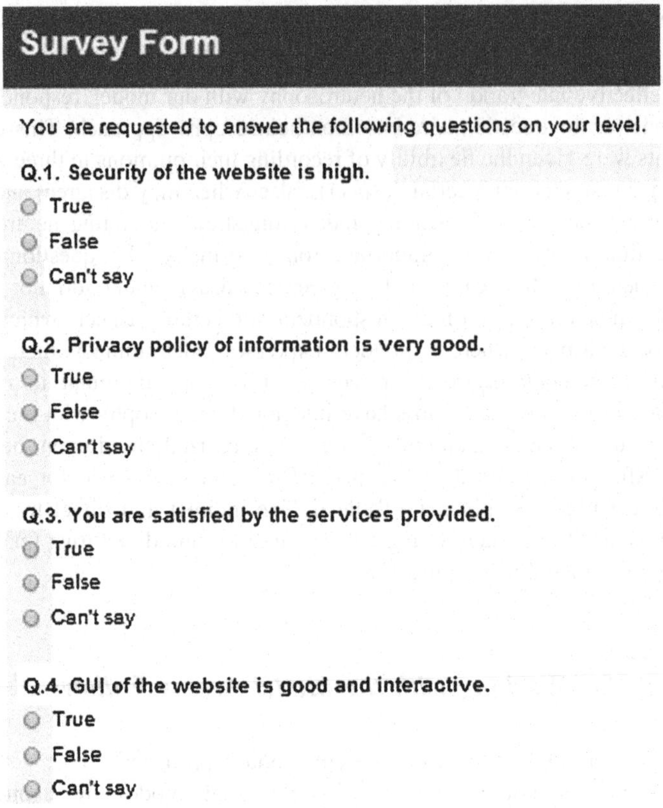

Survey Form

You are requested to answer the following questions on your level.

Q.1. Security of the website is high.

○ True

◉ False

○ Can't say

Q.2. Privacy policy of information is very good.

○ True

◉ False

○ Can't say

Q.3. You are satisfied by the services provided.

○ True

◉ False

○ Can't say

Q.4. GUI of the website is good and interactive.

○ True

◉ False

○ Can't say

Fig. 8 Snapshot of survey form

indeterminacy) for every aspect under study. Final answer is in the triplet form, like security (s_t, s_f, s_i).

5. Generate final value for each attribute using neutrosophic logic equation (Eq. 3) (like Table 6: column 6).

   To finally work further, a consolidated value for every attribute is required instead of corresponding triplet value. This computes one summarized value for every attribute, taking into account truthness, indeterminacy and falsity related for the attribute using neutrosophic Eq. (3).

6. Calculate indirect trust using FIS_indirect_trust that uses the input data from Step 5 (like Table 8).

   Crisp values are computed for the attributes (as shown in Fig. 1) that contribute for computing indirect trust are computed in Step 5, and are fed as an input to FIS_indirect_trust. This step generates indirect trust for each corresponding input.

7. Calculate direct trust using FIS_direct_trust that uses the input data from Step 5 (like Table 9).

Table 4 Summarized survey data (D_1) for website (W_1)

Attributes		Number of votes (out of 50)		
		True (t)	Indeterministic (i)	False (f)
Indirect trust	Security (S)	44	3	3
	Privacy (P)	44	4	2
	Satisfaction (S)	41	6	3
	GUI (G)	44	3	3
	Category (C)	43	4	3
	Navigation (N)	40	5	5
	Word of mouth (W)	39	7	4
	Trust certification (tc)	42	6	2
	Feedback (F)	38	7	5
Direct trust	Pricing scheme (Pr)	43	4	3
	Promotional discounts (Po)	46	2	2
	Personal discounts (Pd)	38	5	7
	Behaviour (B)	44	3	3
	Quality (Q)	44	3	3
	Delivery (D)	43	3	4

Crisp values are computed for the attributes (as shown in Fig. 1) that contribute for computing direct trust are computed in Step 5, and are fed as an input to FIS_direct_trust. This step generates direct trust for each corresponding input.

8. Calculate final trust using FIS_final_trust which depends on indirect and direct trust values received from Step 6 and Step 7 (like Table 13).

Steps 6 and 7, gives two trust values: direct and indirect trust values. These values are fed to FIS_final_trust that gives one consolidated trust value.

6 Experimental Results

The above-discussed model was tested over a population size of 50. Survey was conducted for two Indian E-Commerce websites for the data collection. Generally, the survey questionnaire captures the respondent's agreement or disagreement for a particular fact due to which the indeterminacy of the respondent for a particular fact is not captured. Hence, the underlying advantage of this proposed approach over other fuzzy based models is that it provides room for capturing the indecisiveness of the respondents about certain aspects, which is quite natural in real-world executions.

Step 1: In this study, 50 users who are actively doing online shopping were asked to answer the NL based questionnaire which was designed that not only capture the agreement or disagreement of the respondents but also capture the indeterminacy for particular fact. Questionnaire snapshot of survey is shown in Fig. 8. To answer a question, a user can only select one of the three options (True, False or Can't say). If a user agrees with the statement in the question, he/she would select "true", if he/she disagrees with the statement then option "false" can be selected and if he/she is confused or not sure then "can't say" option can be selected.

Step 2 and 3: The data collected D_1 and D_2 for all the users for two online shopping websites W_1 and W_2 are summarized in Tables 4 and 5, respectively. Data collected clearly reflects the number of votes as true for good/high value, indeterminacy for votes that reflect: can't say for users who are not sure about their opinion and the number of votes as false who thinks the value for an attribute is not high/good.

Tabulated results in Tables 4 and 5 shows the perception of users for each attribute affecting trust. Each attribute is captured along three dimensions: truthness, falsity and indeterminacy. For example, Table 4 shows that out of 50 only 43 users believe that delivery service is very good.

Step 4 and 5: As discussed above that neutrosophic components are real standard or non-standard subsets of $]^-0, 1^+[$, so the summarized data for both websites W_1 and W_2 are normalized in the range of 0–1 by dividing each value by 50 as shown in Tables 6 and 7.

This normalization of the values generates the three-dimensional view of each attribute and gives a better understanding about the perception of the customers about various attributes that governs trust. Like in Tables 6 and 7, it is quite clear to analyze proportion of agreement versus disagreement versus indecisiveness for all the attributes. This three-dimensional view is certainly quite important from the analyst perspective for better detailed understanding of the perception of various

Table 5 Summarized survey data (D_2) for website 2 (W_2)

Attributes		Number of voting (out of 50)		
		True (t)	Indeterminism (i)	False (f)
Indirect trust	Security (S)	38	6	6
	Privacy (P)	38	7	5
	Satisfaction (S)	35	6	9
	GUI (G)	40	5	5
	Category (C)	43	4	3
	Navigation (N)	40	5	5
	Word of mouth (W)	39	7	4
	Trust certification (tc)	36	10	4
	Feedback (F)	41	6	3
Direct trust	Pricing scheme (Pr)	39	7	4
	Promotional discounts (Po)	41	2	7
	Personal discounts (Pd)	39	5	6
	Behaviour (B)	41	4	5
	Quality (Q)	37	5	8
	Delivery (D)	38	5	7

Table 6 Normalized data (D_1') with new value set for website 1 (W_1)

Attributes		True (t)	Indeterminism (i)	False (f)	New value
Indirect trust	Security (S)	0.88	0.06	0.06	0.77
	Privacy (P)	0.88	0.08	0.04	0.78
	Satisfaction (S)	0.82	0.12	0.06	0.74
	GUI (G)	0.88	0.06	0.06	0.77
	Category (C)	0.86	0.08	0.06	0.76
	Navigation (N)	0.8	0.1	0.1	0.72
	Word of mouth (W)	0.78	0.14	0.08	0.72
	Trust certification (tc)	0.84	0.12	0.04	0.76
	Feedback (F)	0.76	0.14	0.1	0.70
Direct trust	Pricing scheme (Pr)	0.86	0.08	0.06	0.76
	Promotional discounts (Po)	0.92	0.04	0.04	0.80
	Personal discounts (Pd)	0.76	0.1	0.14	0.69
	Behaviour (B)	0.88	0.06	0.06	0.77
	Quality (Q)	0.88	0.06	0.06	0.77
	Delivery (D)	0.86	0.06	0.08	0.76

Table 7 Normalized data (D_2') with new value set for website 2 (W_2)

Attributes		True (t)	Indeterministic (i)	False (f)	New value
Indirect trust	Security (S)	0.76	0.12	0.12	0.69
	Privacy (P)	0.76	0.14	0.1	0.70
	Satisfaction (S)	0.7	0.12	0.18	0.65
	GUI (G)	0.8	0.1	0.1	0.72
	Category (C)	0.86	0.08	0.06	0.76
	Navigation (N)	0.8	0.1	0.1	0.72
	Word of mouth(W)	0.78	0.14	0.08	0.72
	Trust certification (tc)	0.72	0.2	0.08	0.69
	Feedback (F)	0.82	0.12	0.06	0.74
Direct trust	Pricing scheme (Pr)	0.78	0.14	0.08	0.72
	Promotional discounts (Po)	0.82	0.04	0.14	0.72
	Personal discounts (Pd)	0.78	0.1	0.12	0.70
	Behaviour (B)	0.82	0.08	0.1	0.73
	Quality (Q)	0.74	0.1	0.16	0.67
	Delivery (D)	0.76	0.1	0.14	0.69

attributes; which is certainly missing in the other existing fuzzy based models. To work further, it is now essential that once 3D breakdown for the attributes once understood, a consolidated value should now be constructed for each attribute.

So now neutrosophic triplet of the form (t, i, f) is given one consolidated value.

Here the authors give the steps to convert the neutrosophic information about the various attributes in Tables 6 and 7 into its valuation that is a single real number representation. Function f is applied on all the attributes A_i (listed in Tables 6 and 7) to get a cumulative value (new value) 'c' of the neutrosophic information and is represented as:

$$c = f(A) = f(T_A, I_A, F_A) : [0, 1] \times [0, 1] \times [0, 1] \to [0, 1] \tag{2}$$

$$c = f(A) = f(T_A, I_A, F_A) = w_1.T_A + w_2.(1 - F_A) + w_3.I_A/2 + w_4.(1 - I_A)/2 \tag{3}$$

The computation of cumulative value 'c' relates to the overall truthness of the neutrosophic representation, to which each component T, I and F contributes. The truth component gives direct information for the degree of truthness, so in Eq. (3), the truth component contribution is represented as $w_1.T_{A'}$. As the falsity component contributes indirectly to the computation of truthness, so it is represented as $w_2.(1 - F_{A'})$. Indeterminacy values can oscillate between the two extremes of 0 and 1. If indeterminacy value is 0, then either the falsity or truthness component is 100 %. For this case of indeterminacy = 0, the indeterminacy component contribution would not be used in the final evaluation, as it would be considered by the truth or falsity component. But, if the indeterminacy value is 1, then truth and falsity each have an equal chance of comprising 50 % of the final interpretation. Considering both the possibilities, indeterminacy contribution in the overall truthness would be represented as: $w_3.I_{A'}/2 + w_4.(1 - I_{A'}/2)$, as shown in Eq. (3).

Depending on the application domain and the experts input, weights w_1, w_2, w_3, w_4 can be assigned to the three parameters, truth, indeterminacy and falsity, in such a way that:

$0 \le w_1, w_2, w_3, w_4 \le 1$, $w_1 + w_2 + w_3 + w_4 = 1$ and normally, $w_1 > w_2 > w_3, w_4$. These weight values are application-specific and can be fine-tuned by using existing methods of neural networks or genetic algorithms for fine tuning the effective confidence computation.

Though there could be varied combination of the weight values, in this paper authors have given highest weightage to the truth value (50 %), 30 % weightage to falsity value and 20 % weightage to indeterminacy aspect. Here, for this problem, the authors have taken values for weights defined in Eq. (3) as:

$$w_1 = 0.5, w_2 = 0.3, w_3 = w_4 = 0.1$$

So, Eq. (3) can be rewritten as shown in Eq. (4).

Then, new value corresponding to each attribute is calculated from the normalized data using Eq. (4).

$$\text{New\_value} = 0.5t + 0.3(1 - f) + 0.1(i/2) + 0.1[(1 - i)/2] \tag{4}$$

Table 8 Indirect trust calculation for website 1 (W_1)

Policies						Web design				
Security (S)	Privacy (Pr)	Satisfaction (St)	Accumulated value, Pa = S + Pr + St	Final crisp value, P_final = Pa/3		GUI (G)	Category (C)	Navigation (N)	Accumulated value, Rw = G + C + N	Final crisp value, W_final = Rw/3
0.77	0.78	0.74	2.29	0.76		0.77	0.76	0.72	2.25	0.75

Recommendation					Mamdani
Words of mouth(W)	Trust certification (Tc)	Feedback (F)	Accumulated value, Ra = W + Tc + F	Final crisp value, R_final = Ra/3	Indirect Trust
0.72	0.76	0.7	2.18	0.73	0.66

For example in Table 6, corresponding to security: true (t) value is 0.88, false (f) value is 0.06 and indeterminism (i) value is 0.06. So new_value for security is:

$$0.5*0.88 + 0.3(1 - 0.06) + 0.1(0.06/2) + 0.1[(1 - 0.06)/2] = 0.77$$

The new values for all attributes have been shown in the last column of the Tables 6 and 7 corresponding to both the websites W_1 and W_2, respectively.

Step 6, 7 and 8: Based on values from normalized data tables D_1' and D_2', the values of indirect trust and direct trust is calculated using Mamdani-based indirect and direct trust model discussed in above section.

In Table 8, accumulated value is the addition of all the values of attributes corresponding to an element and crisp value for an element is obtained by dividing the accumulated value by 3.

For example, accumulated value for policy in case of website 1 is 2.29 and crisp value is 2.29/3 = 0.76. The crisp value for all the three elements acts as input in case of indirect trust calculation using FIS_indirect_trust. The value of indirect trust for website 1 is 0.66.

Similarly, computation of indirect trust value for website 2 is shown in Table 10.

For calculation of direct trust there are two inputs so crisp value for price and past experience is found out. One important thing that needs to be taken into consideration is that trust is indirectly proportional to price; so, for calculation of crisp value for price, its accumulated value is divided by three and then subtracted from one.

The values of direct trust for both websites are shown in Tables 9 and 11 computed using FIS_direct_trust.

This study conducted is a pilot study, which captured opinions of 50 customers indulging in online transactions. To validate the results generated by the proposed approach, analyst entirety view was captured regarding the two websites under study. Expert's voting website 1 (W_1) is considered in high category of trust and website 2 (W_2) is considered in medium category of trust as shown in Table 12. These experts are the people who are involved in designing, working and analyzing of online shopping from E-Commerce sites from last 4 to 5 years.

Table 9 Direct trust calculation for website 1 (W_1)

Price				
Pricing scheme (Pr)	Promotional discounts (Po)	Personal discounts (Pd)	Accumulated value, Pa = Pr + Po + Pd	Final crisp value, P_final = 1 − (Pa/3)
0.76	0.8	0.69	2.25	0.25

Past experience					Mamdani
Behaviour (B)	Quality (Q)	Delivery (D)	Accumulated value, PEa = B + Q + D	Final crisp value, PE_final = PEa/3	Direct trust
0.77	0.77	0.76	2.3	0.77	0.75

Table 10 Indirect trust calculation for website 2 (W_2)

Policies				
Security (S)	Privacy (Pr)	Satisfaction (St)	Accumulated value, Pa = S + Pr + St	Final crisp value, P_final = Pa/3
0.69	0.7	0.65	2.04	0.68

Web design				
GUI (G)	Category (C)	Navigation (N)	Accumulated value, Rw = G + C + N	Final crisp value, W_final = Rw/3
0.72	0.76	0.72	2.2	0.73

Recommendation					Mamdani
Words of mouth (W)	Trust Certification (Tc)	Feedback (F)	Accumulated value, Ra = W + Tc + F	Final crisp value, R_final = Ra/3	Indirect trust
0.72	0.69	0.74	2.15	0.72	0.58

Table 11 Direct trust calculation for website 2 (W_2)

Price				
Pricing scheme (Pr)	Promotional discounts (Po)	Personal discounts (Pd)	Accumulated value, Pa = Pr + Po + Pd	Final crisp value, P_final = 1 − (Pa/3)
0.72	0.72	0.7	2.14	0.29

Past experience					Mamdani
Behaviour (B)	Quality (Q)	Delivery (D)	Accumulated value, PEa = B + Q + D	Final crisp value, PE_final = PEa/3	Direct trust
0.73	0.67	0.69	2.09	0.70	0.63

Table 12 Expert's voting

Trust level	Website 1 (W_1)	Website 2 (W_2)
High	✓	×
Medium	×	✓
Low	×	×

Table 13 Final trust for website 1 and website 2

	Indirect trust	Direct trust	Final trust	Level
Website 1 (W_1)	0.66	0.75	0.71	High
Website 2 (W_2)	0.58	0.63	0.54	Medium

Final trust of both the websites is calculated using their corresponding direct and indirect trust values. For website 1, indirect trust value is 0.66 and direct trust value is 0.75, the final trust value obtained is 0.71 that is shown in Table 13. The final value of trust for website 1 lies in high category and matches with the expert's voting decision. Similarly, final trust value for website 2 is obtained as 0.54 which lies in medium level category and expert's voting decision for website 2 also lies in this category as visualized by comparing the Tables 12 and 13.

Since this was a pilot study, the number of respondents was quite limited, but reconciliation of the results given by the proposed approach with the experts view can be seen. Thus, this study motivates the authors to further experiment this approach on the larger scale, and validate further whether it generates an acceptable value of trust that takes into account indeterminacy of the consumers also.

7   Conclusion

In this chapter, the authors have presented a model that calculates the value of trust for a B2C E-Commerce website based on its user's perception, a perception that need not only be good or bad; it also accommodates the indeterminacy/neutral stance of the user. The proposed model concentrated mainly on the structure and important elements of E-Commerce companies. This paper suggests the model that consolidates various prominent attributes that govern trust, along with capturing the indecisiveness/indeterminacy of the customers for certain aspects. The price which is very affective for trust development and maintenance and adds value to this model. If an item is offered at a considerably low price, then certainly the customer would like to take the risk. However, some customers may give their priority to some other attribute. But the authors have considered most of the important aspects in their model, i.e indeterminacy in the data that can arise from inherent gaps in the perceived values, unsurety or unpredictable external factors that influence decision making capabilities. The advantage of using neutrosophic logic in the reasoning system is that it is very uncommon that the data acquired by the system would be 100 % complete and determinate. Though humans can take intelligent decisions in such situations, this knowledge is difficult to express in precise terms; an imprecise linguistic description of the manner of control can usually be articulated by the operator with relative ease.

The proposed neutrosophic representation and reasoning system here would be a special system which would be more generalized and indeterminacy tolerant in its working as compared to other existing fuzzy counterparts. Neutrosophic reasoning systems similar to their fuzzy counterparts would be capable of utilizing knowledge obtained from human operators. So to deal with such situations wherein there is a possibility of indeterminacy and incompleteness in the data acquired, the neutrosophic representation and reasoning system is proposed. This chapter proposes to extend the capabilities of fuzzy representation and reasoning systems by introducing the neutrosophic representation and reasoning system.

From the comparison of the calculated value of trust with expert's perception, it is concluded that this model gives an acceptable value of trust. Furthermore, the value of trust calculated from survey data would help the vendor to make improvements in the offered services which reflect low value of trust. So, definitely neutrosophic logic holds its chance to be experimented and utilized for real-world executions and human psychology simulations.

References

1. Ba S, Whinston AB, Zhang H (1999) Building trust in the electronic market through an economic incentive mechanism. In: Proceedings of the 20th international conference on information systems, Charlotte, North Carolina, Omnipress, pp 208–213
2. Ba S, Pavlou PA (2002) Evidence of the effect of trust building technology in electronic markets: price premiums and buyer behavior. MIS Q 26:243–268
3. Hoffman DL, Novak TP, Peralta M (1999) Building consumer trust online. Commun of the ACM 42:4
4. Benoit J, John I (2006) Consumer TRUST in E-commerce. Copyright © 2006, Idea Group Inc
5. Andrea O, Jana D (2006) Trust in E-technologies. Copyright © 2006, Idea Group Inc., distributing in print or electronic forms without written permission of IGI is prohibited

6. Jarvenpaa SL, Tractinsky N, Vitale, M (1999) Consumer trust in an internet store. Inf Technol Manage 1(1/2):45–72. 5 Gefen D (2000) E-Commerce: the role of familiarity and trust. Omega 28(6):724–737

7. Zadeh LA (1965) Fuzzy sets. Inf Control 8(3):338–353. ISSN 0019-9958

8. Clyde WH, Sharath S (2005) The dynamics of trust in B2C e-commerce: a research model and agenda. IseB 3(4):377–403

9. Kim DJ, Song YI, Braynov SB, Rao HR (2005) Multidimensional trust formation model in B-to-C e-commerce: a conceptual framework and content analyses of academia/ practitioner perspectives. Decis Support Syst 40(2):143–165

10. Krauter SG, Kaluscha EA (2003) Empirical research in online trust: a review and critical assessment. Int J Hum Comput Stud 58(6):783–812

11. McKnight DH, Choudhury V, Kacmar C (2002) The impact of initial consumer trust on intentions to transact with a web site: a trust building model. J Strateg Inf Syst 11(3):297–323

12. Zhuang H, Wongsoontorn S, Zhao Y (2003) A fuzzy-logic based trust model and its optimization for e-commerce. In: F Florida conference on the recent advances in robotics (FCRAR 2003)

13. Akhter F, Hobbs D, Maamar Z (2005) A fuzzy logic-based system for assessing the level of business-to-consumer (B2C) trust in electronic commerce. Expert Syst Appl 28(4):623–628. ISSN 0957-4174

14. Nilashi M, Fathian M, Gholamian MR, bin Ibrahim O (2010) Offering a model of evaluation of trust suggesting between customers and E-stores (B2C) based on approaches of fuzzy logic. Int J Bus Res Manage (IJBRM) 1(2):46

15. Qin Z, Tian B (2007) A trust evaluation model for B2C E-commerce. In: IEEE International Conference on IEEE Service Operations and Logistics, and Informatics, 2007, SOLI 2007

16. Nefti S, Meziane F, Kasiran K (2005) A fuzzy trust model for E-Commerce. In: Proceedings of the Seventh IEEE International Conference on E-Commerce Technology (CEC '05). IEEE Computer Society, Washington, DC, USA, pp 401–404. doi:10.1109/ICECT.2005

17. Kim D, Benbasat I (2003) Trust-related arguments in internet stores: a framework for evaluation. J Electron Commer Res 4(2):49–64

18. Dodds WB (1991) In search of value: how price and store name information influence buyers' product perceptions. J. Consum Mark 8(2):15–24

19. Ludwig SA, Pulimi V, Hnativ A (2009) Fuzzy approach for the evaluation of trust and reputation of services. In: Proceedings of the 18th international conference on fuzzy systems (FUZZ-IEEE'09)

20. Nafi KW, Kar TS, Hossain M, Hashem MMA (2013) A fuzzy logic based certain trust model for E-commerce. In: 2013 International Conference on Informatics, Electronics & Vision (ICIEV), IEEE, pp 1–6

21. Nafi KW, Hossain A, Hashem MM (2013) An advanced certain trust model using fuzzy logic and probabilistic logic theory

22. Ba Sulin (2001) Establishing online trust through a community responsibility system. Decis Support Syst 31(3):323–336

23. Jøsang A, Ismail R, Boyd C (2007) A survey of trust and reputation systems for online service provision. Decis Support Syst 43(2):618–644

24. Kracher B, Corritore CL, Wiedenbeck S (2005) A foundation for understanding online trust in electronic commerce. J Inf Commun Ethics Soc 3(3):131–141

25. Sabater J, Sierra C (2005) Review on computational trust and reputation models. Artif Intell Rev 24(1):33–60

26. Van der Heijden H, Verhagen T, Creemers M (2003) Understanding online purchase intentions: contributions from technology and trust perspectives. Eur J Inf Syst 12(1):41–48

27. Wang YD, Emurian HH (2005) An overview of online trust: concepts, elements, and implications. Comput Hum Behav 21(1):105–125

28. Yoon SJ (2002) The antecedents and consequences of trust in online-purchase decisions. J Interact Mark 16(2):47–63

29. Anurag B, Aggarwal S (2014) Fuzzy based trust model to evaluate and analyse trust in B2C E-Commerce. In: 2014 IEEE international advance computing conference (IACC), pp 1300, 1306, 21–22 Feb 2014. doi:10.1109/IAdCC.2014.6779515
30. Head M, Hassanein K (2002) Trust in e-Commerce: evaluating the impact of third-party seals. Q J Electron Commer 3(3):307–325
31. McCole P, Ramsey E, Williams J (2010) Trust considerations on attitudes towards online purchasing: the moderating effect of privacy and security concerns. J Bus Res 63(9–10):1018–1024
32. Rashad Y, Abu Tabik MAS, Seyedi AP (2011) Security and trust in electronic commerce-finding the safe side. In: Information communication and management–international proceedings of computer science and information technology (2011)
33. Pennanen K, Kaapu T, Paakki M-K (2006) Trust, risk, privacy, and security in ecommerce. In: Proceedings of the ICEB + eBRF Conference, 2006
34. Ganguly B et al (2010) The effects of website design on purchase intention in online shopping: the mediating role of trust and the moderating role of culture. Int J Electro Bus 8(4):302–330
35. Cheskin Research and Studio Archetype Deliver E-Commerce Trust Study (1999). The Free Library. Research and Studio Archetype Deliver E-Commerce Trust Study.-a053541794. http://www.thefreelibrary.com/Cheskin. Accessed 2 Apr 2014
36. Manchala DW (2000) E-commerce trust metrics and models. IEEE Internet Comput 4(2):36–44
37. Weisberg J, Te'eni D, Arman L (2011) Past purchase and intention to purchase in e-commerce: the mediation of social presence and trust. Internet Res 21(1):82–96
38. Smarandache F (1999) Linguistic paradoxists and tautologies, vol XIX. Libertas Mathematica, University of Texas at Arlington, Arlington, pp 143–154
39. Smarandache F (2002a) A unifying field in logics: neutrosophic logic, multiple-valued logic. Int J 8(3):385–438
40. Smarandache F (ed) (2002c) Proceedings of the first international conference on neutrosophy, neutrosophic logic, neutrosophic set, neutrosophic probability and statistics. University of New Mexico, Gallup Campus, Xiquan, Phoenix, 147 pp
41. Smarandache F (2002b) Neutrosophy, a new branch of philosophy, in multiple-valued logic. Int J 8(3):297–384
42. Smarandache F (2003) Definition of neutrosophic logic: a generalization of the intuitionistic fuzzy logic, In: Proceedings of the third conference of the european society for fuzzy logic and technology, EUSFLAT 2003, University of Applied Sciences at Zittau/Goerlitz, Zittau, Germany, pp 141–146, 10–12 Sept 2003
43. Wang L-X (1999) A course in fuzzy systems. Prentice-Hall Press, Upper Saddle River

Immune-Based Feature Selection in Rigid Medical Image Registration Using Supervised Neural Network

Joydev Hazra, Aditi Roy Chowdhury and Paramartha Dutta

Abstract Different radiological images like computed tomography (CT) and magnetic resonance (MR) are increasingly being used in medical science research for diagnosis and treatment. This article presents an automatic image registration technique to register MR–MR images using gray-level co-occurrence matrix (GLCM) and neural networks. This technique identifies different features of a brain image and its transformational counterpart. GLCM-based image feature extraction is a co-occurrence-based method by which different feature parameters are obtained. These parameters are calculated from the co-occurrence matrix along four directions, namely $0°, 45°, 90°$, and $135°$. Six features are selected from a set of features using an artificial immune system-based optimized feature selection technique and these six parameters are fed into the proposed neural network. Based on the principle of backpropagation algorithm, transformation parameters between the referenced and the sensed images are estimated. To demonstrate the effectiveness of the proposed method, experiment is carried out on MR T1, T2 datasets, and the results are compared with two other existing medical image registration techniques. The proposed method shows convincing results compared to others with respect to the estimation of underlying transformation parameters.

Keywords Image registration · GLCM · Artificial immune system · Feature extraction · Backpropagation

J. Hazra (✉)
Department of Information Technology, Heritage Institute of Technology,
Kolkata, West Bengal, India
e-mail: joydev.hazra@heritageit.edu

A.R. Chowdhury
Department of Computer Science and Technology, Bipradas Pal Chowdhury
Institute of Technology, Krishnagar, West Bengal, India
e-mail: aditihi2007@gmail.com

P. Dutta
Department of Computer and System Sciences, Visva-Bharati University,
Santiniketan, West Bengal, India
e-mail: paramartha.dutta@gmail.com

© Springer India 2016
S. Bhattacharyya et al. (eds.), *Hybrid Soft Computing Approaches*,
Studies in Computational Intelligence 611,
DOI 10.1007/978-81-322-2544-7_15

1 Introduction

Medical images are nowadays widely used for diagnosis, treatment, and supervising disease progression. The term 'medical image' spreads over a vast area of different types of images, with different applications. Medical researchers use medical images to investigate disease processes and to understand different developments as well as aging. Multiple images are acquired from common subjects at different times, or from different imaging modalities. Medical image registration primarily deals with the technique to align two or more images of intermodality or intramodality.

Different tomographic images like computed tomography (CT), magnetic resonance imaging (MRI), single-photon emission computed tomography (SPECT), and positron emission tomography (PET) are popular medical imaging techniques used to analyze and register. Over the past decade, researches on automatic rigid registration methods of medical images have been widely developed. Medical image registration can be manual, landmark-based, surface-based, and intensity-based.

Again according to the nature of matching, image registration can be categorized as intensity based and feature based [1]. According to the nature of transformation, image registration can also be grouped into several categories like rigid body, affine [2], linear elastic [3, 4], viscous fluid [5, 6], radial basis function [7, 8], etc. Medical image applications [9] provide the utility of manual registration to align images from monomodality or multimodality, but it depends on the user. Landmark registration [10–13] involves identification of the locations of corresponding points (internal or external) within different images identified by users manually. Error in rigid landmark-based registration can be analyzed [14]. According to the author, errors can be decomposed into three parts: (1) Fiducial localization error (FLE), (2) Target registration error (TRE), (3) Fiducial registration error (FRE). Surface-based registration [15–18] is a technique to reconstruct the surface of an image from a stack of contours generated by segmentation. Determination of transformation parameters can be done by minimizing the distance between the corresponding surface models. In [19] Besl and Mckay presented a technique called iterative closest point (ICP) to register the closest point in one surface to all the points relative to another surface. Intensity-based methods mainly deal with image intensities to estimate the transformation parameters like rotation, scaling, translation, etc. [20, 21] of both gray and color images [22]. Feature matching techniques use image features to determine corresponding feature pairs from the pair of images and then compute the transformation parameters to register them. Recently, mutual information (MI) [23], cross-correlation [24] are commonly used registration techniques. In [25], the authors demonstrate that MI can be effectively utilized to solve the correspondence problem in feature-based registration.

Monomodal image registration is a feature-based registration method used to identify neuropsychiatric disorders, depression, and Huntington's disease [26]. Brain imaging plays an important role in image registration. In [27], the authors present a monomodal image registration technique based on nonsubsampled coutourlet transform (NSCT), MI and particle swarm optimization (PSO).

Traditionally, soft computing has four technical disciplines and they fall into two categories, namely knowledge-driven and data-driven. Probabilistic reasoning (PR) and fuzzy logic (FL) reasoning systems are based on knowledge-driven reasoning. The other two technical disciplines, neurocomputing (NC) and evolutionary computing (EC), are data-driven reasoning. These techniques such as neural network, genetic algorithm, fuzzy logic, etc., are recently being explored. An artificial neural network (ANN) is a mathematical or computational model primarily nonlinear based on the principles of biological neural networks. They can be used to model complex relationships between inputs and outputs or to establish patterns in data. Radial basis functions [28], self-organizing maps [29] can be used for different computational aspects in image registration. Registration methods using neural networks have been reported [30, 31]. In [32], principal component analysis (PCA) has been used in the neural network for CT–MR and MR–MR registration. A genetic algorithm (GA) [33] is a search technique used in computing to find exact or approximate solutions in optimization and search problems. Each technique can be used separately, but a powerful advantage of soft computing is the complementary nature of the techniques rather than competitive. Hybrid soft computing models have been applied to a large number of classification, prediction, and control problems.

In this chapter, we combine NC with EC. We use a feature-based ANN technique to identify the transformational parameters. GLCM is a statistical feature extraction method through which different features of an image can be obtained. Six features are selected from a set of features using an artificial immune system based-optimized feature selection technique and these six parameters are fed into the proposed neural network. The selected features are calculated from the co-occurrence matrix of an image along four directions ($0°$, $45°$, $90°$, and $135°$). Backpropagation algorithm is used for training the neural network using the selected features obtained from GLCM through clonal selection algorithm (CSA) [34].

The organization of this article is as follows: Sect. 2 revisits concepts of relevant techniques involved in the proposed work. The proposed methodology is described in Sect. 3. Section 4 describes performance study based on different parameters; Sect. 5 presents description of datasets, experimental values and result analysis. Section 6 contains concluding remarks.

2 Prerequisite

In this section, a brief description of affine transformation, feature extraction mechanism, and neural network is provided.

2.1 Affine Transformation

A rigid transformation is one where the preimage and the image after transformation both preserve distance between every pair of points. Here, the original image is called the preimage and the new (copied) image is called the image after transformation. Transformations such as rotation, translation, reflection are rigid in nature. The relationship between the two sets of image coordinates (x', y') and (x, y) is often modeled using affine transformation as

$$\begin{pmatrix} x' \\ y' \\ 1 \end{pmatrix} = \begin{pmatrix} m_1 & m_2 & m_3 \\ m_4 & m_5 & m_6 \\ 0 & 0 & 1 \end{pmatrix} \cdot \begin{pmatrix} x \\ y \\ 1 \end{pmatrix} \tag{1}$$

Affine transformations like rotation, scaling, translation can be represented mathematically as

$$T' = RST \tag{2}$$

where, order may be different and R is the image rotation about the image center by an angle θ as represented by

$$R = \begin{pmatrix} \cos\theta & \sin\theta & 0 \\ -\sin\theta & \cos\theta & 0 \\ 0 & 0 & 1 \end{pmatrix} \tag{3}$$

S is the scaling effect with S_x and S_y representing scaling factors along X and Y axes, respectively, as

$$S = \begin{pmatrix} S_x & 0 & 0 \\ 0 & S_y & 0 \\ 0 & 0 & 1 \end{pmatrix} \tag{4}$$

and T_r is the translation effect with T_x and T_y representing translation factors along X and Y coordinate axes, respectively, as

$$T = \begin{pmatrix} 1 & 0 & T_x \\ 0 & 1 & T_y \\ 0 & 0 & 1 \end{pmatrix} \tag{5}$$

2.2 Feature Extraction

Feature extraction can be performed using statistical or syntactic methods. One of the simplest approaches for describing gray value of an image is to use statistical moments of the intensity histogram of an image or region. However, histograms will only carry information about distribution of intensities, but not about the relative position of pixels with respect to each other. Use of statistical approach such as co-occurrence matrix will help to provide valuable information about the relative position of the neighboring pixels in an image. Statistical methods analyze the spatial distribution of gray values and compute local features at each point in the image. The statistical features could be based on first-order (one pixel), second-order (two pixel), or higher-order (three or more pixels) statistics of gray level of an image. The textural features of an image are derived by using the (1) first-order statistics and (2) second-order statistics computed from spatial gray-level co-occurrence matrices (GLCMs) [35]. It is defined as "A two dimensional histogram of gray levels for a pair of pixels, which are separated by a fixed spatial relationship" [36]. Basically, GLCM calculates how often a pixel with gray level value i occurs either horizontally, vertically, or diagonally to adjacent pixels with the gray level value j. For any image GLCM can be calculated using displacement vector v. The displacement vector can be defined based on radius and orientation.

Various research activities have been carried out to find out the radius of the displacement vector. Very large value cannot detect detail information about an image. Generally its value ranges between 1–10. For best results, value of radius is limited to 1 or 2. Every pixel has eight neighboring pixels at angles 0°, 45°, 90°, 135°, 180°, 225°, 270°, or 315°. The calculation of angle θ in 0° and 180° is equal and so for others. So, in GLCM calculation one has to consider only 0°, 45°, 90°, and 135° as shown in Fig. 1a.

Pixels a_{23} and a_{21}, a_{13} and a_{31}, a_{12} and a_{32}, a_{11} and a_{33} are 0°, 45°, 90°, and 135° neighbors, respectively, with pixel a_{22} where radius is 1.

Sometimes, when the image is isotropic, or directional information is not required, one can obtain isotropic GLCM by integration over all angles. The basic difference is that first-order statistics estimates properties (e.g. average and variance) of individual pixel values, ignoring the spatial interaction between the image pixels, whereas second and higher-order statistics estimate properties of two or more pixel values occurring at specific locations relative to each other. The data thus generated contain normalized feature vectors computed around each pixel.

Fig. 1 **a** Neighbor pixels at different angles. **b** Sample image grid

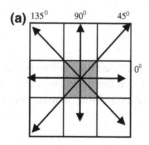
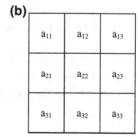

The feature vector includes three features derived from first-order statistics namely: (i) mean, (ii) standard deviation, and (iii) entropy. Fourteen features derived from second-order statistics including (i) contrast, (ii) angular second moment, (iii) inverse difference moment, (iv) correlation, (v) variance, (vi) inertia computed over a distance 1 and 2, etc. The features of GLCM [35, 37], used in this article, are briefly described in the following section.

Let an image to be analyzed is a rectangle which consists of 4×4 cells. Each cell represents a gray tone value 0–3. General form of the gray tone spatial-dependence matrix with gray tone values 0–3 is represented in Fig. 1c, where #(i, j) stands for number of times gray tone i and j have been neighbors.

Consider a 4×4 test images as shown in Fig. 2a, b. Calculation of gray tone spatial-dependence matrix in $0°$ with radius 1 is as follows (assuming pixel co-occurrence as symmetric):

#(0, 0) = {(a11, a12), (a12, a11), (a21, a22), (a22, a21)}
#(0, 1) = {(a12, a13), (a22, a23)}
#(0, 2) = {(a31, a32)}
#(0, 3) = { }
#(1, 0) = {(a13, a12), (a23, a22)}
#(1, 1) = {(a13, a14), (a14, a13), (a23, a24), (a24, a23)}
#(1, 2) = { }
#(1, 3) = { }
#(2, 0) = {(a32, a31)}
#(2, 1) = { }
#(2, 2) = {(a32, a33), (a33, a32), (a33, a34), (a34, a33), (a41, a42), (a42, a41)}
#(2, 3) = {(a42, a43)}
#(3, 0) = { }
#(3, 1) = { }
#(3, 2) = {(a43, a42)}
#(3, 3) = {(a43, a44), (a44, a43)}

Similarly, we can calculate gray value spatial-dependence matrix or gray tone co-occurrence matrix (radius 1) for $45°$, $90°$, and $135°$. Calculated GLCM values for angles $0°$, $45°$, $90°$, and $135°$ with radius 1 are shown in Fig. 2c–f, respectively.

2.2.1 GLCM Features

$p(i, j)$ is (i, j)th entry in a normalized gray color spatial-dependence matrix, G is the number of distinct gray levels in the quantized image, and $p_i(i)$ and $p_j(j)$ are the ith and jth entry in the marginal probability matrix, respectively.

Fig. 2 a Test image. **b** Test image grid. **c–f** Calculated GLCM values for angle 0°, 45°, 90°, and 135° with radius 1

(a)

0	0	1	1
0	0	1	1
0	2	2	2
2	2	3	3

(b)

a11	a12	a13	a14
a21	a22	a23	a24
a31	a32	a33	a34
a41	a42	a43	a44

(c)

	0	1	2	3
0	4	2	1	0
1	2	4	0	0
2	1	0	6	1
3	0	0	1	2

(d)

	0	1	2	3
0	4	1	0	0
1	1	2	2	0
2	0	2	4	1
3	0	0	1	0

(e)

	0	1	2	3
0	6	0	2	0
1	0	4	2	0
2	2	2	2	2
3	0	0	2	0

(f)

	0	1	2	3
0	2	1	3	0
1	1	2	1	0
2	3	1	0	2
3	0	0	2	0

The following notations are used to explain the various textural features: $p_i(i)$ and $p_j(j)$ are the marginal probabilities.

$$p_i(i) = \sum_{j=1}^{G} p(i,j) \tag{6}$$

$$p_j(j) = \sum_{i=1}^{G} p(i,j) \tag{7}$$

μ_i and μ_j are the mean of p_i and p_j

$$\mu_i = \sum_{i=1}^{G} p_i(i), \tag{8}$$

$$\mu_j = \sum_{j=1}^{G} p_j(j) \tag{9}$$

σ_i and σ_j are the standard deviation of p_i and p_j

$$\sigma_i^2 = \sum_{i=1}^{G} (i - \mu_i)^2 p_i(i), \tag{10}$$

$$\sigma_j^2 = \sum_{j=1}^{G} (j - \mu_j)^2 p_j(j) \tag{11}$$

$$p_{x+y} = -\sum_{i=1}^{G} \sum_{j=1}^{G} p(i,j), i+j = k \text{ and } k = 2,3, \ldots, 2G \tag{12}$$

$$p_{x-y} = -\sum_{i=1}^{G} \sum_{j=1}^{G} p(i,j), |i+j| = k \text{ and } k = 0,1, \ldots, G-1 \tag{13}$$

Important features (Haralick) of GLCM can be represented as follows [35]:

i. *Energy or angular second moment*: Homogeneous property of any image can be detected by this feature. It has a normalized range where maximum value is one. For any homogeneous image that has a uniform or periodic gray level distribution reaches the maximum value.

$$\text{So, Energy} = \sum_{i,j} p(i,j)^2 \tag{14}$$

ii. *Variance or sum of squares*: It is correlated with first-order statistics, namely standard deviation. If gray value of any pixel differs from their mean, then variance becomes large.

$$\text{Variance} = \sum_{i,j} \left(p(i,j) \frac{1}{N} \right)^2 \tag{15}$$

iii. *Entropy*: Complexity property of any image can be measured by entropy property. Basically, entropy is a complement of energy. So, for any non-uniform or a periodic gray level distribution entropy value is large.

$$\text{Entropy} = -\sum_{i,j} p(i,j) \log(p(i,j)) \tag{16}$$

iv. *Inverse difference moment or homogeneity*: For any pair of pixels small gray value differences can be detected easily by this property. It attains the maximum value for the same gray value of all pixels. GLCM contrast and homogeneity are strongly, but inversely, correlated. This means homogeneity decreases if contrast increases; while energy is kept constant.

$$\text{Homogeneity} = \sum_{i=0}^{G-1} \sum_{j=0}^{G-1} \frac{1}{1 + (i-j)^2} p(i,j) \tag{17}$$

v. *Correlation*: Measures the linear dependency among gray values of an image.

$$\text{Correlation} = \sum_{i=0}^{G-1} \sum_{j=0}^{G-1} p(i,j) \frac{(i - \mu_x)(j - \mu_y)}{\sigma_x \sigma_y} \tag{18}$$

vi. *Contrast*: Spatial frequency of any image can be measure by this feature. Through this feature we can measure the amount of variations present among pixels in an image.

$$\text{Contrast} = \sum_{n=0}^{G-1} n^2 \left\{ \sum_{i=1}^{G} \sum_{j=1}^{G} p(i,j) \right\}, |i - j| = n \tag{19}$$

The rest of the features are derived from the above-mentioned features.

vii. $$\text{Sum Average} = \sum_{i=2}^{2G} i p_{x+y}(i) \tag{20}$$

viii. $$\text{Sum Entropy} = -\sum_{i=2}^{2G} p_{x+y}(i) \log\{p_{x+y}(i)\} \tag{21}$$

ix. $$\text{Sum Variance} = \sum_{i=2}^{2G} (i - \text{vii})^2 p_{x+y}(i) \tag{22}$$

x. $$\text{Difference Variance} = \text{Variance of } p_{x+y} \tag{23}$$

xi. $$\text{Difference Entropy} = -\sum_{i=0}^{G-1} p_{x-y}(i) \log\{p_{x-y}(i)\} \tag{24}$$

xii. $$\text{Information measure of correlation} = \frac{HXY - HXY1}{\max\{Hx, Hy\}} \tag{25}$$

xiii. $$\text{Information measure of correlation} = (1 - e^{[-2(HXY2 - HXY)]})^{1/2} \tag{26}$$

$$\text{Where, } HXY = -\sum_{i} \sum_{j} p(i,j) \log(p(i,j)), \tag{26a}$$

$$HXY1 = -\sum_{i} \sum_{j} p(i,j) \log\{p_x(i) p_y(j)\} \tag{26b}$$

$$HXY2 = -\sum_{i} \sum_{j} p_x(i) p_y(j) \log\{p_x(i) p_y(j)\} \tag{26c}$$

xiv. Maximal correlation coefficient = (Second largest eigenvalue of A)$^{1/2}$

$$\text{where, } A = \sum_k \frac{p(i,k)p(j,k)}{p_x(i)p_y(k)} \tag{27}$$

We can calculate the features of the image given in Fig. 1 using the above-mentioned GLCM method.

Energy $= 0.2502 + 0.0002 + 0.0832 + 0.0002 + 0.0002 + 0.1672 + 0.0832 + 0.0002$
$\qquad + 0.0832 + 0.0832 + 0.0832 + 0.0832 + 0.0002 + 0.0002 + 0.0832 + 0.0002$
$\qquad = 0.1386$

Entropy $= 0.250 * \ln(0.250) + 0.000 + 0.083 * \ln(0.083) + 0.000 + 0.000$
$\qquad + 0.167 * \ln(0.167) + 0.083 * \ln(0.083) + 0.000 + 0.083 * \ln(0.083) + 0.083 * \ln(0.083)$
$\qquad + 0.083 * \ln(0.083) + 0.083 * \ln(0.083) + 0.000 + 0.000 + 0.083 * \ln(0.083) + 0.000$
$\qquad = 2.0915$

Similarly, other GLCM features are calculated.

2.3 Artificial Immune System

The artificial immune systems are composed of intelligent methodologies, inspired by the natural immune system, for the solution of real-world problems [38]. Nature is one of the main sources of inspiration for the development of various computational algorithms. The biological immune system (IS) is a robust and adaptive system that defends the body from foreign pathogens. The ultimate target of all immune response is an antigen (Ag) which is usually a foreign molecule. The cells that originally belong to our body and are harmless to its functioning are termed self (or self-antigens), while the disease causing elements are named nonself (or nonself antigens). The immune system has to be capable of distinguishing between self and nonself antigens. This process is called self/nonself discrimination [34].

All living organisms are exposed to many different microorganisms and viruses that cause illness. These microorganisms are called pathogens. Pathogens usually act as antigens. After getting stimulated, the immune system generates a number of antibodies that respond to the foreign antigens. The basic mechanism of the biological immune system is a complex process, described in Fig. 3.

- Antigens are ingested and digested by phagocytes (white blood cells). The phagocytes that present antigens to lymphocytes are called antigen-presenting cells or APC.
- APC fragments those antigens into antigenic peptides and they are joined with major histocompatibility complex or MHC molecules and form peptide–MHC combinations. Then they are displayed on the cell surface.

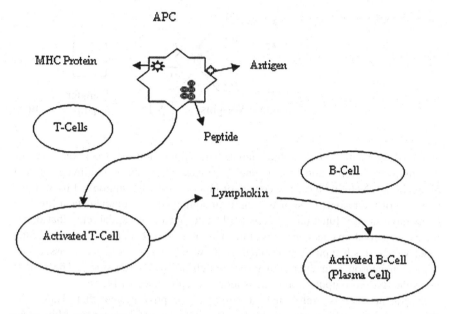

Fig. 3 Basic biological immune system

- Phagocytes that are not APC are called T cells (T lymphocytes). T cells can recognize different peptide–MHC combinations.
- After recognition, activated T cells send lymphokines or chemical signals that mobilize different components of the immune system.
- The signal is received by B cells or B lymphocytes. B cells divide and mature into plasma cells that secrete antibody proteins.
- These antibodies can neutralize the antigens or reduce their destruction capability by producing complement enzymes.
- Some B cells and T cells are memory cells. If the same antigen presents itself in future, then these memory cells boost the immune system to prevent it.
- Antibodies in B cell frequently suffer from affinity maturation, i.e., mutation and modification. This helps them to improve response against antigens.

The immune clonal algorithm inspired by biological immune system is an evolutionary algorithm that can be used in optimization problems.

2.4 Neural Network

An ANN is basically a mathematical model of biological neural networks [39, 40]. This network mainly consists of some artificial neurons that are interconnected together. Every neuron can be seen as a computational unit. Neurons receive inputs

Fig. 4 A sample neuron

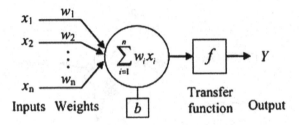

and process them to obtain the output. The connections between the neurons determine the information flow among the nodes. They can be unidirectional and bidirectional. Neurons basically consist of inputs, which are multiplied by weights (strength of the respective signals), and then computed by a mathematical function (activation/transfer function) such as hard limit, linear, sigmoidal, etc., that determines the activation characteristics of the neuron. In Fig. 4, x_i, $i = 1, 2, \ldots, n$ are the input values. Each input is multiplied by its weight w_i and a bias b is associated with each neuron, and their sum goes through a transfer function f. The neuron gives the desired output Y, after activation, as represented in Fig. 4.

Neural network is useful since it acts as an adaptive system that changes its structure during its learning phase. In ANN, three major learning paradigms are supervised learning, unsupervised learning, and reinforcement learning. In supervised learning, a network is trained using a set of inputs and outputs (targets). A learning network estimates an unknown function from representative observations of the relevant variables [31]. One of the well known and frequently used learning techniques is backpropagation learning (BPL).

Multilayer perceptron uses BPL for training purposes. It is basically a simple iterative gradient descent algorithm designed to minimize the mean squared error (MSE) between the desired response output and the actual output. The weights in batch BPL can be updated only after the presentation of the complete set of training data. So, a training iteration incorporates one complete pass through all the training patterns. On the other hand, the sequential BPL adjusts the network parameters as and when one training pattern is provided.

3 Proposed Method

The proposed algorithm consists of three stages: feature extraction, feature selection, and detection of transformation parameters as shown in Fig. 5. In feature extraction, a set of features is extracted from image sets using GLCM method. In feature selection step, we select d number of features from the extracted D number of features ($d \ll D$) using clonal selection algorithm (CSA) [34]. Using supervised learning method, we trained the neural network to detect unknown transformation parameters. The technical specification of clonal selection algorithm is as follows: initial antibody population size is 50 and each antibody consists of 6 random

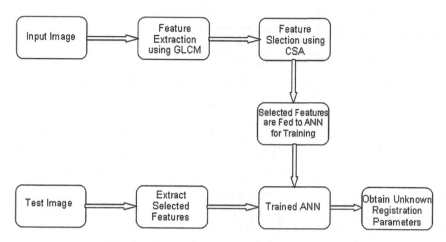

Fig. 5 Proposed registration technique

features among 14 features extracted using GLCM. The length of each antibody is 24-bits. Proposed ANN constitutes one input layer, one output layer, and one hidden layer. We use multilayer perceptron with batch BPL as the learning algorithm. The selected (using CSA) GLCM features of both the original image and its transformed counterpart are used as input dataset to the ANN. We fix target outputs based on the angle of rotation (R), scaling along x-axis (S_x), scaling along y-axis (S_y), translation along x-axis (T_x) and translation along y-axis (T_y). Then we train the network for each pair of original and transformed image. In this training phase, backpropagation algorithm is used to minimize the error between the accepted output and the actual output. After completion of the training process, adjusted weights are used for testing the rigid transformation parameters (Fig. 6).

3.1 Architecture of the Immune Clonal System

As stated above, initial antibody population size is 50 and the length of the antibody is 24-bits since each antibody consists of 6 random features and each feature is represented by 4 bits. Among the 50 antibodies, 4 antibodies are selected as memory cells based on the highest affinity value. Working principle of the clonal selection algorithm [34] is given below.

Algorithm 1:

1. N antibodies are divided into m cell (memory) and r cell (remaining).
2. Loop:

   - Calculate affinity of N antibodies.
   - Selection: b ($b < n$) antibodies with the highest affinity value are selected.

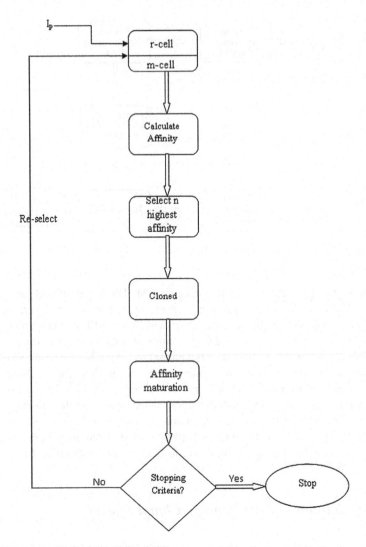

Fig. 6 Flowchart of clonal selection algorithm

- Mutation: Clone the selected antibodies and perform affinity maturation on clonal set.
- Calculate affinity of these newly generated antibodies.
- m antibodies having the highest affinity values are selected for m cell.

3. When the stopping criteria (maximum number of iteration, no change in m cell in two successive iteration, etc.) are met, the antibodies in m cell are the solution.

Here, we consider the set of unique feature that classify the image uniquely and calculate the classification error. The classification error acts as an affinity function. The lowest classification error means highest affinity value. Flowchart of the above mentioned algorithm is given in Fig. 6.

3.2 Architecture of the Neural Network

In this chapter, the following notations are used:

i_k kth node for input layer
h_i ith node for hidden layer
o_j ith node for output layer
d_j jth target vector component
W_{ij} Weight from hidden to output layer across the edge h_i to o_j
v_{ki} Weight from input to hidden layer across the edge i_k to h_i
μ Pattern μ
d_j^μ jth target vector component for training pattern μ
$f(.)$ Activation function
v Learning parameter
ξ Momentum parameter

Consider a multilayer neural network with l' units in the input layer, l units in the hidden layer, p units in the output layer using backpropagation learning algorithm. There is only one hidden layer.

3.2.1 Input Layer

Input layer consists of different features of sensed reference pair images based on GLCM method. We use six features of this pair of images using different angles. First n nodes of this layer are feature vectors obtained from reference image and the remaining n nodes are from sensed image. So, l' is equal to $2n$.

3.2.2 Hidden Layer

The hidden layer is used to determine the weight of neurons from its previous input layer and again to generate signal applying sigmoidal activation function for the next output layer. Mathematically,

$$n_i^\mu = \sum_{k=1}^{l'} v_{ki}\delta_k^\mu + b_i \tag{28}$$

and

$$\delta_k^\mu = f(n_i^\mu) \tag{29}$$

3.2.3 Output Layer

The final output of the proposed neural network is obtained in the output layer. The output layer consists of p nodes. Each node represents one transformation parameter such as rotation, scaling (along both x and y axes), translation (along both x and y axes).

Mathematically,

$$N_j^\mu = \sum_{i=1}^{l} W_{ij}\delta_i^\mu + b_j = \sum_{i=1}^{l} W_{ij} f(\sum_{k=1}^{l'} v_{ki}\delta_k^\mu + b_i) + b_j \tag{30}$$

and

$$\delta_j^\mu = f(\sum_{i=1}^{l} W_{ij} f(\sum_{k=1}^{l'} v_{ki}\delta_k^\mu + b_i) + b_j) \tag{31}$$

For error calculation

$$E = \frac{1}{2}\sum \left[d_j^\mu - f\left(\sum_{i=1}^{l} W_{ij} f\left(\sum_{k=1}^{l'} v_{ki}\delta_k^\mu + b_i\right) + b_j\right)\right]^2 \tag{32}$$

3.2.4 Backpropagation Rule

We use backpropagation learning algorithm for the supervised training of multi-layer networks. For a connection from p to q, weight change can be written as

$$\Delta W_{pq} = v\alpha_q x_{qp} \tag{33}$$

Here, α is the backpropagation error term, x is the value of the input in the weight.

Now, learning rate and momentum rate play crucial role in training of the ANN. To obtain faster learning without oscillation, the weight change is related to the previous weight change to ensure a smooth effect. The momentum coefficient determines the proportion of the last weight change that is added to the new weight change. Effect of these two parameters is offered in detail in Sect. 4.

$$\Delta W_{pq}(t+1) = \xi \Delta W_{pq}(t) + \Delta W_{pq} \tag{34}$$

So,

$$W_{pq}(t+1) = W_{pq}(t) + \xi(W_{pq}(t) - W_{pq}(t-1)) + \Delta W_{pq} \tag{35}$$

for output neuron

$$\alpha_j^\mu = (d_j^\mu - O_j^\mu)O_j^\mu(1 - O_j^\mu) \tag{36}$$

for hidden neuron

$$\alpha_i^\mu = O_i^\mu(1 - O_i^\mu)\sum W_{ij}\alpha_j^\mu \tag{37}$$

3.3 Algorithm of the Proposed Method

The step-by-step process of our proposed methodology is described in Algorithm 2. The whole process is continued until stopping criterion is attained.

Algorithm 2

(a) Initialization:

   1. Set the number of feature neurons in the input layer, l'.
   2. Set the number of neurons in the hidden layer, l.
   3. Set the number of output to be p.
   4. Initialize the weights v_{ki}.
   5. Initialize the weights W_{ij}.
   6. For both reference and sense images, create a feature list using GLCM algorithm.
   7. The input of the neural network consists of the vector containing the feature list of both reference and sense images.

(b) Apply the inputs to the network and workout the outputs from hidden layer.
(c) Feed the output of hidden layer as the input to the output layer and again compute the output.
(d) Calculate the errors in the output layer.
(e) Backpropagate the errors from the output layer to hidden layer neurons.
(f) After having obtained the error for the hidden layer neurons, now proceed to change weights from hidden layer to output layer.
(g) Repeat steps (b) to (f) until convergence.

For the given architecture, number of nodes in the input layer and output layer are 48 and 5, respectively. Justification of parameter selection for hidden layer is provided in Sect. 4.

4 Performance Study Based on Different Parameters

While designing a multilayer neural network based on backpropagation learning algorithm, different parameters such as initial weights and biases, learning rate, number of hidden nodes, momentum rate, and activation functions play a vital role. Improper choice of any of these parameters may lead to slow convergence and poor performance of the network. In this study, the best parameter choice has been identified by varying number of hidden layer neurons ranging over 6, 12, 24, 48, 96. For each network, different combinations of learning rate 0, 0.05, 0.1, 0.15, 0.2, 0.25, 0.3, 0.4, 0.5 and momentum rate 0, 0.1, 0.2, 0.3, 0.4, 0.5, 0.6, 0.7, 0.8, 0.9 have been explored to fix final setting of the parameters.

4.1 Influence of Number of Neurons in Hidden Layer

The network topology is based on the varying number of neurons in the hidden layer. From the comparison among different sizes shown in Fig. 7, it is clear that for our dataset optimal size is 24.

4.2 Influence of Learning Rate

The learning rate is important for accurate classification of the testing examples. The learning rate is an adjustable factor that regulates the speed of the learning process. With a faster learning rate, the ANN model will learn faster. However, if the learning rate is too high, the oscillations of weight changes can impede the

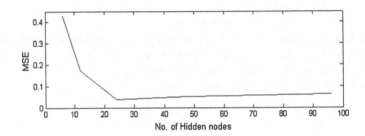

Fig. 7 Influence of hidden layer neuron member over mean squared error

convergence of the error surface, and may lead to overshooting of a near-optimal weight factor. In contrast, if the learning rate is too slow, the ANN model may be caught in a local error minimum instead of achieving the global minimum. So, learning rate selection is a crucial step in finding global minima. The best choice of learning rate is interestingly data-dependent and can be found using some trial and error.

From Fig. 8, it is clear that MSE is minimum at learning rate 0.05.

4.3 Influence of Momentum Rate

The momentum rate is used to avoid the network getting stuck at any local minima. The error surface is surrounded by a positive and negative gradient. So, it can incorrectly converge upon a solution that may not correspond to the minimum error. Momentum in backpropagation algorithm helps to speedup the convergence, to avoid getting trapped into local minima and also to reduce oscillation of weight change. We compute MSE for different momentum rate and the error is minimum at 0.9. The graph depicted in Fig. 9 clearly shows this.

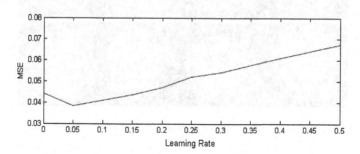

Fig. 8 Influence of learning rate over MSE

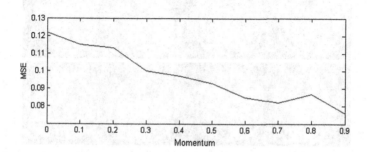

Fig. 9 Influence of momentum rate over MSE

5 Experimental Results

To evaluate the performance of the proposed method, several experiments were conducted on different MRI images of human brain. Registration of MR T1 and T2 [41, 42] weighted images is probably the best test case for a gradient-based method. We carry out different experiments on 20 monomodal intrasubject MRI brain image datasets. In this article, we only present four such datasets as indicated in Figs. 10, 11, 12 and 13. Prior to the experiment, normalization of data has been done. The training of the network is performed by presenting the network, a set of 50 artificial affine transformation applied on the original image. For the whole experiments, as well as comparison, the initial transformation parameters were bounded within some range. Rotational angle θ is maintained within $-30°$ to $30°$. Scaling parameters along both X and Y coordinate axes are maintained within 0.5–3 while translational parameters along X and Y axes are kept within 1–10.

Table 1 shows result of the proposed algorithm on the dataset 2 given in Fig. 11.

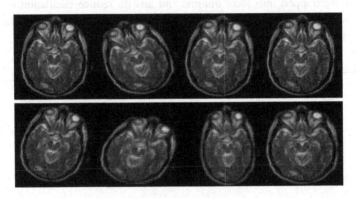

Fig. 10 A T2-weighted MR image slice of brain undergoing different transformations

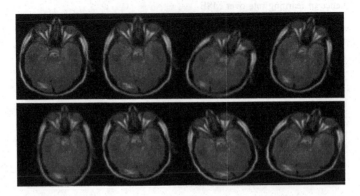

Fig. 11 A T1-weighted MR image slice of brain with different transformations

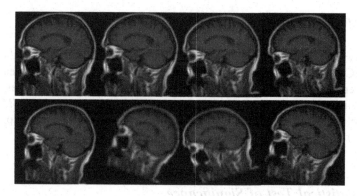

Fig. 12 A T1-weighted MR image corpus of brain with different transformations

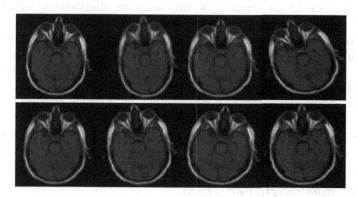

Fig. 13 A T1-weighted MR image corpus of brain with different transformations

Table 1 MRI–MRI rigid registration: experimental parameters and final values from proposed algorithm

Dataset 2 (Fig. 11)	Experimental parameters					Result obtained from proposed method				
	R	S_x	S_y	T_x	T_y	R	S_x	S_y	T_x	T_y
Sub2	3	1.0	2.0	5	2	2.952	1.079	2.033	4.982	1.850
Sub3	−18	0.5	2.0	7	9	−18.124	0.521	1.995	6.991	9.137
Sub4	−12	1.5	1.0	10	1	−12.156	1.537	1.019	9.888	0.996
Sub5	4	2.5	0.5	2	8	3.972	2.552	0.558	2.138	7.945
Sub6	8	1.0	2.0	3	6	7.928	1.066	2.049	2.784	6.356
Sub7	20	1.5	2.5	10	1	19.916	1.490	2.502	10	1.030
Sub8	−10	1.0	3.0	5	10	−10.008	1.000	2.999	5.144	10

5.1 Analysis of Results

We carry out MR–MR registration using our proposed method, PCA method [32] and NSCT [27], and show the comparative study in Table 2. As mentioned, experiments are performed on four datasets described earlier. The best results are marked in bold. From Table 2, it is observed that the proposed GLCM-based methodology outperforms all the other two methods used in our experiment (Table 3).

5.2 Statistical Test of Significance

We used Kolmogorov-Smirnov (K-S) two-sample nonparametric tests on two datasets for testing the equality of two unknown distributions yielding the respective result sets. The K-S test [43] tries to determine if two datasets differ significantly. The K-S-test is nonparametric and devoid of any assumption on distribution. The null and alternative hypotheses for the K-S test relate to the equality of the two distribution estimates corresponding to the result sets available. Depending on the nature of the alternative hypothesis, the test will offer, whether there is any stochastic ordering among the result sets X and Y obtained from two separate methods and, if there be any such ordering, which of the two is stochastically smaller than the other. $X_1, X_2, \ldots , X_m$ and $Y_1, Y_2, \ldots ,Y_n$ are independent random samples from populations representing unknown distributions F_X and F_Y.

Null hypothesis H_0: $F_X(t) = F_Y(t) \; \forall \; t$,
Alternative hypothesis H_1: $F_X(t) < F_Y(t) \; \forall \; t$

We have three sets of data, obtained from (i) our algorithm (ii) PCA algorithm, and (iii) NSCT method. These datasets represent MSE between the actual value and the obtained value. Table 4 shows the p-values of the data between each pair of registration methods after applying K-S test. p-values report if the datasets differ significantly statistical. The null hypothesis is rejected if p is statistically "insignificant".

We apply K-S test on dataset 3 and dataset 1. Maximum difference between the cumulative distributions D is 0.2024, i.e., $D^- = \max(F1 - F3) = 0.2024 > 0.1484$ (tabulated critical value). The test rejects the null hypothesis at the 5 % level of significance, i.e., this test accepts the alternative hypothesis that the dataset 3 is stochastically smaller than dataset 1 generated by PCA method.

Again, we apply K-S test on dataset 3 and dataset 2. Maximum difference between the cumulative distributions D is 0.2738 i.e. $D^- = \max (F2 - F3) = 0.2738 > 0.1484$ (tabulated critical value). The test rejects the null hypothesis at the 5 % level of

Table 2 Overall error estimation of PCA, NSCT based, and the proposed methods for different datasets given in Figs. 10, 11, 12 and 13

	PCA			NSCT					Proposed				
	R	T_x	T_y	R	S_x	S_y	T_x	T_y	R	S_x	S_y	T_x	T_y
Dataset 1 (Fig. 10)													
Sub2	**0.329**	0.020	0.011	0.571	0.151	0.391	0.032	0.053	0.624	**0.013**	**0.009**	**0.000**	**0.002**
Sub3	**0.258**	0.039	0.056	0.478	0.037	0.410	0.055	0.071	0.404	**0.000**	**0.000**	**0.011**	**0.042**
Sub4	0.207	0.047	0.007	0.412	0.024	0.120	0.050	0.046	**0.036**	**0.000**	**0.005**	**0.022**	**0.000**
Sub5	0.331	0.030	0.033	0.513	0.019	0.020	0.041	0.036	**0.148**	**0.018**	**0.007**	**0.029**	**0.021**
Sub6	**0.293**	0.009	0.020	0.337	0.080	0.011	0.009	0.011	0.308	**0.008**	**0.001**	**0.002**	**0.017**
Sub7	0.179	0.050	0.033	0.376	0.059	0.061	0.071	0.042	**0.160**	**0.023**	**0.03**	**0.004**	**0.006**
Sub8	0.311	0.028	0.039	0.471	0.024	0.008	0.032	0.049	**0.096**	**0.014**	**0.002**	**0.012**	**0.036**
Dataset 2 (Fig. 11)													
Sub2	0.312	0.333	0.451	0.461	0.021	0.077	0.417	0.773	**0.048**	**0.079**	**0.033**	**0.018**	**0.150**
Sub3	0.172	0.164	0.080	0.377	0.071	0.051	0.131	0.090	**0.124**	**0.021**	**0.004**	**0.009**	**0.137**
Sub4	0.217	0.007	0.009	0.452	0.020	0.010	0.006	0.008	**0.156**	**0.037**	**0.019**	**0.112**	**0.004**
Sub5	0.111	0.021	0.050	0.301	0.034	0.073	0.034	0.066	**0.028**	**0.052**	**0.057**	**0.138**	**0.055**
Sub6	0.303	0.051	**0.271**	0.591	**0.007**	0.027	0.069	0.312	**0.072**	0.066	**0.049**	0.216	0.356
Sub7	0.512	0.003	0.052	0.663	0.020	0.007	0.006	0.042	**0.084**	**0.009**	**0.002**	**0.000**	**0.030**
Sub8	0.010	0.213	0.008	0.015	0.005	0.007	0.212	0.007	**0.008**	**0.000**	**0.001**	**0.144**	**0.000**
Dataset 3 (Fig. 12)													
Sub2	0.312	0.301	0.422	0.512	0.005	0.079	0.391	0.501	**0.232**	**0.000**	**0.069**	**0.238**	**0.217**
Sub3	0.211	0.129	0.097	0.126	0.067	0.069	0.221	0.098	0.328	0.051	0.044	0.321	0.076
Sub4	0.203	0.007	0.011	0.213	0.029	0.017	0.018	0.009	**0.136**	**0.015**	**0.002**	**0.001**	**0.005**
Sub5	0.106	0.015	0.041	0.257	0.031	0.072	0.021	0.044	**0.076**	**0.012**	**0.051**	**0.002**	**0.039**
Sub6	**0.215**	0.026	**0.081**	0.312	0.010	0.021	0.027	0.211	0.428	**0.001**	**0.016**	**0.015**	0.105
Sub7	0.091	0.013	0.021	0.073	0.311	0.039	0.042	0.019	**0.060**	0.208	**0.012**	**0.007**	**0.003**
Sub8	0.097	0.021	0.057	0.271	0.051	0.077	0.051	0.077	**0.088**	**0.011**	**0.069**	**0.002**	**0.042**

(continued)

Table 2 (continued)

	PCA			NSCT					Proposed				
	R	T_x	T_y	R	S_x	S_y	T_x	T_y	R	S_x	S_y	T_x	T_y
Dataset 4 (Fig. 13)													
Sub2	0.371	0.022	0.251	0.444	0.121	0.059	0.102	0.211	**0.232**	**0.081**	**0.036**	**0.012**	**0.109**
Sub3	0.217	**0.056**	0.023	0.414	0.031	0.040	0.317	0.051	**0.156**	**0.012**	**0.006**	0.226	**0.019**
Sub4	0.090	0.472	0.211	0.176	0.122	0.318	0.445	0.216	**0.068**	**0.002**	**0.126**	0.249	**0.148**
Sub5	0.071	0.077	0.069	0.139	0.221	**0.072**	0.092	0.026	**0.052**	**0.057**	0.098	0.044	**0.017**
Sub6	0.211	0.151	0.176	0.312	0.066	0.064	0.221	0.052	**0.188**	**0.026**	**0.053**	0.133	**0.022**
Sub7	0.315	0.244	0.069	0.294	0.077	0.008	0.413	0.079	**0.104**	**0.023**	**0.002**	0.187	**0.050**
Sub8	**0.269**	0.421	0.337	0.312	0.006	0.009	0.315	0.211	0.276	**0.000**	**0.000**	0.255	**0.148**

The best results are marked in bold

Table 3 MSE statistics of the registration methods

PCA (dataset 1)		NSCT (dataset 2)		Proposed (dataset 3)	
$\mu_m$se	$\sigma_m$se	$\mu_m$se	$\sigma_m$se	$\mu_m$se	$\sigma_m$se
0.1461	0.135	0.2039	0.189	0.1070	0.119

Table 4 Comparison among different registration methods using K-S test (p-values)

Registration methods	p-value
Proposed method and PCA	0:055
Proposed method and NSCT	0:003

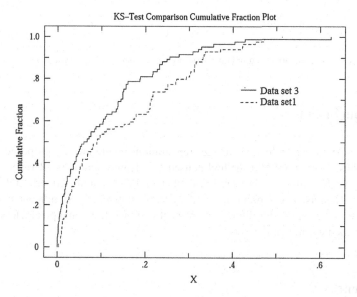

Fig. 14 K-S test comparison cumulative fraction plot of dataset 3 (proposed) versus dataset 1 (PCA)

significance. In other words, this test accepts the alternative hypothesis with 95 % confidence that the proposed algorithm or dataset 3 is stochastically smaller than dataset 2 generated by NSCT method (Figs. 14 and 15).

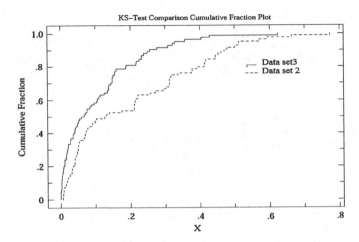

Fig. 15 K-S test comparison cumulative fraction plot of dataset 3 (proposed) versus dataset 2 (NSCT)

6 Conclusions

In this article, an effective image registration technique is proposed using feature-based ANN. GLCM method is used as a feature extractor. Performance of the proposed method is compared with that of PCA neural network and NSCT methods. Results of the proposed method show statistically significant improvement over the existing algorithms with respect to transformational parameters for all the four medical image datasets.

References

1. Goshtaby A (2005) 2-D and 3-D image registration for medical, remote sensing, and industrial applications. Wiley, New Jersy
2. Collins DL, Neelin P, Peters TM, Evans AC (1994) Automatic 3D intersubject registration of MR volumetric data in standardized Talairach space. J Comput Assis Tomog 18(2):192–205
3. Gee J, Reivich M, Bajacsy R (1993) Elastically deforming 3D atlas to match anatomical brain images. J Comput Assist Tomogr 17(2):225–236
4. Bajcsy R, Kovacic S (1989) Multiresolution elastic matching. Comput Vision Graphics Image Process 46(1):1–21
5. Christensen GE, Rabitt RD, Miller MI (1996) Deformable templates using large deformation kinematics. IEEE Trans Med Imag 5(10):1435–1447
6. Christensen GE (1996) Individualizing neuroanatomical atlases using a massively parallel computer. IEEE Comput 29(1):32–38
7. Bookstein FL (1989) Principal warps: thin-plate splines and the decomposition of deformations. IEEE Trans Pattern Anal Mach Intell 11(6):567–585

8. Rueckert D, Sonoda LI, Hayes C, Hill DLG, Leach MO, Hawkes DJ (1999) Non-rigid registration using free-form deformation: application to breast MR images. IEEE Trans Med Imag 18(8):712–721

9. Habboush IH, Mitchell KD, Mulkern RV, Barnes PD, Treves ST (1996) Registration and alignment of three-dimensional images: an interactive visual approach. Radiology 199 (2):573–578

10. Maurer CR, Fitzpatrick JM, Wang MY, Galloway RL, Maciunas RJ, Allen GG (1997) Registration of head volume images using implantable fiducial markers. IEEE Trans Med Imag 16(4):447–462

11. Fox PT, Perlmutter JS, Raichle ME (1985) A stereotactic method of anatomical localization of positron emission tomography. J Comput Assist Tomogr 9:141–153

12. Evans AC, Marrett S, Collins L, Peters TM (1989) Anatomical-functional correlative analysis of the human brain using three dimensional imaging systems. Med Imag Process 1092:264–274

13. Strother SC, Anderson JR, Xu X, Liow J, Bonar DC, Rottenberg DA (1994) Quantitative comparisons of image registration techniques based on high-resolution MRI of the brain. J Comput Assist Tomogr 18(6):954–962

14. Fitzpatrick JM, West JB, Maurer CR (1998) Predicting error in rigid-body point based registration. IEEE Trans Med Imag 17(5):694–702

15. Grimpson WEL, Ettinger GJ, White SJ, Lozano-Perez T, Wells WM, Kikinis R (1996) An automatic registration method for frameless stereotaxy, image guided surgery, and enhanced reality visualization. IEEE Trans Med Imag 15(2):129–140

16. Herring JL, Dawant BM, Maurer CR Jr, Muratore DM, Galloway RL, Fitzpatrick JM (1998) Surface-based registration of CT images to physical space for image-guided surgery of the spine: a sensitivity study. IEEE Trans Med Imag 17(5):743–752

17. Kanatani K (1994) Analysis of 3-D rotation fitting. IEEE Trans Pattern Analysis Machine Intell 16(5):543–549

18. Declerc J, Feldmar J, Betting F, Goris ML (1997) Automatic registration and alignment on a template of cardiac stress and rest SPECT images. IEEE Trans Med Imag 16:727–737

19. Besl PJ, McKay ND (1992) A method for registration of 3-D shapes. IEEE Trans Pattern Recogn Mach Intell 14(2):239–256

20. Kim J, Fessler Jeffrey A (2004) Intensity-based image registration using robust correlation coefficients. IEEE Trans Med Imag 23(11):98–104

21. Hazra J, Roy Chowdhury A, Dutta P (2013) An approach for determining angle of rotation of a gray image using weighted statistical regression. Int J Sci Eng Res 4(8):1006–1013

22. Hazra J, Roy Chowdhury A, Dutta P (2014) Statistical regression based rotation estimation technique of color image. Int J Comput Appl 102(15):1–4

23. Pluim JPW, Maintz JBA, Viergever MA (2003) Mutual-information based registration of medical images: a survey. IEEE Trans Med Imag 22(6):986–1004

24. Langevin F, Didon JP (1995) Registration of MR images: from 2D to 3D, using a projection based cross correlation method. In: IEEE 17th Annual Conference, vol 1. Engineering in medicine and biology society, Canada

25. Rangarajan A, Chui H, Duncan JS (1999) Rigid point feature registration using mutual information. Med Image Anal 3(4):425–440

26. Puri BK (2004) Monomodal rigid-body registration and applications to the investigation of the effects of eicosapentaenoic acid intervention in neuropsychiatric disorders. Prostaglandins, Leukotrienes and Essential Fatty Acids (PLEFA) 3:137–200

27. AI-Azzawi N, Abdullah WAKW (2012) MRI monomodal feature based registration based on the efficiency of multiresolution representation and mutual information. American J Biomed Eng 2(3):98–104

28. Davis MH, Khotanzad A, Flaming DP (1996) 3D image matching using radial basis function neural network. In: WCNN'96: World congress on neural networks, pp 1174–1179

29. Sabisch T, Ferguson A, Bolouri H (1998) Automatic registration of complex images using a self organizing neural system. In: Proceedings of 1998 international joint conference on neural networks, pp 165–170

30. Elhanany I, Sheinfeld M, Beck A, Kadmon Y, Tal N, Tirosh D (2000) Robust image registration based on feedforward neural networks. In: Proceedings of 2000 IEEE international conference on systems, man and cybernetics, TN, USA, pp 1507–1511

31. Mostafa MG, Farag AA, Essock E (2000) Multimodality image registration and fusion using neural network. Information Fusion. FUSION 2000, vol 2, pp 10–13

32. Shang L, Cheng Lv J, Yi Z (2006) Rigid medical image registration using pca neural network. Neurocomputing 69(13–15):1717–1722

33. Goldberg DE (1989) Genetic algorithm in search, optimization and machine learning. Addison-Wesley Professional, Reading

34. Castro L, Zuben F (1999) Artificial immune systems: Part I—Basic theory and applications. Technical Report TR—DCA 01/99

35. Haralick RM, Shanmugam K, Dinstein I (1973) Textural features for image classification. IEEE Trans Syst Man Cybern SMC-3(6):610–621

36. Renzetti FR, Zortea L (2011) Use of a gray level cooccurrence matrix to characterize duplex stainless steel phases microstructure. Frattura ed Integrita Strutturale 16:43–51

37. Tuan Anh Pham (2010) Optimization of texture feature extraction algorithm. Thesis Paper, The Netherlands

38. Dasgupta D (1999) Artificial immune systems and their applications. Springer, Berlin

39. Haykin S (1999) Neural networks: a comprehensive foundation, 2nd edn. Prentice-Hall

40. Meva DT, Kumbharana CK, Kothari AD (2012) The study of adoption of neural network approach in fingerprint recognition. Int J Comput Appl 40(11):8–11

41. http://www.med.wayne.edu/diagRadiology/Anatomy_Modules/brain/brain.html

42. http://www.med.harvard.edu/aanlib/home.html

43. Goon AM, Gupta MK, Dasgupta B (2008) An outline of statistical theory. The World Press Private Limited, Kolkata

Author Index

© Springer India 2016
S. Bhattacharyya et al. (eds.), *Hybrid Soft Computing Approaches*,
Studies in Computational Intelligence 611,
DOI 10.1007/978-81-322-2544-7

Printed in the United States
By Bookmasters